WITHDRAWN

ADVANCED ROBOTICS
Redundancy and Optimization

Yoshihiko Nakamura
University of California
Santa Barbara

Addison-Wesley Publishing Company
Reading, Massachusetts • Menlo Park, California
New York • Don Mills, Ontario • Wokingham, England
Amsterdam • Bonn • Sydney • Singapore • Tokyo • Madrid • San Juan

Library of Congress Cataloging-in-Publication Data

Nakamura, Yoshihiko.
    Advanced robotics : redundancy and optimization / Yoshihiko Nakamura.
       p.   cm.
    Includes bibliographical references and index.
    ISBN 0-201-15198-7
    1. Robotics.   I. Title.
TJ211.N34 1991
629.8'92–dc20   90-931
                                                                                                     CIP

Copyright © 1991 by Addison-Wesley Publishing Company, Inc.

All rights reserved. No part of this publication may be reproduced, stored in a retrieval system, or transmitted, in any form or by any means, electronic, mechanical, photocopying, recording, or otherwise, without the prior written permission of the publisher. Printed in the United States of America.

123456789-MA-93210

In memory of Koichi Nakamura

This book is in the **Addison-Wesley Series in Electrical and Computer Engineering: Control Engineering.**

Consulting Editor: John J. Craig, Robotic Systems

Adaptive Control, 09720	Karl J. Åström and Björn Wittenmark
Introduction to Robotics, Second Edition, 09528	John J. Craig
Modern Control Systems, Fifth Edition, 14278	Richard C. Dorf
Digital Control of Dynamic Systems, Second Edition, 11938	Gene F. Franklin, J. David Powell, and Michael L. Workman
Computer Control of Machines and Processes, 10645	John G. Bollinger and Neil A. Duffie
Feedback Control of Dynamic Systems, 11540	Gene F. Franklin, J. David Powell, and Michael L. Workman
Adaptive Control of Mechanical Manipulators, 10490	John J. Craig
Modern Control System Theory and Application, Second Edition, 07494	Stanley M. Shinners

Preface

This book is intended as a second-level graduate course textbook in robotics and as a research monograph. Although the book is intended to be self-contained, I suggest that the reader has completed an introductory course in robot kinematics and dynamics.

In graduate curricula, there has been a recent need for an advanced robotics course for those students who have already taken an introductory robotics course and have started graduate research. There are several good textbooks available for introductory robotics, such as the one by John J. Craig (*Introduction to Robotics*, second edition, Addison-Wesley, 1989). However, there is a substantial gap between the cutting edge of research and the contents of an introductory course. *Advanced Robotics: Redundancy and Optimization* bridges that gap. It is my hope that this book will invite students and engineers to view one of the forefronts of recent robotics research.

The specific field in robotics to be covered in this book is *redundancy*. If we are to provide machines with autonomy and dexterity, the next generation of robots must be equipped with more actuators and more sensors. The aim of research on redundancy is to develop the technologies that utilize redundancy efficiently. Specific research on kinematics, dynamics, control, and sensing has been done in an isolated manner. The common features are found in the mathematic techniques to optimize or utilize the redundancy.

The first goal of this book is to establish a comprehensive methodological perspective for the mathematical optimization techniques commonly used in the field. Chapter 2 serves this purpose. Numerous mathematical optimization techniques, as common and important as linear algebraic methods, variational methods, and Pontryagin's maximum principle, are covered here and form the basis of discussions throughout the book. Chapter 3 is a summary of robot kinematics, with which the reader should already be familiar from an introductory robotics course.

The second goal is to provide a view of research in the field as examples of how the mathematical optimization techniques are applied. Chapters 4 through 11 serve this purpose. They are compiled from research publications of myself and my colleagues. It is not correct in any sense to say that these chapters provide a balanced perspective of research in this field.

The draft of this book was used in ME270A, *Advanced Robotics*, at the University of California, Santa Barbara (UCSB), in the spring quarter, 1989. Since I had taught an introductory course in robot kinematics and dynamics, ME170A, Chapter 3 was omitted. Chapters 2, 4 through 7, 10, and 11 were fitted for the 10-weeks' quarter period. Chapters 8 and 9 were included after the course. Many examples in Chapter 2 were added afterward, based on feedback from the students.

This book is a result of the invaluable benefits that I reaped from many people. I was fortunate enough to study with Professor Hideo Hanafusa, who was my advisor for my bachelor's, master's, and doctoral theses. From him, I learned more than just robotics, which I believe has been the basis of my academic life. Chapters 4, 5, and 9 are based on my doctoral work, which was supervised by him at the Automation Research Laboratory at Kyoto University. At Kyoto University, I was also indebted to Professor Tsuneo Yoshikawa. Discussions with him were the driving force of my research, and my admiration of his strength in mathematics encouraged me to learn mathematics for robotics. Chapters 4, 6, and 10 describe work in which I benefited from him as a coworker. Chapters 6, 7, 8, and 11 are based on the research I conducted at the Center for Robotic Systems in Microelectronics (CRSM), at UCSB. Professors Susan Hackwood and Gerardo Beni deserve special thanks for providing me with this ideal environment and opportunity.

I would like to thank Dr. John J. Craig for recommending that I write this book. My thanks are extended to the reviewers who read the manuscript thoroughly and offered many valuable and constructive suggestions.

I also appreciated the contributions of the following individuals who helped me in coauthoring papers and setting up experiments: Ichiro Futamata, Modjtaba Ghodoussi, Steve Jordan, Heng Liu, Kiyoshi Nagai, Katsumi Nakashima, Frank Reveles, Timo Ropponen, Yuzo Sawada, Toru Tsuchida, Hirotoshi Yamamoto, Yasuyoshi Yokokohji, and Yingti Xu. I am grateful to the students enrolled in ME270A, spring 1989, UCSB, who suffered from reading the very first version of the manuscript, among whom Modjtaba Ghodoussi and Ranjan Mukherjee spent a significant amount of time and made a number of critical editorial comments that helped me improve the manuscript.

I would like to acknowledge the financial support provided by the National Science Foundation, through the CRSM, UCSB General Research Fund, and the Grant in Aid for Scientific Research, Ministry of Culture and Education, Japan.

I wish to thank Tom Robbins, Don Fowley, Laurie McGuire, and Laurie Petrycki at Addison-Wesley.

Finally, my wife Kyoko was always a source of inspiration and moral support. My daughter Salena and son Hugh, who came into being while I was writing this book, provided me with silent or sometimes noisy encouragement. Their support and encouragement played an essential role in completing this book.

Santa Barbara, California Y. N.

Contents

1	**Introduction**	**1**
1.1	Redundancy for Autonomy and Dexterity	1
1.2	Examples of Redundant Robots	3
1.3	Scope of this Book	11
	References	12
2	**Mathematical Toolbox**	**15**
2.1	Introduction	15
2.2	Derivatives of Vectors, Matrices, and their Functions	16
2.3	Singular Value Decomposition	30
2.4	Generalized Inverse and Pseudoinverse	41
2.5	Variational Methods	68
2.6	Pontryagin's Maximum Principle	81
	Summary	101
	References	102
3	**Differential Kinematics and Redundancy**	**105**
3.1	Introduction	105
3.2	Jacobian Matrix	106
3.3	Manipulability and Redundancy	114

3.4	Measure of Manipulability	119
	Summary	122
	References	122

4 Local Optimization of Kinematic Redundancy — 125

4.1	Introduction	125
4.2	Tasks with the Order of Priority	127
4.3	Inverse Kinematics Considering the Order of Priority	128
4.4	Numerical Simulations	132
4.5	Experiments	140
	Summary	148
	References	150

5 Global Optimization of Kinematic Redundancy — 153

5.1	Introduction	153
5.2	Optimal Control Problem of Redundancy	154
5.3	Application of Pontryagin's Maximum Principle	155
5.4	Boundary Conditions	156
5.5	An Example of Performance Index	159
5.6	Consideration of Dynamics	160
5.7	Numerical Examples	162
	Summary	169
	References	170

6 Redundancy in Multirobot Coordination — 173

6.1	Introduction	173
6.2	Coordinative Manipulation by Multiple Robotic Mechanisms	176
6.3	Coordinative Manipulability	182
6.4	Computation of Minimum Forces	184
6.5	Contact Stability and Optimal Forces	191
6.6	Examples of Minimum Force Computation	199
	Summary	200
	References	201

7 Actuation Redundancy of Closed-Link Mechanisms — 205

7.1	Introduction	205
7.2	Inverse Dynamics of Closed–Link Robots	207
7.3	Redundant Actuation Systems	213
7.4	Examples	217
	Summary	224
	References	225

8 A Manipulator with Kinematic and Actuation Redundancy — 229

- 8.1 Introduction — 229
- 8.2 A Closed-Link Mechanism with Kinematic Redundancy — 230
- 8.3 Redundant Actuation of the Closed-Link Mechanism — 232
- 8.4 Singularity-Free Parameterization of Actuation Redundancy — 232
- 8.5 Numerical Analysis — 239
- Summary — 248
- References — 250

9 Singularity-Robust Inverse of Jacobian Matrix — 253

- 9.1 Introduction — 253
- 9.2 Singularity and Pseudoinverse — 255
- 9.3 SR-Inverse of the Jacobian Matrix — 257
- 9.4 Properties of the SR-Inverse — 259
- 9.5 Computational Complexity of the SR-Inverse — 264
- 9.6 Variable Scale Factor — 267
- 9.7 Simulations — 273
- Summary — 279
- References — 279

10 Redundancy in Multiaxis Force Sensing — 283

- 10.1 Introduction — 283
- 10.2 Force Sensing — 285
- 10.3 Structure Evaluation of Elastic Components — 289
- 10.4 Use of Redundancy in Force Sensing — 293
- 10.5 An Example of Force-Sensor Design — 295
- Summary — 298
- References — 299

11 Geometric Optimization for Sensor Fusion — 303

- 11.1 Introduction — 303
- 11.2 Background — 304
- 11.3 Sensing Model and Uncertainty Ellipsoid — 306
- 11.4 Geometric Fusion Method — 311
- 11.5 Fusion of Partial Information — 319
- Summary — 323
- References — 324

Index — 327

CHAPTER 1

Introduction

1.1 Redundancy for Autonomy and Dexterity

Redundancy has been incorporated into various systems that need high reliability, such as aircraft, spacecraft, and military systems. For example, in a spacecraft, so that the reliability of the computer system will be improved, several computers do the same computation simultaneously. If all the results do not coincide, then the result reached by the majority of the machines is chosen as the solution. Improving reliability has been the main purpose of adopting redundancy.

In traditional machine design, the best design synthesizes a mechanism that functions with the minimum complexity of the structure. Most industrial robots have been designed based on this commonsense rule. Industrial robots that must be able arbitrarily to position and orient end-effectors in a three-dimensional workspace have six actuators. Welding robots, which usually do not need rotation about the welding torch, have five actuators. SCARA[1.1]

[1.1] SCARA stands for "selectively compliant assembly robot arm."

type robots are designed to perform simple assembly tasks. Many simple assembly tasks require that the robot pick up a part on a horizontal plane and place that part onto another horizontal plane. The orientation may be changed only about the axis perpendicular to the planes. Hence, SCARA-type robots have only four actuators.

Robots so far have been used primarily for applications that have well-arranged, rather artificial environments. The part that the robot is to grasp is placed at an *easy*[1.2] location, and the path and the goal are both in an *easy* area within the workspace that is basically free of obstacles. There should be no unexpected disturbance. Satisfying these assumptions is the responsibility of the engineers or scientists who set up the total system. The best-designed robots can work in only such environments. Setting up environments that meet these assumptions costs more than does building the robots themselves. This fact has limited robotic applications in both industrial and nonindustrial areas.

To ease or avoid these limitations, we must build a robot that has functions that allow it to understand its situation, and to plan and control its motion consistent with the situation. Such a robot is said to be *autonomous* and *dexterous*. The need for an autonomous and dexterous robot system is becoming important for the microelectronics manufacturing in clean or vacuum environments. Such robots also could play an important role in space and undersea missions, where they would allow us to automate the hazardous jobs of human crews and to make the missions cost effective.

If it is to be autonomous, a robot system should have many sensors. Although these sensors are necessary to identify the total situation, they tend to be redundant for a particular measurement. For example, the position and orientation of the end-effector may be computed either from joint sensors or from a stereo vision, although the two sensors have different roles for other measurements. Therefore, as in systems that require high reliability, we can improve the reliability or accuracy of measurements by actively using this sensing redundancy. High-level motion planning based on a reliable and accurate understanding of the situation is a major component of autonomy.

Dexterity implies the mechanical ability to carry out various kinds of tasks in various situations. If it is to have dexterity, a robot system should have two or more arms, rather than one; three or more fingers, rather than a simple jaw-type gripper; and seven or more joints for each arm, rather than six or less. Consistent management of this high degree of freedom is required if the system is to satisfy a given goal. This type of mechanical redundancy is different from the sensing redundancy. Mechanical redundancy is divided into *kinematic* and *actuation* redundancies.

In a system with kinematic redundancy, we are able to change the internal

---

[1.2]Here, *easy* implies that the robot is not forced to be close to or at a singular point. Singularity will be discussed in detail in Chapters 3, 4, 5, and 9.

structure or configuration of the mechanisms without changing the position and orientation of the end-effector or of the object. A typical example is a human arm, which has 7 degrees of freedom (DOF) from the shoulder to the wrist. Because there is kinematic redundancy, it is possible to change the position of the elbow while the hand grasps a fixed object. Kinematic redundancy can be used to avoid joint limits, obstacles, and singularity, as well as to minimize energy consumption. Since the decision at $t = t_0$ to change the internal structure affects the internal structure at $t > t_0$, kinematic redundancy has causality and, therefore, forms a global problem that can be optimized by viewing the total motion.

Actuation redundancy is found in only closed-link mechanisms. When two arms or multiple fingers hold a single object, they form a closed-link structure. When a multilegged robot walks, the robot and the ground form a closed-link structure. These are examples of closed-link mechanisms formed by multilimb mechanisms. For example, if an object is grasped by three fingers with 3 DOF each, the system has freedom to adjust how hard to squeeze the object, since the total closed-link mechanism has nine actuators, whereas the object motion is specified by only six independent variables. This is a typical actuation redundancy. By utilizing this redundancy, we can change the quality of the grasp. The actuation redundancy of multilimb systems is represented in the relationship between the resultant force and moment applied to the object and the forces and moments that the individual mechanisms exert at the contact points.

General nonmultilimb closed-link mechanisms can have actuation redundancy. A general closed-link mechanism usually has actuated joints and unactuated joints. The number of actuated joints is commonly equal to the dimension of the output variables. Adding actuators at some of the unactuated joints provides the mechanism with another kind of actuation redundancy, which is distinct from that in multilimb systems in that the redundancy is not represented in terms of the forces and moments at the ends of the individual limbs. Hence, the parametric representation of actuation redundancy is different depending on whether the closed-link mechanism is a multilimb system or a nonmultilimb system.

Actuation redundancy concerns only the determination of forces and moments. It has nothing to do with the internal structure or configuration of the mechanisms. The determination of forces and moments at $t = t_0$ does not affect the determination at $t > t_0$. Accordingly, actuation redundancy does not have causality and, therefore, is intrinsically a local problem that can be optimized by viewing the relationship at $t = t_0$.

In summary, the next generation of robot systems must be autonomous and dexterous, and the robots must have more sensors and more actuators. These sensors and actuators provide the system with both sensing and mechanical redundancies. The management of these redundancies is a central technology to be established for building real autonomous and dexterous robot systems.

4   Chapter 1   Introduction

Figure 1.1 MELARM, a system with dual 7-DOF anthropomorphic manipulators powered by rotary hydraulic actuators. (Courtesy of Dr. Tatsuo Arai, Mechanical Engineering Laboratory, MITI, Japan.)

1.2   Examples of Redundant Robots

Many mechanically redundant robots have been developed so far. Some were developed for research purposes. Some have already been used in real applications. In this section, we shall examine these examples.

Figure 1.1 shows a system, named MELARM, with two 7-DOF arms. MELARM was developed at the Mechanical Engineering Laboratory, the Ministry of International Trade and Industry (MITI), Japan (Nakano and Ozaki 1974). Each arm has an anthropomorphic structure.[†1.3] The joints are driven by electrohydraulic actuators. The figure shows an experiment in master–slave control.

Figure 1.2 shows a 7-DOF manipulator named UJIBOT that has a nonanthropomorphic structure. UJIBOT was developed in 1979 at Kyoto University

---

[†1.3] A 7-DOF manipulator with three joints intersecting at the shoulder, three joints intersecting at the wrist, and an elbow joint, is called an *anthropomorphic arm*, since the structure is similar to that of a human arm.

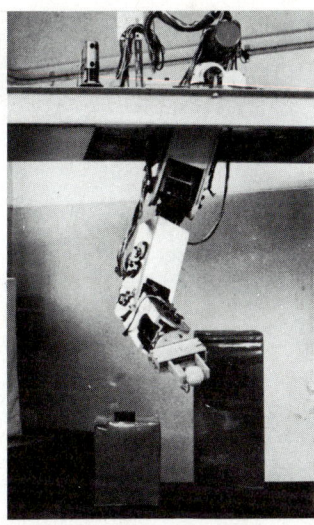

Figure 1.2 UJIBOT, a system with a 7-DOF nonanthropomorphic arm.

(Hanafusa, Yoshikawa, and Nakamura 1981). This robot was used for experiments in obstacle avoidance. The details of this robot and of the experiments performed with it will be described in Chapter 4.

Figure 1.3 shows a 7-DOF direct-drive manipulator that has an anthropomorphic structure; it was also developed at the Mechanical Engineering Laboratory, MITI, Japan (Arai, Yano, Hashimoto, and Nakano 1987). This robot is intended for the automation of sewing tasks in garment industries.

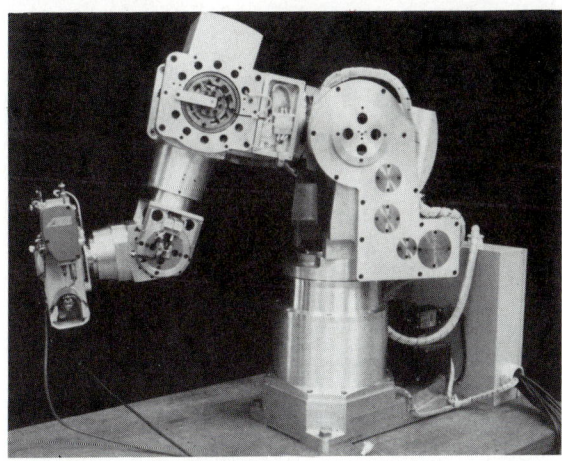

Figure 1.3 A 7-DOF direct-drive manipulator applied in sewing automation. (Courtesy of Dr. Tatsuo Arai, Mechanical Engineering Laboratory, MITI, Japan.)

6   Chapter 1   Introduction

Figure 1.4 A welding robot system with 7 DOF (5-DOF arm and 2-DOF table). (Courtesy of Shin Meiwa Industries Co. Limited.)

The three robots in Figures 1.1 through 1.3 are research robots. Now, let's look at redundant robots designed for practical use. There are several commercial welding robot systems consisting of a 5- or 6-DOF arm and a 2-DOF positioning table. Figure 1.4 shows one of these welding robots. Although

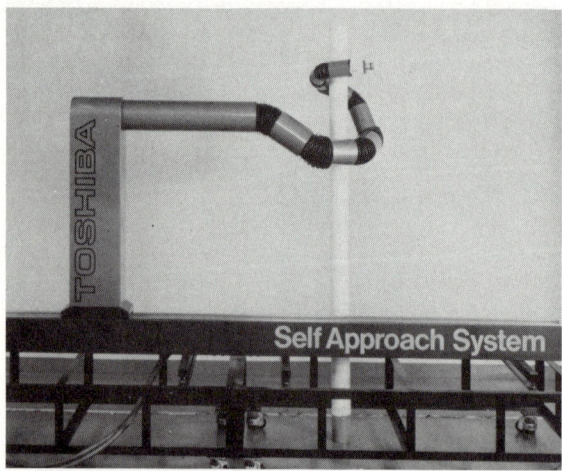

Figure 1.5 A 17-DOF inspection robot. (Courtesy of Toshiba Corp.)

their redundancy has been used in a rather limited way, these robots are the typical example of mechanically redundant robot systems.

Figure 1.5 shows a 17-DOF robot designed to inspect the inside of nuclear plants; the system was developed by Toshiba Co. (Asano, Obama, Arimura, Kondo, and Hitomi 1983). This robot makes snakelike motions. Redundancy is used to avoid collision with the environment.

Figure 1.6 shows another 17-DOF robot system for research of space, telerobotics developed by Robotics Research Co. (Vold, Karlen, Thompson, Farrell, and Eismann 1989). The robot system consists of two 7-DOF arms and one 3-DOF torso. The arms look anthropomorphic, but are not in the strict sense since the three joints at the shoulder and the wrist do not intersect.

Figure 1.6 A 17-DOF robot system; dual 7-DOF manipulators and one 3-DOF torso. (Courtesy of Robotic Research Corp.)

8 Chapter 1 Introduction

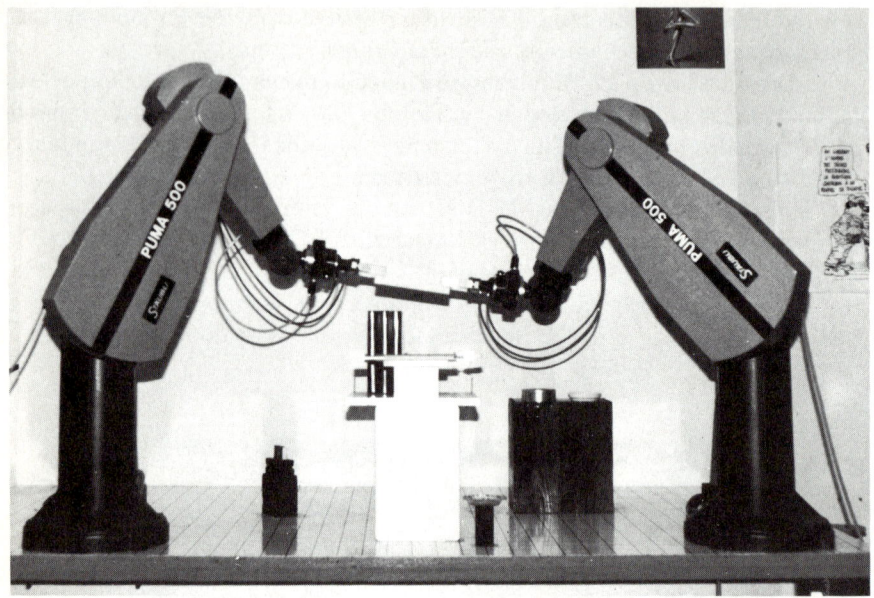

Figure 1.7 Two 6-DOF arms manipulating a rigid object. (Courtesy of Dr. Pierre Dauchez, Laboratoire d'Automatique et de Microélectronique de Montpellier.)

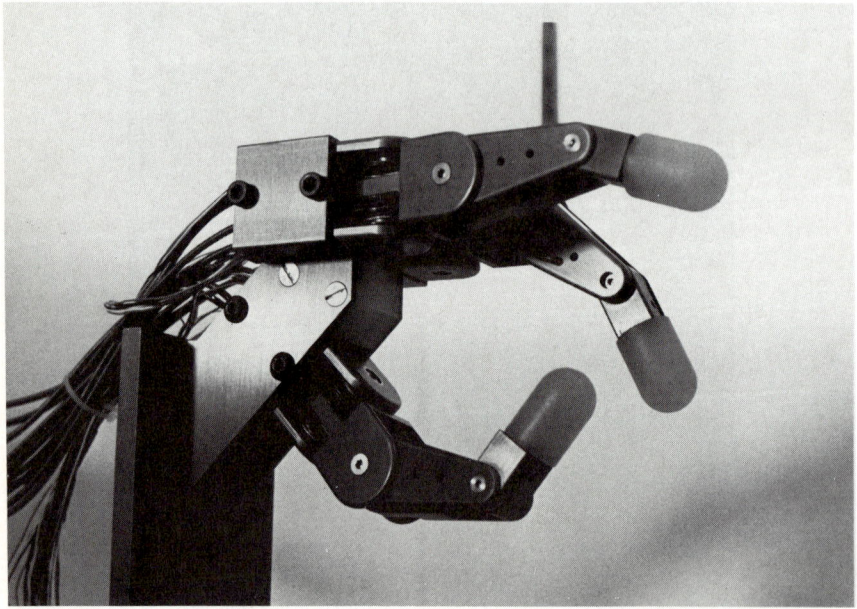

Figure 1.8 Salisbury hand, three 3-DOF fingers. (Courtesy of Dr. Kenneth Salisbury, Massachusetts Institute of Technology, photo credit: David Lampe.)

For research on multiarm coordination, two 6-DOF arms are often used. Figure 1.7 shows an example of such an experimental system.

Several multifingered-hand systems have been developed for research on multifinger coordination. Figures 1.8 through 1.11 show examples of such sys-

Figure 1.9 UTAH/MIT dextrous hand; four fingers with 4-DOF each. (Courtesy of Center for Engineering Design, University of Utah, photo credit: Ed Rosenberger.)

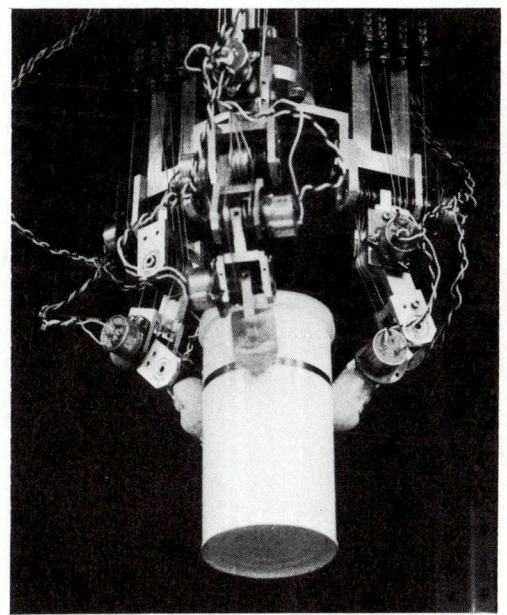

Figure 1.10 A three-fingered hand; a system with three 4-DOF fingers. (Courtesy of Professor Hiroaki Kobayashi, Meiji University.)

10   Chapter 1 Introduction

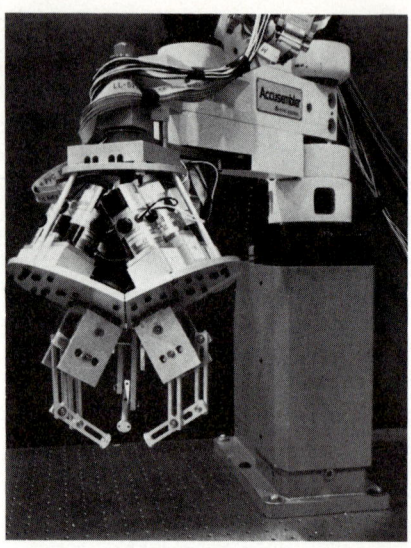

Figure 1.11 A three-fingered hand; a system with three 3-DOF closed link fingers. (University of California, Santa Barbara.)

tems (Salisbury 1982; Jacobsen, Wood, Knutti, Biggers, and Iversen 1985; Hanafusa, Kobayashi, and Terasaki 1983; Nakamura and Ghodoussi 1988).

Finally, Fig. 1.12 shows a six-legged walking machine developed at Ohio State University (Waldron 1986). Force distribution among legs was a central problem of the research (Orin and Oh 1981).

Figure 1.12 The adaptive suspension vehicle; a six-legged walking machine. (Courtesy of Professor Kenneth J. Waldron, Ohio State University.)

1.3 Scope of This Book

Various optimization techniques have been applied to *redundant robots*. In Chapters 2 and 3, we overview the mathematical and kinematic backgrounds of these optimization methods. Then, we discuss the specific issues of kinematic redundancy, actuation redundancy, and sensing redundancy in Chapters 4 through 11. The reader is expected to have a basic knowledge of robot kinematics and dynamics—subjects usually covered in an introductory robotics course (see, for example, Yoshikawa 1990; Craig 1989; Spong and Vidyasagar 1989; Koivo 1989; Fu, Gonzalez, and Lee 1987; Wolovich 1987; Asada and Slotine 1986; Paul 1981). A brief summary of each chapter follows.

Chapter 2 provides an overview of the basic mathematical methods. Linear algebraic methods, classical variational methods, and Pontryagin's maximum principle are covered as typical mathematical tools for redundancy analysis and optimization. This chapter forms a firm ground for the rest of the book: The mathematical techniques established in this chapter are used extensively in the following chapters.

Chapter 3 is a short summary of robot kinematics, with which the reader should already be familiar from an introductory robotic course. The Jacobian matrix and its computation, the definition of manipulability and redundancy, and the measure of manipulability are explained.

Chapters 4 and 5 discuss kinematic redundancy. In Chapter 4, the local approach based on the concept of task priority is introduced. Chapter 5 describes the global approach to kinematic redundancy. The problem is formulated in the framework of Pontryagin's maximum principle.

Chapter 6 discusses the actuation redundancy of multilimb-type closed-link mechanisms. The dynamics and stability of grasping by multifingered hand are the subjects of this chapter. The focus is the determination and optimization of finger forces. The type of contact condition (such as point contact, surface contact, soft contact, and rolling contact), which determines the nature of grasping, is an important research topic. In this chapter, we assume point contact.

The actuation redundancy of nonmultilimb-type closed-link mechanisms is the subject of Chapters 7 and 8. Chapter 7 addresses the dynamics computation of general closed-link mechanisms and a parameterization of actuation redundancy. In Chapter 8, a closed-link mechanism with both kinematic and actuation redundancy is proposed. The algorithmic singularity of the parameterization in Chapter 7 is pointed out, and a new singularity-free parameterization is established.

In Chapter 9, we reformulate the inverse kinematics problem as an optimization problem, in order to cope with kinematic singularity. The proposed singularity-robust inverse can be used in place of the pseudoinverse in kinematic redundancy control.

Chapters 10 and 11 discuss examples of optimization in sensing redundancy. A multiaxis force sensor uses redundant strain gauges to form

Wheatstone bridges. In Chapter 10, we show how redundancy of strain gauges is used to perform analog signal processing to calculate force and moment. The criteria used in force-sensor design also are discussed. In multisensor fusion, we extract reliable and accurate information by using redundancy in sensing. In Chapter 11, a mathematical method of multisensor fusion is derived based on the geometric optimization method.

References

Arai, T., Yano, T., Hashimoto, R., and Nakano, E. 1987. Development of a direct-drive human-like manipulator. *J. Japan Robotics Society* 5 (1): 27-35 *(in Japanese)*.

Asada, H., and Slotine, J. J. 1986. *Robot analysis and control.* New York: Wiley.

Asano, K., Obama, M., Arimura, Y., Kondo, M., and Hitomi, Y. 1983. Multijoint inspection robot. *IEEE Trans. Industrial Electronics* IE-30 (3): 277-281.

Craig, J. J. 1989. *Introduction to robotics: Mechanics and control*, second edition. Reading, MA: Addison-Wesley.

Fu, K., Gonzalez, R., and Lee, C. S. G. 1987. *Robotics: Control, sensing, vision, and intelligence.* New York: McGraw-Hill.

Hanafusa, H., Kobayashi, H., and Terasaki, N. 1983 (Tokyo). Fine control of the object with articulated multi-finger robot hands. *Proc. 1983 Int. Conf. Advanced Robotics*, pp. 245-251.

Hanafusa, H., Yoshikawa, T., and Nakamura, Y. 1981 (Kyoto). Analysis and control of articulated robot arms with redundancy. *Control Science and Technology for the Progress of Society (Proc. 8th Triennial World Congress of IFAC*, ed. H. Akashi, Vol. 4: pp. 1927-1932.

Jacobsen, S. C., Wood, J. E., Knutti, D. F., Biggers, K. B., and Iversen, E. K. 1985. The version I Utah/MIT dextrous hand. In *Robotics research 2*, eds. H. Hanafusa and H. Inoue, pp. 301-308. Cambridge: MIT Press.

Koivo, A. J. 1989. *Fundamentals for control of robotic manipulators.* New York: Wiley.

Nakamura, Y., and Ghodoussi, M. 1988 (Philadelphia). A computational scheme of closed link robot dynamics derived by d'Alembert principle. *Proc. 1988 IEEE Int. Conf. Robotics and Automation*, pp. 1354-1360.

Nakano, E., and Ozaki, S. 1974 (Tokyo). Coorperative control of a pair of anthropomorphous manipulators — MELARM. *Proc. 4th Int. Symp. Industr. Robots*, pp. 250-260.

Orin, D. E., and Oh, S. Y. 1981. Control of force distribution in robotic mechanisms containing closed kinematic chains. *ASME J. Dyn. Sys., Meas. Contr.* 102 (2): 134–141.

Paul, R. P. 1981. *Robot manipulators: Mathematics, programming, and control.* Cambridge: MIT Press.

Salisbury, J. K. 1982. Kinematic and force analysis of articulated hands. Ph.D. thesis, Stanford University, Dept. of Mechanical Engineering, Stanford, CA.

Spong, M. W., and Vidyasagar, M. 1989. *Robot dynamics and control.* New York: Wiley.

Vold, H. I., Karlen, J. P., Thompson, J. M., Farrell, J. D., and Eismann, P. H. 1989. A 17 degree of freedom anthropomorphic manipulator. *Proc. NASA Conf. Space Telerobotics*, Vol. 1: pp. 19–28.

Waldron, K. J. 1986. Force and motion management in legged locomotion. *IEEE J. Robotics and Automat.* 2 (4): 214–220.

Wolovich, W. 1987. *Robotics: Basic analysis and design.* New York: Holt, Rinehart, and Winston.

Yoshikawa, T. 1990. *Foundations of robotics: Analysis and control.* Cambridge: MIT Press.

CHAPTER 2

Mathematical Toolbox

2.1 Introduction

The mathematical methods to be used in the following chapters are classified into three categories: *linear algebraic methods, variational methods,* and *Pontryagin's maximum principle.* Since each of these forms a broad subject, it is not the goal of this chapter to cover systematically the three methods. We focus on the specific topics that are necessary to understand the following chapters, or are relevant to them. So that we can overview the mathematical tools concisely, we often describe the important results without rigorous proofs. For readers interested in such proofs, we provide references. The mathematical techniques summarized in this chapter not only are necessary for work in the following chapters, but also are powerful tools for further study and research on robotics.

Derivatives of vectors, matrices, and their functions are summarized in Section 2.2. In Sections 2.3 and 2.4, two important linear algebraic tools—*singular value decomposition* and *pseudoinverses*—are overviewed. In Section 2.5, the basic techniques of classical variational methods are summarized. Section 2.6 covers *Pontryagin's maximum principle,* which is an extension of the classical variational methods.

In Chapters 3 through 11, various optimization problems concerning robotic redundancy are solved using the established mathematical tools. The relevance of each chapter and the mathematical tools are: Chapter 3, singular value decomposition and pseudoinverses; Chapter 4, pseudoinverses; Chapter 5, Pontryagin's maximum principle; Chapter 6, pseudoinverses and variational methods; Chapter 7, variational methods; Chapter 8, pseudoinverses; Chapter 9, singular value decomposition and pseudoinverses; Chapter 10, singular value decomposition and pseudoinverses; Chapter 11, singular value decomposition and variational methods.

2.2 Derivatives of Vectors, Matrices, and their Functions

2.2.1 Derivatives of Vectors

Throughout this book, vectors are represented in column forms. For example, a vector $\boldsymbol{x} \in R^n$ is an n-dimensional column vector, and its row form is represented by $\boldsymbol{x}^T$, where $*^T$ indicates the transpose of vector or matrix $*$. Hence,

$$\boldsymbol{x} = \begin{pmatrix} x_1 \\ x_2 \\ \vdots \\ x_n \end{pmatrix} \in R^n \qquad 2.1$$

$$\boldsymbol{x}^T = (\, x_1 \quad x_2 \quad \cdots \quad x_n \,) \qquad 2.2$$

Let a vector $\boldsymbol{y} \in R^m$ be a function of $\boldsymbol{x}$. Namely,

$$\boldsymbol{y} = \boldsymbol{y}(\boldsymbol{x})$$

$$= \begin{pmatrix} y_1(\boldsymbol{x}) \\ y_2(\boldsymbol{x}) \\ \vdots \\ y_m(\boldsymbol{x}) \end{pmatrix}$$

$$= \begin{pmatrix} y_1(x_1, x_2, \cdots, x_n) \\ y_2(x_1, x_2, \cdots, x_n) \\ \vdots \\ y_m(x_1, x_2, \cdots, x_n) \end{pmatrix} \qquad 2.3$$

The derivative of $\boldsymbol{y}$ with respect to $\boldsymbol{x}$ is defined by

$$\frac{\partial \boldsymbol{y}}{\partial \boldsymbol{x}} \triangleq \begin{pmatrix} \partial y_1/\partial \boldsymbol{x} \\ \partial y_2/\partial \boldsymbol{x} \\ \vdots \\ \partial y_m/\partial \boldsymbol{x} \end{pmatrix}$$

2.2 Derivatives of Vectors, Matrices, and their Functions

$$= \begin{pmatrix} \partial y_1/\partial x_1 & \partial y_1/\partial x_2 & \cdots & \partial y_1/\partial x_n \\ \partial y_2/\partial x_1 & \partial y_2/\partial x_2 & \cdots & \partial y_2/\partial x_n \\ \vdots & \vdots & \ddots & \vdots \\ \partial y_m/\partial x_1 & \partial y_m/\partial x_2 & \cdots & \partial y_m/\partial x_n \end{pmatrix} \in R^{m \times n} \qquad 2.4$$

Note that the order of elements of the numerator of $\partial \boldsymbol{y}/\partial \boldsymbol{x}$ is maintained column-wise and that of the denominator appears row-wise. Also note that the derivative of a scalar with respect to a vector—that is, the case of $m = 1$ in Eq. 2.4—is defined as a row vector as follows:

$$\frac{\partial y}{\partial \boldsymbol{x}} = (\frac{\partial y}{\partial x_1} \ \frac{\partial y}{\partial x_2} \ \cdots \ \frac{\partial y}{\partial x_n}) \qquad 2.5$$

When a scalar $y(\boldsymbol{x})$ is a potential function of $\boldsymbol{x}$ and we want to find the change in $\boldsymbol{x}$, namely $\Delta \boldsymbol{x}$, that most effectively increases the potential function, the common method is to take the gradient of $y(\boldsymbol{x})$. Namely,

$$\text{grad}\{y(\boldsymbol{x})\} \overset{\triangle}{=} \begin{pmatrix} \partial y/\partial x_1 \\ \partial y/\partial x_2 \\ \vdots \\ \partial y/\partial x_n \end{pmatrix} = (\frac{\partial y}{\partial \boldsymbol{x}})^T \qquad 2.6$$

The normalized gradient

$$\frac{\text{grad}\{y(\boldsymbol{x})\}}{\| \text{grad}\{y(\boldsymbol{x})\} \|} \qquad 2.7$$

implies the unit vector indicating the direction of $\Delta \boldsymbol{x}$ that maximizes the change in $y(\boldsymbol{x})$; namely, Δy. $\| \text{grad}\{y(\boldsymbol{x})\} \|$, on the other hand, means the magnification ratio from $\| \Delta \boldsymbol{x} \|$ to Δy if $\Delta \boldsymbol{x}$ is chosen in the direction of Eq. 2.7. These characteristics will be derived in Example 2.6, Section 2.5.3, using a Lagrange multiplier. Note that the gradient is represented by the transposed form of the derivative of $y(\boldsymbol{x})$ with respect to $\boldsymbol{x}$, as shown in Eq. 2.6. In this book, we shall instead use $(\partial y/\partial \boldsymbol{x})^T$ whenever the gradient is needed. In addition, if a potential function or penalty function $y(\boldsymbol{x})$ must be reduced, the most effective unit vector $\Delta \boldsymbol{x}$ is

$$\Delta \boldsymbol{x} = -\frac{\text{grad}\{y(\boldsymbol{x})\}}{\| \text{grad}\{y(\boldsymbol{x})\} \|} \qquad 2.8$$

which is obtained as the best unit vector $\Delta \boldsymbol{x}$ that increase $-y(\boldsymbol{x})$.

If the elements of $\boldsymbol{x}$ are the functions of a parameter t, the derivative of $\boldsymbol{y}$ with respect to t is given by

$$\frac{d\boldsymbol{y}}{dt} = \begin{pmatrix} dy_1/dt \\ dy_2/dt \\ \vdots \\ dy_m/dt \end{pmatrix}$$

$$= \begin{pmatrix} \partial y_1/\partial x_1\, dx_1/dt + \partial y_1/\partial x_2\, dx_2/dt + \cdots + \partial y_1/\partial x_n\, dx_n/dt \\ \partial y_2/\partial x_1\, dx_1/dt + \partial y_2/\partial x_2\, dx_2/dt + \cdots + \partial y_2/\partial x_n\, dx_n/dt \\ \vdots \\ \partial y_m/\partial x_1\, dx_1/dt + \partial y_m/\partial x_2\, dx_2/dt + \cdots + \partial y_m/\partial x_n\, dx_n/dt \end{pmatrix}$$

$$= \begin{pmatrix} \partial y_1/\partial x_1 & \partial y_1/\partial x_2 & \cdots & \partial y_1/\partial x_n \\ \partial y_2/\partial x_1 & \partial y_2/\partial x_2 & \cdots & \partial y_2/\partial x_n \\ \vdots & \vdots & \ddots & \vdots \\ \partial y_m/\partial x_1 & \partial y_m/\partial x_2 & \cdots & \partial y_m/\partial x_n \end{pmatrix} \begin{pmatrix} dx_1/dt \\ dx_2/dt \\ \vdots \\ dx_n/dt \end{pmatrix}$$

$$= \frac{\partial \boldsymbol{y}}{\partial \boldsymbol{x}} \frac{d\boldsymbol{x}}{dt} \qquad 2.9$$

Example 2.1

Figure 2.1 shows a 2-DOF manipulator in $x_1 x_2$ plane. The position of the end-effector $\boldsymbol{r} \in R^2$ is represented as a function of joint angle vector $\boldsymbol{\theta} = (\theta_1\ \theta_2)^T \in R^2$ as follows:

$$\begin{aligned}\boldsymbol{r} &= \begin{pmatrix} r_1(\boldsymbol{\theta}) \\ r_2(\boldsymbol{\theta}) \end{pmatrix} \\ &= \begin{pmatrix} l_1 \cos\theta_1 + l_2 \cos(\theta_1 + \theta_2) \\ l_1 \sin\theta_1 + l_2 \sin(\theta_1 + \theta_2) \end{pmatrix}\end{aligned} \qquad 2.10$$

When θ_1 and θ_2 change as functions of time t, the velocity of the end effector is calculated by

$$\begin{aligned}\frac{d\boldsymbol{r}}{dt} &= \frac{\partial \boldsymbol{r}}{\partial \boldsymbol{\theta}} \frac{d\boldsymbol{\theta}}{dt} \\ &= \begin{pmatrix} \partial r_1/\partial \theta_1 & \partial r_1/\partial \theta_2 \\ \partial r_2/\partial \theta_1 & \partial r_2/\partial \theta_2 \end{pmatrix} \begin{pmatrix} d\theta_1/dt \\ d\theta_2/dt \end{pmatrix} \\ &= \begin{pmatrix} -l_1 \sin\theta_1 - l_2 \sin(\theta_1 + \theta_2) & -l_2 \sin(\theta_1 + \theta_2) \\ l_1 \cos\theta_1 + l_2 \cos(\theta_1 + \theta_2) & l_2 \cos(\theta_1 + \theta_2) \end{pmatrix} \begin{pmatrix} d\theta_1/dt \\ d\theta_2/dt \end{pmatrix}\end{aligned} \quad 2.11$$

The coefficient matrix $\partial \boldsymbol{r}/\partial \boldsymbol{\theta}$ is called the *Jacobian matrix*. Note that we sometimes use $\dot{*} \stackrel{\Delta}{=} d*/dt$ to denote the derivative with respect to time.

2.2 Derivatives of Vectors, Matrices, and their Functions   19

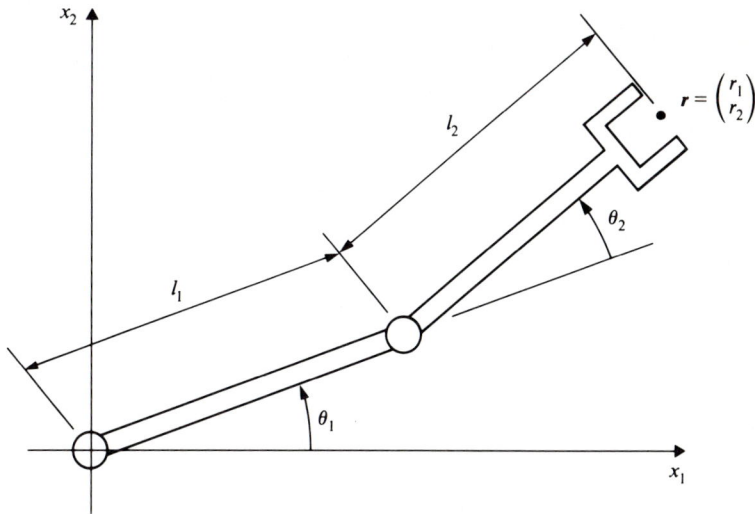

Figure 2.1 Planar 2-DOF manipulator.

2.2.2 Derivatives of Matrices

When a matrix $A \in R^{l \times m}$ is a function of $x \in R^n$, and x is a function of t, their relationship is represented as follows:

$$A = \begin{pmatrix} a_{11} & a_{12} & \cdots & a_{1m} \\ a_{21} & a_{22} & \cdots & a_{2m} \\ \vdots & \vdots & \ddots & \vdots \\ a_{l1} & a_{l2} & \cdots & a_{lm} \end{pmatrix} \in R^{l \times m} \qquad 2.12$$

$$a_{ij} = a_{ij}(\boldsymbol{x}) = a_{ij}(x_1, x_2, \cdots, x_n)$$

$$\boldsymbol{x} = \boldsymbol{x}(t) = \begin{pmatrix} x_1(t) \\ x_2(t) \\ \vdots \\ x_n(t) \end{pmatrix}$$

The derivative of A with respect to t is calculated as follows:

$$\begin{aligned} \frac{d\boldsymbol{A}}{dt} &= \frac{\partial \boldsymbol{A}}{\partial x_1} \frac{dx_1}{dt} + \frac{\partial \boldsymbol{A}}{\partial x_2} \frac{dx_2}{dt} + \cdots + \frac{\partial \boldsymbol{A}}{\partial x_n} \frac{dx_n}{dt} \\ &= \sum_{i=1}^{n} \frac{\partial \boldsymbol{A}}{\partial x_i} \frac{dx_i}{dt} \end{aligned} \qquad 2.13$$

$$\frac{\partial \boldsymbol{A}}{\partial x_i} = \begin{pmatrix} \partial a_{11}/\partial x_i & \partial a_{12}/\partial x_i & \cdots & \partial a_{1m}/\partial x_i \\ \partial a_{21}/\partial x_i & \partial a_{22}/\partial x_i & \cdots & \partial a_{2m}/\partial x_i \\ \vdots & \vdots & \ddots & \vdots \\ \partial a_{l1}/\partial x_i & \partial a_{l2}/\partial x_i & \cdots & \partial a_{lm}/\partial x_i \end{pmatrix} \qquad 2.14$$

It is possible to represent Eq. 2.13 using the form similar to that of Eq. 2.9, as follows:

$$\frac{d\boldsymbol{A}}{dt} = \frac{\partial \boldsymbol{A}}{\partial \boldsymbol{x}} \frac{d\boldsymbol{x}}{dt} \qquad 2.15$$

Note, however, that $\partial \boldsymbol{A}/\partial \boldsymbol{x}$ no longer maintains a two-dimensional matrix form, and should be treated as a tensor. Although it is difficult to see the structure of $\partial \boldsymbol{A}/\partial \boldsymbol{x}$ as being similar to that of Eq. 2.4, $\partial \boldsymbol{A}/\partial \boldsymbol{x}$ could be considered an $\ell \times m \times n$ three-dimensional matrix. The structure of Eqs. 2.9 and 2.15 would be visually represented as shown in Fig. 2.2.

Example 2.2

In Example 2.1, we can obtain the acceleration of $\boldsymbol{r}$ by differentiating Eq. 2.11 with respect to time. Namely,

$$\frac{d^2 \boldsymbol{r}}{dt^2} = \frac{\partial \boldsymbol{r}}{\partial \boldsymbol{\theta}} \frac{d^2 \boldsymbol{\theta}}{dt^2} + \frac{d}{dt}\left(\frac{\partial \boldsymbol{r}}{\partial \boldsymbol{\theta}}\right) \frac{d\boldsymbol{\theta}}{dt} \qquad 2.16$$

The first term of the right-hand side of Eq. 2.16 represents the effect of joint acceleration. In the second term, the Jacobian matrix, $\partial \boldsymbol{r}/\partial \boldsymbol{\theta} \triangleq \boldsymbol{J}$, is differentiated. From Eq. 2.13, we have

$$\frac{d}{dt}\left(\frac{\partial \boldsymbol{r}}{\partial \boldsymbol{\theta}}\right) = \frac{d\boldsymbol{J}}{dt}$$

$$= \sum_{i=1}^{2} \frac{\partial \boldsymbol{J}}{\partial \theta_i} \frac{d\theta_i}{dt} \qquad 2.17$$

Since the Jacobian matrix is given as the coefficient of Eq. 2.11, Eq. 2.17 is computed as follows:

$$\begin{aligned}\frac{d\boldsymbol{J}}{dt} &= \frac{\partial \boldsymbol{J}}{\partial \theta_1}\dot{\theta}_1 + \frac{\partial \boldsymbol{J}}{\partial \theta_2}\dot{\theta}_2 \\ &= \begin{pmatrix} -l_1\cos\theta_1 - l_2\cos(\theta_1+\theta_2) & -l_2\cos(\theta_1+\theta_2) \\ -l_1\sin\theta_1 - l_2\sin(\theta_1+\theta_2) & -l_2\sin(\theta_1+\theta_2) \end{pmatrix}\dot{\theta}_1 \\ &\quad + \begin{pmatrix} -l_2\cos(\theta_1+\theta_2) & -l_2\cos(\theta_1+\theta_2) \\ -l_2\sin(\theta_1+\theta_2) & -l_2\sin(\theta_1+\theta_2) \end{pmatrix}\dot{\theta}_2 \end{aligned} \qquad 2.18$$

2.2 Derivatives of Vectors, Matrices, and their Functions

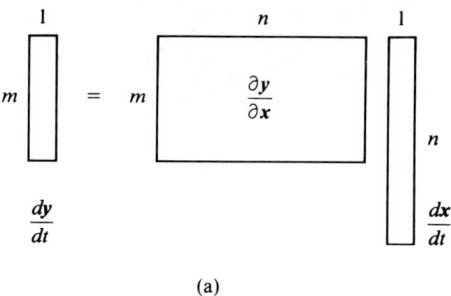

(a)

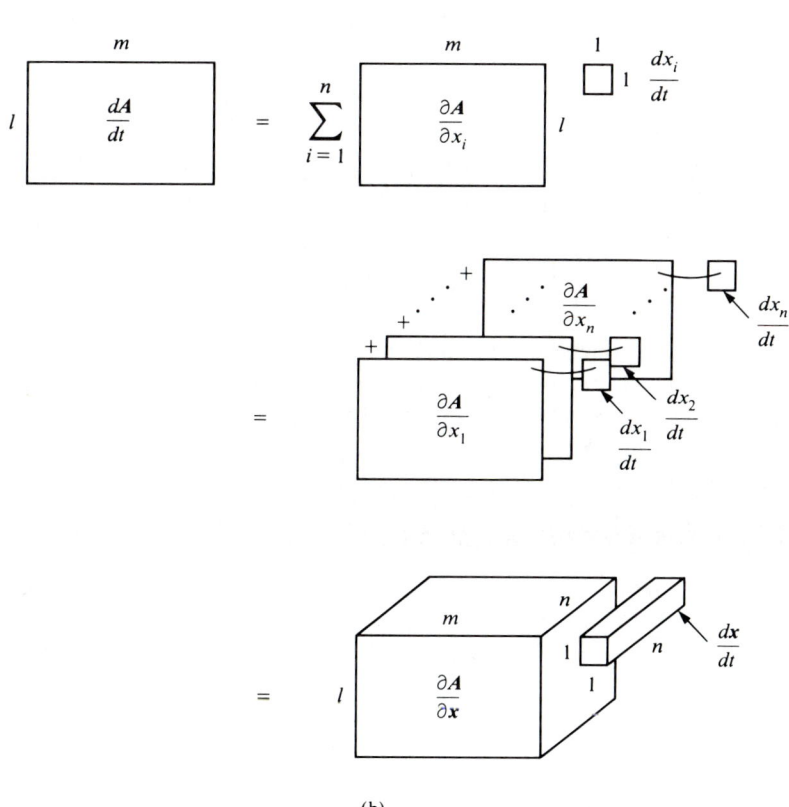

(b)

Figure 2.2 Structure of Eqs. (a) 2.9 and (b) 2.15.

From Eq. 2.15, Eq. 2.18 is now represented as

$$\frac{d\boldsymbol{J}}{dt} = \frac{\partial \boldsymbol{J}}{\partial \boldsymbol{\theta}}\frac{d\boldsymbol{\theta}}{dt} \qquad 2.19$$

Consequently, the second term of Eq. 2.16 is obtained as follows:

$$\frac{d}{dt}\left(\frac{\partial \boldsymbol{r}}{\partial \boldsymbol{\theta}}\right)\frac{d\boldsymbol{\theta}}{dt}$$

$$= \left(\frac{\partial \boldsymbol{J}}{\partial \theta_1}\dot{\theta}_1 + \frac{\partial \boldsymbol{J}}{\partial \theta_2}\dot{\theta}_2\right)\frac{d\boldsymbol{\theta}}{dt}$$

$$= \dot{\theta}_1 \frac{\partial \boldsymbol{J}}{\partial \theta_1}\frac{d\boldsymbol{\theta}}{dt} + \dot{\theta}_2 \frac{\partial \boldsymbol{J}}{\partial \theta_2}\frac{d\boldsymbol{\theta}}{dt}$$

$$= \dot{\theta}_1 \begin{pmatrix} -l_1\cos\theta_1 - l_2\cos(\theta_1+\theta_2) & -l_2\cos(\theta_1+\theta_2) \\ -l_1\sin\theta_1 - l_2\sin(\theta_1+\theta_2) & -l_2\sin(\theta_1+\theta_2) \end{pmatrix} \begin{pmatrix} \dot{\theta}_1 \\ \dot{\theta}_2 \end{pmatrix}$$

$$+ \dot{\theta}_2 \begin{pmatrix} -l_2\cos(\theta_1+\theta_2) & -l_2\cos(\theta_1+\theta_2) \\ -l_2\sin(\theta_1+\theta_2) & -l_2\sin(\theta_1+\theta_2) \end{pmatrix} \begin{pmatrix} \dot{\theta}_1 \\ \dot{\theta}_2 \end{pmatrix}$$

$$= \begin{pmatrix} -l_1\cos\theta_1 - l_2\cos(\theta_1+\theta_2) \\ -l_1\sin\theta_1 - l_2\sin(\theta_1+\theta_2) \end{pmatrix} \dot{\theta}_1^2$$

$$+ \begin{pmatrix} -l_2\cos(\theta_1+\theta_2) \\ -l_2\sin(\theta_1+\theta_2) \end{pmatrix} (2\dot{\theta}_1\dot{\theta}_2 + \dot{\theta}_2^2) \qquad 2.20$$

Therefore, we conclude that the second term of Eq. 2.16 implies the effect of joint velocity to the end-effector acceleration. Note that the effect appears as quadratic functions of joint velocity.

2.2.3 Derivatives of Functions of Vectors and Matrices

Derivatives of Quadratic Functions. Let q be a quadratic function of the component of $\boldsymbol{x} \in R^n$ such that

$$q = \sum_{i=1}^{n}\sum_{j=1}^{n} a_{ij} x_i x_j$$

$$= \boldsymbol{x}^T \boldsymbol{A} \boldsymbol{x} \qquad 2.21$$

where $\boldsymbol{A} \in R^{n \times n}$ is a coefficient matrix having a_{ij} as the (i,j) entry. Without loss of generality, we can assume that $\boldsymbol{A}$ is symmetric, because if $\boldsymbol{A}$ is not symmetric in Eq. 2.21, we can represent the same quadratic function by using a symmetric matrix as follows:

$$q = \frac{1}{2}(\boldsymbol{x}^T \boldsymbol{A} \boldsymbol{x} + \boldsymbol{x}^T \boldsymbol{A}^T \boldsymbol{x})$$

$$= \boldsymbol{x}^T \boldsymbol{B} \boldsymbol{x} \qquad 2.22$$

2.2 Derivatives of Vectors, Matrices, and their Functions

$$B = \frac{1}{2}(A + A^T) \in R^{n \times n} \qquad 2.23$$

where B is a symmetric matrix. Note that, to obtain Eq. 2.22, we used

$$x^T A^T x = (x^T A x)^T$$
$$= x^T A x \qquad 2.24$$

because $x^T A x$ is a scalar.

The derivative of q with respect to x is represented from Eqs. 2.5 and 2.21 as follows:

$$\frac{\partial q}{\partial x} = \left(\frac{\partial q}{\partial x_1} \; \frac{\partial q}{\partial x_2} \; \cdots \; \frac{\partial q}{\partial x_n} \right) \qquad 2.25$$

$$\frac{\partial q}{\partial x_k} = \sum_{j=1}^{n} a_{kj} x_j + \sum_{i=1}^{n} a_{ik} x_i \qquad 2.26$$

Equations 2.25 and 2.26 are summarized in the following matrix form:

$$\frac{\partial q}{\partial x} = x^T (A^T + A) \qquad 2.27$$

Equation 2.27 is used for the case where A is not symmetric. If A is symmetric, we have

$$\frac{\partial q}{\partial x} = 2 x^T A \qquad 2.28$$

Example 2.3

For an n-DOF manipulator, the total kinetic energy of links is represented by the quadratic form, as follows:

$$T = \frac{1}{2} \dot{\theta}^T A(\theta) \dot{\theta} \qquad 2.29$$

where $\theta \in R^n$, $A(\theta) \in R^{n \times n}$, and $A(\theta)$ is symmetric. If the potential energy U is constant with respect to θ, as in the case of a manipulator constrained in a horizontal plane or zero-gravity environments, the dynamics of the manipulator are calculated by the following Lagrange equation:

$$\frac{d}{dt}\left(\frac{\partial L}{\partial \dot{\theta}}\right) - \frac{\partial L}{\partial \theta} = \frac{d}{dt}\left(\frac{\partial T}{\partial \dot{\theta}}\right) - \frac{\partial T}{\partial \theta} = \tau^T \qquad 2.30$$

where $\tau \in R^n$ is the joint torque vector. The first term of the left-hand side of Eq. 2.30 is computed from Eqs. 2.28 and 2.29 as follows:

$$\frac{d}{dt}\left(\frac{\partial T}{\partial \dot{\theta}}\right) = \frac{d}{dt}\{\dot{\theta}^T A(\theta)\}$$

$$= \ddot{\boldsymbol{\theta}}^T \boldsymbol{A}(\boldsymbol{\theta}) + \dot{\boldsymbol{\theta}}^T \frac{d\boldsymbol{A}(\boldsymbol{\theta})}{dt} \qquad 2.31$$

If $\boldsymbol{A}(\boldsymbol{\theta})$ is represented by

$$\boldsymbol{A}(\boldsymbol{\theta}) = (\boldsymbol{a}_1 \quad \boldsymbol{a}_2 \quad \cdots \quad \boldsymbol{a}_n) \qquad 2.32$$

where $\boldsymbol{a}_i$ is the ith column vector of $\boldsymbol{A}(\boldsymbol{\theta})$, then the second term of Eq. 2.31 becomes

$$\dot{\boldsymbol{\theta}}^T \frac{d\boldsymbol{A}(\boldsymbol{\theta})}{dt} = \dot{\boldsymbol{\theta}}^T \left(\frac{\partial \boldsymbol{a}_1}{\partial \boldsymbol{\theta}} \dot{\boldsymbol{\theta}} \quad \frac{\partial \boldsymbol{a}_2}{\partial \boldsymbol{\theta}} \dot{\boldsymbol{\theta}} \quad \cdots \quad \frac{\partial \boldsymbol{a}_n}{\partial \boldsymbol{\theta}} \dot{\boldsymbol{\theta}}\right)$$

$$= (\dot{\boldsymbol{\theta}}^T \frac{\partial \boldsymbol{a}_1}{\partial \boldsymbol{\theta}} \dot{\boldsymbol{\theta}} \quad \dot{\boldsymbol{\theta}}^T \frac{\partial \boldsymbol{a}_2}{\partial \boldsymbol{\theta}} \dot{\boldsymbol{\theta}} \quad \cdots \quad \dot{\boldsymbol{\theta}}^T \frac{\partial \boldsymbol{a}_n}{\partial \boldsymbol{\theta}} \dot{\boldsymbol{\theta}}) \qquad 2.33$$

The second term of Eq. 2.30 becomes

$$\frac{\partial T}{\partial \boldsymbol{\theta}} = \frac{1}{2} \frac{\partial}{\partial \boldsymbol{\theta}} \{\dot{\boldsymbol{\theta}}^T \boldsymbol{A}(\boldsymbol{\theta}) \dot{\boldsymbol{\theta}}\}$$

$$= \frac{1}{2}(\dot{\boldsymbol{\theta}}^T \frac{\partial \boldsymbol{A}(\boldsymbol{\theta})}{\partial \theta_1} \dot{\boldsymbol{\theta}} \quad \dot{\boldsymbol{\theta}}^T \frac{\partial \boldsymbol{A}(\boldsymbol{\theta})}{\partial \theta_2} \dot{\boldsymbol{\theta}} \quad \cdots \quad \dot{\boldsymbol{\theta}}^T \frac{\partial \boldsymbol{A}(\boldsymbol{\theta})}{\partial \theta_n} \dot{\boldsymbol{\theta}}) \qquad 2.34$$

Hence, from Eqs. 2.30, 2.31, 2.33, and 2.34, the dynamics of the manipulator become

$$\ddot{\boldsymbol{\theta}}^T \boldsymbol{A}(\boldsymbol{\theta}) + (\dot{\boldsymbol{\theta}}^T \frac{\partial \boldsymbol{a}_1}{\partial \boldsymbol{\theta}} \dot{\boldsymbol{\theta}} \quad \cdots \quad \dot{\boldsymbol{\theta}}^T \frac{\partial \boldsymbol{a}_n}{\partial \boldsymbol{\theta}} \dot{\boldsymbol{\theta}})$$

$$- \frac{1}{2}(\dot{\boldsymbol{\theta}}^T \frac{\partial \boldsymbol{A}(\boldsymbol{\theta})}{\partial \theta_1} \dot{\boldsymbol{\theta}} \quad \cdots \quad \dot{\boldsymbol{\theta}}^T \frac{\partial \boldsymbol{A}(\boldsymbol{\theta})}{\partial \theta_n} \dot{\boldsymbol{\theta}}) = \boldsymbol{\tau}^T \qquad 2.35$$

Equation 2.35 can be represented in the following form:

$$\boldsymbol{A}(\boldsymbol{\theta})\ddot{\boldsymbol{\theta}} + \boldsymbol{B}(\boldsymbol{\theta}, \dot{\boldsymbol{\theta}}) = \boldsymbol{\tau} \qquad 2.36$$

$$\boldsymbol{B}(\boldsymbol{\theta}, \dot{\boldsymbol{\theta}}) = \text{col}\{\dot{\boldsymbol{\theta}}^T (\frac{\partial \boldsymbol{a}_i}{\partial \boldsymbol{\theta}} - \frac{1}{2} \frac{\partial \boldsymbol{A}}{\partial \theta_i})^T \dot{\boldsymbol{\theta}}\} \qquad 2.37$$

where col$\{*\}$ represents a column vector whose ith entry is $*$. Equations 2.36 and 2.37 offer important structural information about the dynamics of manipulators. In Eq. 2.36, $\boldsymbol{A}(\boldsymbol{\theta})$ is called the *inertia matrix*. The inertia matrix is necessarily symmetric, because it is derived from the quadratic function of Eq. 2.29. Equation 2.37 implies that each component of vector $\boldsymbol{B}(\boldsymbol{\theta}, \dot{\boldsymbol{\theta}})$ is a quadratic function of $\dot{\boldsymbol{\theta}}$. In each component, the terms of $\dot{\theta}_i^2$ ($i = 1, \cdots, n$) are the effect of the centrifugal forces, and the terms of

$\dot{\theta}_i\dot{\theta}_j$ $(i \neq j)$ are the effect of the Coriolis forces. Hence, $B(\theta,\dot{\theta})$ is called the *centrifugal* and *Coriolis torque*.

Positive Definiteness. The *positive definiteness* of matrices is an important concept to discuss in relation to the structural characteristics of mechanisms and to synthesize control schemes. The concept is formally defined as follows.

Definition 2.1

A matrix $A \in R^{n \times n}$ is said to be *positive definite* if and only if

$$x^T A x > 0 \qquad \text{for } x \neq O \qquad (2.38)$$

A is said to be *nonnegative definite (positive semidefinite)* if and only if

$$x^T A x \geq 0 \qquad \text{for } x \neq O \qquad (2.39)$$

A is said to be *negative definite* if and only if

$$x^T A x < 0 \qquad \text{for } x \neq O \qquad (2.40)$$

A is said to be *nonpositive definite (negative semidefinite)* if and only if

$$x^T A x \leq 0 \qquad \text{for } x \neq O \qquad (2.41)$$

The following two theorems are used to verify whether or not a given symmetric matrix A is positive definite (Kodama and Suda 1978).

Theorem 2.1

A symmetric matrix $A \in R^{n \times n}$ is positive (nonnegative) definite if and only if all the eigenvalues of A are positive (nonnegative).

Example 2.4

A^1, A^2, and $A^3 \in R^{2 \times 2}$ are the following three symmetric matrices:

$$A^1 = \begin{pmatrix} 1 & 1 \\ 1 & 4 \end{pmatrix}$$
$$A^2 = \begin{pmatrix} 1 & 2 \\ 2 & 4 \end{pmatrix} \qquad (2.42)$$
$$A^3 = \begin{pmatrix} 1 & 3 \\ 3 & 4 \end{pmatrix}$$

The eigenvalues of the matrices are

$$\begin{aligned}\boldsymbol{A}^1 &\Rightarrow \lambda = 0.6972,\ 4.3028\\ \boldsymbol{A}^2 &\Rightarrow \lambda = 0,\ 5\\ \boldsymbol{A}^3 &\Rightarrow \lambda = -0.8541,\ 5.8541\end{aligned} \qquad 2.43$$

From Theorem 2.1, it turns out that $\boldsymbol{A}^1$ and $\boldsymbol{A}^2$ are positive definite and nonnegative definite, respectively.

Let $\boldsymbol{x}^1 \triangleq (2\ -1)^T$ and $\boldsymbol{x}^2 \triangleq (0\ -1)^T$. The following equation shows that $\boldsymbol{x}^1$ makes $\boldsymbol{x}^{1^T}\boldsymbol{A}^2\boldsymbol{x}^1 = 0$:

$$\begin{aligned}\boldsymbol{x}^{1^T}\boldsymbol{A}^2\boldsymbol{x}^1 &= (2\ -1)\begin{pmatrix}1 & 2\\ 2 & 4\end{pmatrix}\begin{pmatrix}2\\ -1\end{pmatrix}\\ &= 0\end{aligned} \qquad 2.44$$

The following equations show that $\boldsymbol{A}^3$ is not any one of positive, nonnegative, negative, or nonpositive definite:

$$\begin{aligned}\boldsymbol{x}^{1^T}\boldsymbol{A}^3\boldsymbol{x}^1 &= (2\ -1)\begin{pmatrix}1 & 3\\ 3 & 4\end{pmatrix}\begin{pmatrix}2\\ -1\end{pmatrix}\\ &= -4\\ \boldsymbol{x}^{2^T}\boldsymbol{A}^3\boldsymbol{x}^2 &= (0\ -1)\begin{pmatrix}1 & 3\\ 3 & 4\end{pmatrix}\begin{pmatrix}0\\ -1\end{pmatrix}\\ &= 4\end{aligned} \qquad 2.45$$

We define $\boldsymbol{A}_i$ $(i = 1, \cdots, n)$ as the $i \times i$ square matrix consisting of the first i rows and i columns of $\boldsymbol{A}$. The determinant of $\boldsymbol{A}_i$ is called a *leading principal minor*.

Theorem 2.2 (Sylvester's Criterion)

A symmetric matrix $\boldsymbol{A} \in R^{n \times n}$ is positive definite if and only if all the leading principal minors are positive. Namely,

$$\det \boldsymbol{A}_i > 0 \qquad \text{for } i = 1, \cdots, n \qquad 2.46$$

Example 2.5

For $\boldsymbol{A}^1$, $\boldsymbol{A}^2$, and $\boldsymbol{A}^3$ as defined in Example 2.4, the leading principal minors are as follows:

$$\boldsymbol{A}^1 \Rightarrow \begin{cases}\det \boldsymbol{A}_1^1 = \det(1) = 1 > 0\\ \det \boldsymbol{A}_2^1 = \det\begin{pmatrix}1 & 1\\ 1 & 4\end{pmatrix} = 3 > 0\end{cases} \qquad 2.47$$

2.2 Derivatives of Vectors, Matrices, and their Functions

$$A^2 \implies \begin{cases} \det A_1^2 = \det(1) = 1 > 0 \\ \det A_2^2 = \det \begin{pmatrix} 1 & 2 \\ 2 & 4 \end{pmatrix} = 0 \end{cases} \qquad 2.48$$

$$A^3 \implies \begin{cases} \det A_1^3 = \det(1) = 1 > 0 \\ \det A_2^3 = \det \begin{pmatrix} 1 & 3 \\ 3 & 4 \end{pmatrix} = -5 < 0 \end{cases} \qquad 2.49$$

Theorem 2.2 and Eq. 2.47 verify that only A^1 is positive definite.

From Eq. 2.46 with $i = n$, it can be seen that a symmetric positive definite matrix is nonsingular. Equations 2.29 and 2.36 offer further structural information about the inertia matrix of a manipulator. The kinetic energy never becomes zero or negative for nonzero $\dot{\theta}$. Therefore, the inertia matrix is always symmetric and positive definite. Needless to say, the inertia matrix is nonsingular and, therefore, is invertible. The positive definiteness of the inertia matrix plays an important role in synthesizing various control schemes.

We have the following theorem for nonsymmetric matrices.

Theorem 2.3

A square nonsymmetric matrix $A \in R^{n \times n}$ is positive definite if and only if the symmetric matrix $(A + A^T)/2$ is positive definite.

By examining the positive definiteness of the symmetric matrix $(A + A^T)/2$ using either Theorem 2.1 or Theorem 2.2, we can verify the positive definiteness of nonsymmetric matrix A.

Example 2.6

Let A^1, A^2, and A^3 be the following nonsymmetric matrices:

$$\begin{aligned} A^1 &= \begin{pmatrix} 1 & 1 \\ 0 & 1 \end{pmatrix} \\ A^2 &= \begin{pmatrix} 1 & 2 \\ 0 & 1 \end{pmatrix} \\ A^3 &= \begin{pmatrix} 1 & 3 \\ 0 & 1 \end{pmatrix} \end{aligned} \qquad 2.50$$

For each A^i, $(A^i + A^{iT})/2$ becomes

$$(A^1 + A^{1T})/2 = \begin{pmatrix} 1 & 0.5 \\ 0.5 & 1 \end{pmatrix} = A^{1*}$$

$$(A^2 + A^{2T})/2 = \begin{pmatrix} 1 & 1 \\ 1 & 1 \end{pmatrix} = A^{2*} \qquad 2.51$$

$$(A^3 + A^{3T})/2 = \begin{pmatrix} 1 & 1.5 \\ 1.5 & 1 \end{pmatrix} = A^{3*}$$

The leading principal minors of A^{i*} are computed as follows:

$$A^{1*} \implies \begin{cases} \det A^{1*}_1 = \det(1) = 1 > 0 \\ \det A^{1*}_2 = \det \begin{pmatrix} 1 & 0.5 \\ 0.5 & 1 \end{pmatrix} = 0.75 > 0 \end{cases} \qquad 2.52$$

$$A^{2*} \implies \begin{cases} \det A^{2*}_1 = \det(1) = 1 > 0 \\ \det A^{2*}_2 = \det \begin{pmatrix} 1 & 1 \\ 1 & 1 \end{pmatrix} = 0 \end{cases} \qquad 2.53$$

$$A^{3*} \implies \begin{cases} \det A^{3*}_1 = \det(1) = 1 > 0 \\ \det A^{3*}_2 = \det \begin{pmatrix} 1 & 1.5 \\ 1.5 & 1 \end{pmatrix} = -1.25 < 0 \end{cases} \qquad 2.54$$

From Theorem 2.2, only A^{1*} is positive definite. Therefore, using Theorem 2.3, we can conclude that, among the three nonsymmetric matrices, only A^1 is positive definite.

Derivatives of Determinants. The *determinant* of a square matrix is an important scalar characteristic of the matrix. In robotics, as we shall discuss in Chapter 3, the determinant is often used to evaluate the kinematic characteristics. Here, we examine the derivative of the determinants of square matrices.

Let all the entries of a square matrix $A \in R^{n \times n}$ be represented as functions of a parameter t. That is, $A = A(t)$. The derivative of the determinant with respect to t is obtained by

$$\frac{d}{dt} \det A(t) = \sum_{i=1}^{n} \sum_{j=1}^{n} \frac{\partial}{\partial a_{ij}} \{\det A(t)\} \frac{d\, a_{ij}(t)}{dt} \qquad 2.55$$

where a_{ij} is the (i, j) entry of A.

The cofactor expansion of $\det A$ about the ith row becomes

$$\det A = \sum_{j=1}^{n} a_{ij} \Delta_{ij} \qquad 2.56$$

where Δ_{ij} is the cofactor of $\boldsymbol{A}$. Using an $(n-1) \times (n-1)$ square matrix $\boldsymbol{A}_{ij}$ obtained by taking out the ith row and jth column of $\boldsymbol{A}$, we represent the cofactor by

$$\Delta_{ij} = (-1)^{i+j} \det \boldsymbol{A}_{ij} \qquad 2.57$$

The $n \times n$ square matrix that has Δ_{ij} as the (j, i) entry is called the *adjoint matrix* of $\boldsymbol{A}$. Namely,

$$\operatorname{adj} \boldsymbol{A} \triangleq \begin{pmatrix} \Delta_{11} & \Delta_{21} & \cdots & \Delta_{n1} \\ \Delta_{12} & \Delta_{22} & \cdots & \Delta_{n2} \\ \vdots & \vdots & \ddots & \vdots \\ \Delta_{1n} & \Delta_{2n} & \cdots & \Delta_{nn} \end{pmatrix} \qquad 2.58$$

If $\det \boldsymbol{A} \neq 0$, the inverse matrix of $\boldsymbol{A}$ is given by

$$\boldsymbol{A}^{-1} = \frac{1}{\det \boldsymbol{A}} \operatorname{adj} \boldsymbol{A} \qquad 2.59$$

Since, in Eq. 2.56, a_{ij} does not appear explicitly in Δ_{ij}, Eq. 2.55 can be calculated by

$$\frac{d}{dt} \det \boldsymbol{A}(t) = \sum_{i=1}^{n} \sum_{j=1}^{n} \Delta_{ij} \frac{d\, a_{ij}(t)}{dt}$$

$$= \operatorname{trace} \left\{ \frac{d\, \boldsymbol{A}(t)}{dt} \operatorname{adj} \boldsymbol{A}(t) \right\} \qquad 2.60$$

where trace $\boldsymbol{M}$ is the trace of square matrix $\boldsymbol{M} \in R^{n \times n}$ defined by

$$\operatorname{trace} \boldsymbol{M} \triangleq \sum_{i=1}^{n} M_{ii} \qquad 2.61$$

where M_{ij} is the (i, j) entry of $\boldsymbol{M}$.

Example 2.7

Let's examine a specific example of Eq. 2.60. Let $\boldsymbol{A}(t)$ be as follows:

$$\boldsymbol{A}(t) = \begin{pmatrix} 1 & t \\ 1-t & 2t \end{pmatrix} \qquad 2.62$$

Then, $\det \boldsymbol{A}(t)$, $d\boldsymbol{A}(t)/dt$, and $\operatorname{adj} \boldsymbol{A}(t)$ are obtained as follows:

$$\det \boldsymbol{A}(t) = \det \begin{pmatrix} 1 & t \\ 1-t & 2t \end{pmatrix} = t^2 + t$$

$$\frac{d\boldsymbol{A}(t)}{dt} = \frac{d}{dt} \begin{pmatrix} 1 & t \\ 1-t & 2t \end{pmatrix} = \begin{pmatrix} 0 & 1 \\ -1 & 2 \end{pmatrix} \qquad 2.63$$

$$\operatorname{adj} \boldsymbol{A}(t) = \begin{pmatrix} 2t & -t \\ -1+t & 1 \end{pmatrix}$$

Hence, Eq. 2.60 is verified as follows:

$$\operatorname{trace}\{\frac{d\boldsymbol{A}(t)}{dt}\operatorname{adj}\boldsymbol{A}\}$$

$$= \operatorname{trace}\{\begin{pmatrix} 0 & 1 \\ -1 & 2 \end{pmatrix}\begin{pmatrix} 2t & -t \\ -1+t & 1 \end{pmatrix}\}$$

$$= \operatorname{trace}\begin{pmatrix} t-1 & 1 \\ -2 & t+2 \end{pmatrix}$$

$$= (t-1) + (t+2)$$

$$= 2t + 1$$

$$= \frac{d}{dt}\det \boldsymbol{A}(t) \qquad\qquad 2.64$$

2.3 Singular Value Decomposition

2.3.1 *Concept*

The *singular value decomposition (SVD)* is the most significant decomposition of matrices. The characteristics of matrices are examined from various viewpoints using SVD. SVD is defined by the following theorem (Kodama and Suda 1978).

Theorem 2.4 (Singular Value Decomposition)

If $\boldsymbol{A} \in R^{m \times n}$ and rank $\boldsymbol{A} = k$, then there exist orthogonal matrices

$$\boldsymbol{U} = (\boldsymbol{u}_1 \quad \cdots \quad \boldsymbol{u}_m) \in R^{m \times m} \qquad\qquad 2.65$$

$$\boldsymbol{V} = (\boldsymbol{v}_1 \quad \cdots \quad \boldsymbol{v}_n) \in R^{n \times n} \qquad\qquad 2.66$$

such that $\boldsymbol{A}$ is represented by

$$\boldsymbol{A} = \boldsymbol{U}\boldsymbol{\Sigma}\boldsymbol{V}^T \qquad\qquad 2.67$$

$$\boldsymbol{\Sigma} \triangleq \operatorname{diag}(\sigma_1, \cdots, \sigma_p) \in R^{m \times n}$$

$$p = \min\{m, n\}$$

$$\sigma_1 \geq \sigma_2 \geq \cdots \geq \sigma_k > 0$$

$$\sigma_{k+1} = \cdots = \sigma_p = 0$$

2.3 Singular Value Decomposition

In Eq. 2.67, the σ_i ($i = 1, \cdots, p$) are called the *singular values* of $\boldsymbol{A}$. Particularly, σ_1 and σ_p are referred to as the largest and smallest singular values respectively. Note that the singular values are uniquely determined, although $\boldsymbol{U}$ and $\boldsymbol{V}$ may not be. The computational algorithm of SVD can be found in Golub and Van Loan (1983). The FORTRAN and Pascal codings of the algorithm are given in Press, Flannery, Teukolsky, and Vetterling (1986).

Example 2.8

Let's examine two examples of SVD. $\boldsymbol{A}^1$ and $\boldsymbol{A}^2$ are the following matrices:

$$\boldsymbol{A}^1 = \begin{pmatrix} 1 & 2 & 0 \\ 0 & 1 & 1 \end{pmatrix}$$

$$\boldsymbol{A}^2 = \begin{pmatrix} 1 & 2 & 0 \\ 0 & 0.1 & 0.1 \end{pmatrix} \tag{2.68}$$

The SVD of each matrix is computed as follows:

$$\boldsymbol{A}^1 = \boldsymbol{U}^1 \boldsymbol{\Sigma}^1 \boldsymbol{V}^{1T}$$

$$= \begin{pmatrix} 0.8944 & -0.4472 \\ 0.4472 & 0.8944 \end{pmatrix} \begin{pmatrix} 2.4495 & 0 & 0 \\ 0 & 1.0000 & 0 \end{pmatrix}$$

$$\cdot \begin{pmatrix} 0.3651 & -0.4472 & -0.8165 \\ 0.9129 & 0.0000 & 0.4082 \\ 0.1826 & 0.8944 & -0.4082 \end{pmatrix}^T$$

$$\boldsymbol{A}^2 = \boldsymbol{U}^2 \boldsymbol{\Sigma}^2 \boldsymbol{V}^{2T}$$

$$= \begin{pmatrix} 0.9992 & -0.0401 \\ 0.0401 & 0.9992 \end{pmatrix} \begin{pmatrix} 2.2379 & 0 & 0 \\ 0 & 0.1095 & 0 \end{pmatrix}$$

$$\cdot \begin{pmatrix} 0.4465 & -0.3660 & -0.8165 \\ 0.8948 & 0.1808 & 0.4082 \\ 0.0018 & 0.9129 & -0.4082 \end{pmatrix}^T \tag{2.69}$$

Note that $\boldsymbol{U}$ and $\boldsymbol{V}$ of SVD are orthogonal matrices. Indeed, for $\boldsymbol{A}^1$,

$$\boldsymbol{U}^1 \boldsymbol{U}^{1T} = \begin{pmatrix} 0.8944 & -0.4472 \\ 0.4472 & 0.8944 \end{pmatrix} \begin{pmatrix} 0.8944 & -0.4472 \\ 0.4472 & 0.8944 \end{pmatrix}^T$$

$$= \begin{pmatrix} 1 & 0 \\ 0 & 1 \end{pmatrix}$$

$$\boldsymbol{V}^1 \boldsymbol{V}^{1T} = \begin{pmatrix} 0.3651 & -0.4472 & -0.8165 \\ 0.9129 & 0.0000 & 0.4082 \\ 0.1826 & 0.8944 & -0.4082 \end{pmatrix}$$

$$\cdot \begin{pmatrix} 0.3651 & -0.4472 & -0.8165 \\ 0.9129 & 0.0000 & 0.4082 \\ 0.1826 & 0.8944 & -0.4082 \end{pmatrix}^T$$

$$= \begin{pmatrix} 1 & 0 & 0 \\ 0 & 1 & 0 \\ 0 & 0 & 1 \end{pmatrix}$$

2.3.2 Properties

Condition Number. The *condition number* κ denotes the ratio of the largest and the smallest singular values; namely,

$$\kappa = \frac{\sigma_1}{\sigma_p} \qquad (\kappa \geq 1) \qquad\qquad 2.70$$

A matrix with a small condition number is said to be *well conditioned*, whereas a matrix with a large condition number is said to be *ill conditioned*. These terms are used because the numerical computation of a linear equation with an ill-conditioned coefficient matrix may involve large computational errors, as we shall explain in detail in Section 2.3.3.

Example 2.9

The condition numbers κ_1 and κ_2 of $\boldsymbol{A}^1$ and $\boldsymbol{A}^2$ in Example 2.8 are obtained as follows:

$$\kappa_1 = \frac{2.4495}{1.0000} = 2.4495$$
$$\kappa_2 = \frac{2.2379}{0.1095} = 20.437 \qquad\qquad 2.71$$

Accordingly, we observe that $\boldsymbol{A}^1$ is better conditioned than is $\boldsymbol{A}^2$.

Eigenvalues. The set of nonzero singular values of $\boldsymbol{A} \in R^{m \times n}$ is equivalent to the set of the positive square roots of nonzero eigenvalues of $\boldsymbol{A}^T \boldsymbol{A} \in R^{n \times n}$, and also is equal to that of $\boldsymbol{A}\boldsymbol{A}^T \in R^{m \times m}$. Note that, for arbitrary matrices $\boldsymbol{A} \in R^{m \times n}$ and $\boldsymbol{B} \in R^{n \times m}$, the two sets of nonzero eigenvalues of their products $\boldsymbol{A}\boldsymbol{B} \in R^{m \times m}$ and $\boldsymbol{B}\boldsymbol{A} \in R^{n \times n}$ coincide with each other. Furthermore, for square symmetric matrices, there are the following two results: (1) the set of singular values of a symmetric matrix is equivalent to the set of absolute values of that matrix's eigenvalues, and (2) the set of singular values of a symmetric positive or nonnegative definite matrix is equivalent to the set of that matrix's eigenvalues.

Example 2.10

For A^1 as given by Eq. 2.68, $A^{1^T}A^1$ and $A^1 A^{1^T}$ are computed as follows:

$$A^{1^T} A^1 = \begin{pmatrix} 1 & 0 \\ 2 & 1 \\ 0 & 1 \end{pmatrix} \begin{pmatrix} 1 & 2 & 0 \\ 0 & 1 & 1 \end{pmatrix}$$

$$= \begin{pmatrix} 1 & 2 & 0 \\ 2 & 5 & 1 \\ 0 & 1 & 1 \end{pmatrix} \qquad 2.72$$

$$A^1 A^{1^T} = \begin{pmatrix} 1 & 2 & 0 \\ 0 & 1 & 1 \end{pmatrix} \begin{pmatrix} 1 & 0 \\ 2 & 1 \\ 0 & 1 \end{pmatrix}$$

$$= \begin{pmatrix} 5 & 2 \\ 2 & 2 \end{pmatrix}$$

The eigenvalues of $A^{1^T}A^1$ and $A^1 A^{1^T}$ become

$$\begin{aligned} A^{1^T} A^1 &\Rightarrow \lambda = 0,\ 1,\ 6 \\ A^1 A^{1^T} &\Rightarrow \lambda = 1,\ 6 \end{aligned} \qquad 2.73$$

Note that $6 = (2.4495)^2$. Hence, the positive square roots of nonzero eigenvalues coincide with the singular values of A^1—namely, 1.0000 and 2.4495.

Determinant. The product of singular values of a square matrix $A \in R^{n \times n}$ is equal to the absolute value of that matrix's determinant. This result can be shown readily using Eq. 2.67 and $\det U = \pm 1$, $\det V^T = \pm 1$ as follows:

$$\begin{aligned} \det A &= \det U \det \Sigma \det V^T \\ &= \pm \det \Sigma \\ &= \pm \prod_{i=1}^{n} \sigma_i \end{aligned} \qquad 2.74$$

Example 2.11

Let $A \in R^{3 \times 3}$ be given by

$$A = \begin{pmatrix} 1 & 1 & 0 \\ 0 & 3 & 2 \\ 0 & 1 & 1 \end{pmatrix} \qquad 2.75$$

The SVD of A is computed as

$$A = U\Sigma V^T$$
$$= \begin{pmatrix} 0.2253 & -0.9674 & 0.1156 \\ 0.9095 & 0.1662 & -0.3810 \\ 0.3493 & 0.1910 & 0.9173 \end{pmatrix} \begin{pmatrix} 3.9577 & 0 & 0 \\ 0 & 1.1345 & 0 \\ 0 & 0 & 0.2227 \end{pmatrix}$$
$$\cdot \begin{pmatrix} 0.0569 & -0.8527 & 0.5192 \\ 0.8346 & -0.2448 & -0.4935 \\ 0.5479 & 0.4614 & 0.6978 \end{pmatrix}^T \qquad 2.76$$

Therefore,

$$\prod_{i=1}^{3} \sigma_i = \sigma_1 \sigma_2 \sigma_3 = 3.9577 \cdot 1.1345 \cdot 0.2227 = 1$$
$$\det A = 1 \cdot 3 \cdot 1 + 1 \cdot 2 \cdot 0 + 0 \cdot 1 \cdot 0 \qquad 2.77$$
$$\quad - 1 \cdot 2 \cdot 1 - 1 \cdot 0 \cdot 1 - 0 \cdot 3 \cdot 0$$
$$= 1$$

Pseudoinverse. The *pseudoinverse* of $A \in R^{m \times n}$, rank $A = k$, will be formally defined in Section 2.4. Here, we see the relationship between the pseudoinverse and SVD. This relationship offers a computational scheme of pseudoinverses. When A is decomposed by SVD as in Eq. 2.67, its pseudoinverse $A^\# \in R^{n \times m}$ is represented by (Golub and Van Loan 1983)

$$A^\# = V\Sigma^\# U^T \qquad 2.78$$

$$\Sigma^\# \triangleq \text{diag}\,(\underbrace{\frac{1}{\sigma_1}, \frac{1}{\sigma_2}, \cdots, \frac{1}{\sigma_k}}_{p=\min\,(m,\,n)}, 0, \cdots, 0) \in R^{n \times m}$$

$$\sigma_1 \geq \sigma_2 \geq \cdots \geq \sigma_k > 0$$

Equation 2.78 implies that $1/\sigma_i$ ($i = 1, \cdots, k$) and $p - k$ zeros are the singular values of $A^\#$, among which $1/\sigma_k$ is the largest.

Example 2.12

A is the following 3×4 matrix:

$$A = \begin{pmatrix} 1 & 0 & -1 & -2 \\ 1 & 2 & 1 & 0 \\ 0 & 1 & 1 & 1 \end{pmatrix} \qquad 2.79$$

The SVD of A becomes

$$A = U\Sigma V^T$$
$$= \begin{pmatrix} -0.5774 & 0.7071 & -0.4082 \\ 0.5774 & 0.7071 & 0.4082 \\ 0.5774 & 0 & -0.8165 \end{pmatrix} \begin{pmatrix} 3.0000 & 0 & 0 & 0 \\ 0 & 2.4495 & 0 & 0 \\ 0 & 0 & 0 & 0 \end{pmatrix}$$
$$\cdot \begin{pmatrix} 0 & 0.5774 & 0 & -0.8165 \\ 0.5774 & 0.5774 & -0.4082 & 0.4082 \\ 0.5774 & 0 & 0.8165 & 0 \\ 0.5774 & -0.5774 & -0.4082 & -0.4082 \end{pmatrix}^T \qquad 2.80$$

Using U, Σ, and V of Eq. 2.80, we have

$$V\Sigma^{\#}U^T = \begin{pmatrix} 0 & 0.5774 & 0 & -0.8165 \\ 0.5774 & 0.5774 & -0.4082 & 0.4082 \\ 0.5774 & 0.0000 & 0.8165 & 0.0000 \\ 0.5774 & -0.5774 & -0.4082 & -0.4082 \end{pmatrix}$$
$$\cdot \begin{pmatrix} 0.3333 & 0 & 0 \\ 0 & 0.4082 & 0 \\ 0 & 0 & 0 \\ 0 & 0 & 0 \end{pmatrix}$$
$$\cdot \begin{pmatrix} -0.5774 & 0.7071 & -0.4082 \\ 0.5774 & 0.7071 & 0.4082 \\ 0.5774 & 0 & -0.8165 \end{pmatrix}^T$$
$$= \begin{pmatrix} 0.1667 & 0.1667 & 0 \\ 0.0556 & 0.2778 & 0.1111 \\ -0.1111 & 0.1111 & 0.1111 \\ -0.2778 & -0.0556 & 0.1111 \end{pmatrix} \qquad 2.81$$

On the other hand, the pseudoinverse of A and its SVD are directly computed as follows:

$$A^{\#} = \begin{pmatrix} 0.1667 & 0.1667 & 0 \\ 0.0556 & 0.2778 & 0.1111 \\ -0.1111 & 0.1111 & 0.1111 \\ -0.2778 & -0.0556 & 0.1111 \end{pmatrix}$$
$$= \begin{pmatrix} 0.5774 & 0.0000 & -0.7080 & 0.4067 \\ 0.5774 & 0.5774 & 0.5573 & 0.1506 \\ 0.0000 & 0.5774 & -0.4067 & -0.7080 \\ -0.5774 & 0.5774 & -0.1506 & 0.5573 \end{pmatrix}$$
$$\cdot \begin{pmatrix} 0.4082 & 0 & 0 \\ 0 & 0.3333 & 0 \\ 0 & 0 & 0 \\ 0 & 0 & 0 \end{pmatrix}$$

$$\begin{pmatrix} 0.7071 & -0.5774 & 0.4082 \\ 0.7071 & 0.5774 & -0.4082 \\ 0.0000 & 0.5774 & 0.8165 \end{pmatrix}^T \qquad 2.82$$

Comparing Eq. 2.81 with Eq. 2.82 verifies Eq. 2.78. Note that the right-hand side of Eq. 2.78 does not necessarily imply the SVD of $A^\#$, since the diagonal components of $\Sigma^\#$ may not be in the descending order.

2.3.3 Sensitivity Analysis

We can discuss the sensitivity analysis of a linear square equation

$$y = Ax \qquad 2.83$$

where $x, y \in R^n$ and $A \in R^{n \times n}$, using the singular values (Iri, Kodama, and Suda 1982). Assuming that A is nonsingular, the solution of Eq. 2.83 is computed by

$$\begin{aligned} x &= A^{-1} y \\ &= (U \Sigma V^T)^{-1} y \\ &= V \Sigma^{-1} U^T y \end{aligned} \qquad 2.84$$

Using $x^* \triangleq V^T x$ and $y^* \triangleq U^T y$, Eq. 2.84 is represented as follows:

$$x^* = \Sigma^{-1} y^* \qquad 2.85$$

Therefore, the magnification ratio $\| x^* \| / \| y^* \|$ is limited by the following inequality:

$$\frac{1}{\sigma_1} \leq \frac{\| x^* \|}{\| y^* \|} \leq \frac{1}{\sigma_n} \qquad 2.86$$

Since U and V are orthogonal matrices,

$$\| x^* \| = \sqrt{(V^T x)^T V^T x} = \| x \| \qquad 2.87$$

$$\| y^* \| = \sqrt{(U^T y)^T U^T y} = \| y \| \qquad 2.88$$

Hence, Eq. 2.86 is equivalent to

$$\frac{1}{\sigma_1} \leq \frac{\| x \|}{\| y \|} \leq \frac{1}{\sigma_n} \qquad 2.89$$

When y includes noise or roundoff error Δy, the solution error Δx is represented by

$$x + \Delta x = A^{-1}(y + \Delta y) \qquad 2.90$$

Subtracting Eq. 2.84 from Eq. 2.90, we have

$$\Delta x = A^{-1} \Delta y \qquad 2.91$$

Accordingly, the error magnification ratio becomes

$$\frac{1}{\sigma_1} \leq \frac{\|\Delta x\|}{\|\Delta y\|} \leq \frac{1}{\sigma_n} \qquad 2.92$$

Equation 2.92 implies that the magnification ratio of the absolute error is not smaller than the reciprocal of the largest singular value, and is not larger than the reciprocal of the smallest singular value. Hence, if the smallest singular value of A is too small, the solution may include an extremely magnified error due to the noise or roundoff error of y.

The magnification ratio of relative errors of x and y is defined by

$$\frac{\|\Delta x\| / \|x\|}{\|\Delta y\| / \|y\|} \qquad 2.93$$

From Eqs. 2.89 and 2.92, the magnification ratio of relative error is limited by the following inequality:

$$\kappa^{-1} \leq \frac{\|\Delta x\| / \|x\|}{\|\Delta y\| / \|y\|} \leq \kappa \qquad 2.94$$

Equation 2.94 implies that the magnification ratio of relative error is not smaller than the reciprocal of the condition number, and is not larger than the condition number. The solution of a linear equation that has a coefficient matrix with a large condition number may involve an extremely magnified relative error due to signal noise or computational roundoff error. This is why such a matrix is called *ill conditioned*.

Example 2.13

Matrix A is equal to A^1 of Eq. 2.68 used in Example 2.8. The condition number of A^1 was computed in Example 2.9—namely, $\kappa_1 = 2.4495$. In this case, from Eq. 2.94, the magnification ratio of relative error in solving Eq. 2.83 is limited by the following equation:

$$0.4082 \leq \frac{\|\Delta x\| / \|x\|}{\|\Delta y\| / \|y\|} \leq 2.4495 \qquad 2.95$$

On the other hand, when A is equal to A^2 of Eq. 2.68 in Example 2.8, using $\kappa_2 = 20.437$ computed in Example 2.9, we can show that the magnification ratio of relative error is included in the following range:

$$0.0489 \leq \frac{\|\Delta x\| / \|x\|}{\|\Delta y\| / \|y\|} \leq 20.4374 \qquad 2.96$$

Therefore, the linear algebraic equation with A^2 as a coefficient possibly involves more computational error than does one with A^1.

2.3.4 Elliptic Geometry

The SVD has a clear geometric interpretation, which motivates various geometric optimizations using SVD. Consider a linear equation represented by

$$y = Ax \qquad 2.97$$

where $y \in R^m$, $x \in R^n$, $A \in R^{m \times n}$, and rank $A = k$. Transforming Eq. 2.97 by means of x^* and y^* defined for Eq. 2.85, we have

$$y^* = \Sigma x^* = \begin{pmatrix} \sigma_1 x_1^* \\ \vdots \\ \sigma_k x_k^* \\ 0 \\ \vdots \\ 0 \end{pmatrix} \qquad 2.98$$

Note that the transformations from x to x^* and from y to y^* imply the rotation of the coordinate frames and changes only the direction of x and y, since U and V are orthogonal matrices. From Eq. 2.98, we have

$$x_i^* = \frac{1}{\sigma_i} y_i^* \qquad (i = 1, \cdots, k) \qquad 2.99$$

When x^* is constrained by $\|x^*\| \leq 1$, from Eq. 2.99, y^* satisfies the following inequality:

$$\frac{y_1^{*2}}{\sigma_1^2} + \frac{y_2^{*2}}{\sigma_2^2} + \cdots + \frac{y_k^{*2}}{\sigma_k^2} \leq 1 \qquad 2.100$$

We can see that, when x_i^* takes all the values in a unit sphere of n-dimensional space, y^* takes all the values in the hyperellipsoid defined by Eq. 2.100 within a k-dimensional subspace of the m-dimensional space. Singular values are precisely the length of semiaxes of the ellipsoid. $\|x^*\| \leq 1$ is equivalent to $\|x\| \leq 1$ and Eq. 2.100 becomes a rotated ellipsoid in y space. The rotation is determined by the orthogonal matrix U. Note that, from $y = Uy^*$ and $x = Vx^*$, the direction of the ith longest principal axis in y space is that of the ith column vector of U, and y lies in the ith longest principal axis when

2.3 Singular Value Decomposition

$\boldsymbol{x}$ lies in the direction of the ith column vector of $\boldsymbol{V}$. Figure 2.3 shows the elliptic geometry of Eqs. 2.97 and 2.98.

It is well known that the m-dimensional volume of the hyperellipsoid of Eq. 2.100 is computed by

$$\text{volume} = \frac{\pi^{m/2}}{\Gamma(1+m/2)} \prod_{i=1}^{m} \sigma_i \qquad 2.101$$

where $\Gamma(*)$ is the gamma function. Equation 2.101 implies that volume of the ellipsoid is proportional to the product of all the singular values, and that it becomes zero if rank $\boldsymbol{A} = k < \min(m, n)$.

Example 2.14 (Manipulability Ellipsoid)

We revisit the planar 2-DOF manipulator shown in Fig. 2.1. When $l_1 = l_2 = 1$ and $\theta_1 = \theta_2 = 30°$, the Jacobian matrix obtained in Eq. 2.11 becomes

$$\frac{\partial \boldsymbol{r}}{\partial \boldsymbol{\theta}} = \begin{pmatrix} -1.3660 & -0.8660 \\ 1.3660 & 0.5000 \end{pmatrix} \qquad 2.102$$

The SVD of the Jacobian matrix is computed as follows:

$$\frac{\partial \boldsymbol{r}}{\partial \boldsymbol{\theta}} = \begin{pmatrix} -0.7443 & -0.6678 \\ 0.6678 & -0.7443 \end{pmatrix} \begin{pmatrix} 2.1630 & 0 \\ 0 & 0.2312 \end{pmatrix}$$
$$\cdot \begin{pmatrix} 0.8918 & -0.4524 \\ 0.4524 & 0.8918 \end{pmatrix}^T \qquad 2.103$$

When the joint velocity takes all the values of $\|\dot{\boldsymbol{\theta}}\| \leq 1$, the envelope of the set of end-effector velocities becomes an ellipsoid described by

$$\frac{\dot{r}_1^{*2}}{2.1630^2} + \frac{\dot{r}_2^{*2}}{0.2312^2} \leq 1 \qquad 2.104$$

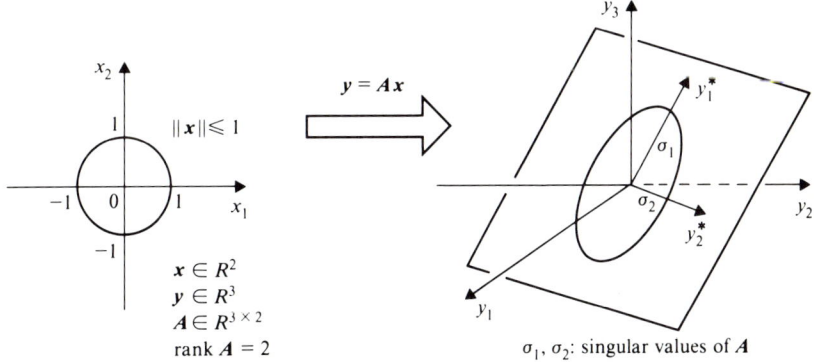

Figure 2.3 Elliptic geometry of mapping by Eqs. 2.97 and 2.98.

where $\dot{r}_1^*$ and $\dot{r}_2^*$ are the components of the end-effector velocity represented in the principal axes. The volume of the ellipsoid is

$$\text{volume} = \pi 2.1630 \cdot 0.2312 = 1.5708 \qquad 2.105$$

Yoshikawa (1984) named the ellipsoid and its volume the *manipulability ellipsoid* and the *measure of manipulability*, respectively.

Example 2.15 (Inertia Ellipsoid)

When an n-DOF manipulator is stationary (that is, when $\dot{\theta} = O$), Eq. 2.36 and SVD of the inertia matrix are as follows:

$$\tau = A(\theta)\ddot{\theta} \qquad 2.106$$

$$A(\theta) = U\Sigma U^T \qquad 2.107$$

Note that $V = U$ because $A(\theta)$ is symmetric. Equations 2.106 and 2.107 imply that, when the joints are accelerated by every $\ddot{\theta}$ that satisfies $\|\ddot{\theta}\| \leq 1$, the set of all the necessary joint torques form an ellipsoid determined by SVD of $A(\theta)$. The first column vector of U indicates the direction of $\ddot{\theta}$ for which the largest joint torque is required. Note that, in this case, the direction of $\ddot{\theta}$ and the direction of the required joint torques τ coincide with each other, and $\|\tau\| / \|\ddot{\theta}\|$ is equal to the largest singular value. Hence, σ_1 is the effective inertia in this direction. The ellipsoid is called the *inertia ellipsoid*.

Example 2.16 (Generalized Inertia Ellipsoid)

The relationship between the joint acceleration $\ddot{\theta} \in R^n$ and the end-effector acceleration $\ddot{x} \in R^6$ is represented by

$$\ddot{x} = J(\theta)\ddot{\theta} + \dot{J}(\theta)\dot{\theta} \qquad 2.108$$

where $J(\theta) \in R^{6 \times n}$ is the Jacobian matrix. Let $F \in R^6$ be the force and moment applied at the end-effector of the manipulator. The equivalent joint torque $\tau \in R^n$ is obtained using the principle of virtual work as follows (Craig 1989):

$$\tau = J^T(\theta)F \qquad 2.109$$

When $\dot{\theta} = O$, the relationship between the end-effector force and the end-effector acceleration can be computed using Eqs. 2.106, 2.108, and 2.109 as follows

$$\ddot{x} = JA^{-1}J^T F \qquad 2.110$$

If F is constrained by $\|F\| \leq 1$, the set of all the possible end-effector accelerations forms an ellipsoid. This ellipsoid is called the *generalized inertia ellipsoid* (Asada 1983). Note that A^{-1} is symmetric and positive definite, and $JA^{-1}J^T$ is also symmetric and positive definite if rank $J = 6$. Hence, Eq. 2.110 can be transformed to

$$\begin{aligned} F &= (JA^{-1}J^T)^{-1}\ddot{x} \quad &\text{(if } n > 6) \\ &= (J^T)^{-1}AJ^{-1}\ddot{x} \quad &\text{(if } n = 6) \end{aligned} \qquad 2.111$$

$(JA^{-1}J^T)^{-1}$ can be interpreted as the inertia matrix that we feel when we accelerate the end-effector by applying direct force to the end-effector. When $\ddot{x}$ takes all the values such that $\|\ddot{x}\| \leq 1$, the necessary end-effector force takes all the values inside the ellipsoid determined by SVD of $(JA^{-1}J^T)^{-1}$; that is,

$$(JA^{-1}J^T)^{-1} = U\Sigma U^T \qquad 2.112$$

The direction of the largest (smallest) inertia is that of the first (nth) column vector of U, and the corresponding inertia is the largest (smallest) singular value.

The coefficient $JA^{-1}J^T$ of Eq. 2.110 is referred to as the *mechanical impedance matrix*. Note that the mechanical impedance matrix is defined even if rank $J < 6$, although the inertia matrix is not. If rank $J = 6$, the mechanical impedance matrix and the inertia matrix are inverse matrices of each other.

2.4 Generalized Inverse and Pseudoinverse

2.4.1 Definitions

For $A \in R^{m \times n}$ and $X \in R^{n \times m}$, the following equations are used to define a generalized inverse, a reflexive generalized inverse, and a pseudoinverse of A (Boullion and Odell 1971):

$$AXA = A \qquad 2.113$$

$$XAX = X \qquad 2.114$$

$$(AX)^T = AX \qquad 2.115$$

$$(XA)^T = XA \qquad 2.116$$

Equations 2.113 through 2.116 are called the *Penrose conditions* (Penrose 1955).

Definition 2.2 (Generalized Inverse)

A *generalized inverse* of a matrix $A \in R^{m \times n}$ is a matrix $X = A^- \in R^{n \times m}$ satisfying Eq. 2.113.

Definition 2.3 (Reflexive Generalized Inverse)

A *reflexive generalized inverse* of a matrix $A \in R^{m \times n}$ is a matrix $X = A_r^- \in R^{n \times m}$ satisfying Eqs. 2.113 and 2.114.

Definition 2.4 (Pseudoinverse)

A *pseudoinverse* of a matrix $A \in R^{m \times n}$ is a matrix $X = A^\# \in R^{n \times m}$ satisfying Eqs. 2.113 through 2.116.

A pseudoinverse is sometimes called the *Moore-Penrose inverse* after the pioneering works by Moore (1920, 1935) and Penrose (1955).

Example 2.17

Let $A \in R^{2 \times 3}$, P, Q, and $R \in R^{3 \times 2}$ be as follows:

$$A = \begin{pmatrix} 1 & -1 & 1 \\ -1 & 1 & -1 \end{pmatrix}$$

$$P = \begin{pmatrix} 1 & 0 \\ 0 & 1 \\ 0 & 1 \end{pmatrix}$$

$$Q = \begin{pmatrix} 1 & 0 \\ 0 & 0 \\ 0 & 0 \end{pmatrix} \qquad 2.117$$

$$R = \frac{1}{6} \begin{pmatrix} 1 & -1 \\ -1 & 1 \\ 1 & -1 \end{pmatrix}$$

Now, we show that P, Q, and R are a generalized inverse, a reflexive generalized inverse, and a pseudoinverse of A, respectively.

$$APA = \begin{pmatrix} 1 & -1 & 1 \\ -1 & 1 & -1 \end{pmatrix} = A$$

$$PAP = \begin{pmatrix} 1 & 0 \\ -1 & 0 \\ -1 & 0 \end{pmatrix} \neq P$$

$$(AP)^T = \begin{pmatrix} 1 & -1 \\ 0 & 0 \end{pmatrix} \neq AP$$

$$(PA)^T = \begin{pmatrix} 1 & -1 & -1 \\ -1 & 1 & 1 \\ 1 & -1 & -1 \end{pmatrix} \neq PA \qquad 2.118$$

Equation 2.118 shows that P satisfies Eq. 2.113 only. Hence, P is a generalized inverse of A.

$$AQA = \begin{pmatrix} 1 & -1 & 1 \\ -1 & 1 & -1 \end{pmatrix} = A$$

$$QAQ = \begin{pmatrix} 1 & 0 \\ 0 & 0 \\ 0 & 0 \end{pmatrix} = Q$$

$$(AQ)^T = \begin{pmatrix} 1 & -1 \\ 0 & 0 \end{pmatrix} \neq AQ \qquad 2.119$$

$$(QA)^T = \begin{pmatrix} 1 & 0 & 0 \\ -1 & 0 & 0 \\ 1 & 0 & 0 \end{pmatrix} \neq QA$$

Equation 2.119 indicates that Q satisfies Eqs. 2.113 and 2.114, but does not satisfy Eqs. 2.115 and 2.116. This result implies that Q is a reflexive generalized inverse.

$$ARA = \begin{pmatrix} 1 & -1 & 1 \\ -1 & 1 & -1 \end{pmatrix} = A$$

$$RAR = \frac{1}{6}\begin{pmatrix} 1 & -1 \\ -1 & 1 \\ 1 & -1 \end{pmatrix} = R$$

$$(AR)^T = \frac{1}{2}\begin{pmatrix} 1 & -1 \\ -1 & 1 \end{pmatrix} = AR \qquad 2.120$$

$$(RA)^T = \frac{1}{3}\begin{pmatrix} 1 & -1 & 1 \\ -1 & 1 & -1 \\ 1 & -1 & 1 \end{pmatrix} = RA$$

R satisfies all four *Penrose conditions*. Therefore, we conclude that R is a pseudoinverse of A.

2.4.2 Properties

Generalized Inverse.[2.1]
(1) For a linear equation

$$y = Ax \qquad 2.121$$

[2.1] Rao and Mitra 1971; Kodama and Suda 1978.

where $A \in R^{m \times n}$, $x \in R^n$, and $y \in R^m$, a necessary and sufficient condition for the existence of solution x is

$$\text{rank}\,[A\ \ y] = \text{rank}\,A \qquad 2.122$$

If Eq. 2.122 is satisfied for Eq. 2.121, then

$$x = A^- y \qquad 2.123$$

is a solution of Eq. 2.121.

Example 2.18

We now discuss the solution of Eq. 2.121 with the A matrix given in Example 2.17, for two different y vectors—namely,

$$y_1 = \begin{pmatrix} -1 \\ 1 \end{pmatrix}$$
$$y_2 = \begin{pmatrix} 1 \\ 0 \end{pmatrix} \qquad 2.124$$

For both y_1 and y_2, Eq. 2.122 becomes

$$\text{rank}\,[A\ \ y_1] = \text{rank}\begin{pmatrix} 1 & -1 & 1 & -1 \\ -1 & 1 & -1 & 1 \end{pmatrix} = 1$$
$$= \text{rank}\,A$$
$$\text{rank}\,[A\ \ y_2] = \text{rank}\begin{pmatrix} 1 & -1 & 1 & 1 \\ -1 & 1 & -1 & 0 \end{pmatrix} = 2 \qquad 2.125$$
$$\neq \text{rank}\,A = 1$$

We compute x_1 and x_2 as follows:

$$x_1 = A^- y_1 = P y_1$$
$$= \begin{pmatrix} 1 & 0 \\ 0 & 1 \\ 0 & 1 \end{pmatrix} \begin{pmatrix} -1 \\ 1 \end{pmatrix} = \begin{pmatrix} -1 \\ 1 \\ 1 \end{pmatrix}$$
$$x_2 = A^- y_2 = P y_2 \qquad 2.126$$
$$= \begin{pmatrix} 1 & 0 \\ 0 & 1 \\ 0 & 1 \end{pmatrix} \begin{pmatrix} 1 \\ 0 \end{pmatrix} = \begin{pmatrix} 1 \\ 0 \\ 0 \end{pmatrix}$$

We verify by the following equations that x_1 is an exact solution of Eq. 2.121 for $y = y_1$, but that x_2 is not an exact solution for $y = y_2$:

$$Ax_1 = \begin{pmatrix} 1 & -1 & 1 \\ -1 & 1 & -1 \end{pmatrix} \begin{pmatrix} -1 \\ 1 \\ 1 \end{pmatrix}$$

$$= \begin{pmatrix} -1 \\ 1 \end{pmatrix} = y_1$$

$$Ax_2 = \begin{pmatrix} 1 & -1 & 1 \\ -1 & 1 & -1 \end{pmatrix} \begin{pmatrix} 1 \\ 0 \\ 0 \end{pmatrix}$$

$$= \begin{pmatrix} 1 \\ -1 \end{pmatrix} \neq y_2$$

2.127

(2) For an arbitrary $A \in R^{m \times n}$, there exists at least one generalized inverse A^- and rank $A^- \geq$ rank A. A^- coincides with a reflexive generalized inverse if and only if rank $A^- =$ rank A.

Example 2.19

Let's compare the ranks of A, $P = A^-$, and $Q = A_r^-$ used in Example 2.17.

$$\text{rank } A = \text{rank} \begin{pmatrix} 1 & -1 & 1 \\ -1 & 1 & -1 \end{pmatrix} = 1$$

$$\text{rank } A^- = \text{rank} \begin{pmatrix} 1 & 0 \\ 0 & 1 \\ 0 & 1 \end{pmatrix} = 2$$

2.128

$$\text{rank } A_r^- = \text{rank} \begin{pmatrix} 1 & 0 \\ 0 & 0 \\ 0 & 0 \end{pmatrix} = 1$$

Therefore,

$$\text{rank } A^- > \text{rank } A$$
$$A^- \neq A_r^-$$

2.129

Moreover,

$$\text{rank } A_r^- = \text{rank } A$$

2.130

(3) Generally A^- and A_r^- are not unique. If A is square and nonsingular, then the generalized inverse A^- and the reflexive generalized inverse A_r^- are unique, and $A^- = A_r^- = A^{-1}$.

(4) AA^- and A^-A are idempotent.[2.2]

Example 2.20

Now, we use A and $P = A^-$ in Example 2.17. AA^- and A^-A become:

$$AA^- = AP$$
$$= \begin{pmatrix} 1 & 0 \\ -1 & 0 \end{pmatrix}$$
$$A^-A = PA \qquad 2.131$$
$$= \begin{pmatrix} 1 & -1 & 1 \\ -1 & 1 & -1 \\ -1 & 1 & -1 \end{pmatrix}$$

Therefore,

$$(AA^-)^2 = AA^-AA^-$$
$$= \begin{pmatrix} 1 & 0 \\ -1 & 0 \end{pmatrix} \begin{pmatrix} 1 & 0 \\ -1 & 0 \end{pmatrix}$$
$$= \begin{pmatrix} 1 & 0 \\ -1 & 0 \end{pmatrix}$$
$$= AA^- \qquad 2.132$$

$$(A^-A)^2 = A^-AA^-A$$
$$= \begin{pmatrix} 1 & -1 & 1 \\ -1 & 1 & -1 \\ -1 & 1 & -1 \end{pmatrix} \begin{pmatrix} 1 & -1 & 1 \\ -1 & 1 & -1 \\ -1 & 1 & -1 \end{pmatrix}$$
$$= \begin{pmatrix} 1 & -1 & 1 \\ -1 & 1 & -1 \\ -1 & 1 & -1 \end{pmatrix}$$
$$= A^-A \qquad 2.133$$

We can readily prove statement 4 rigorously by substituting $AA^-A = A$ into Eqs. 2.132 and 2.133.

(5) Using an arbitrary matrix $U \in R^{n \times m}$ and a generalized inverse A^-, all the generalized inverses of A can be represented by the following X:

$$X = A^- + U - A^-AUAA^- \qquad 2.134$$

[2.2] A square matrix M is called idempotent if $M^2 = M$.

This result can be readily shown by

$$AXA = A(A^- + U - A^-AUAA^-)A$$
$$= AA^-A + AUA - (AA^-A)UAA^-A$$
$$= A + AUA - AUA = A \qquad 2.135$$

where $AA^-A = A$ was used.

Example 2.21

Once again, we use A and $P = A^-$ as in Example 2.17. $U \in R^{3\times 2}$ is chosen as follows:

$$U = \begin{pmatrix} 1 & 0 \\ 1 & 1 \\ 1 & 0 \end{pmatrix} \qquad 2.136$$

The X matrix of Eq. 2.134 is now computed using Eqs. 2.117, 2.131, and 2.136 as follows:

$$X = A^- + U - A^-AUAA^-$$
$$= \begin{pmatrix} 1 & 0 \\ 0 & 1 \\ 0 & 1 \end{pmatrix} + \begin{pmatrix} 1 & 0 \\ 1 & 1 \\ 1 & 0 \end{pmatrix}$$
$$- \begin{pmatrix} 1 & -1 & 1 \\ -1 & 1 & -1 \\ -1 & 1 & -1 \end{pmatrix} \begin{pmatrix} 1 & 0 \\ 1 & 1 \\ 1 & 0 \end{pmatrix} \begin{pmatrix} 1 & 0 \\ -1 & 0 \end{pmatrix}$$
$$= \begin{pmatrix} 0 & 0 \\ 3 & 2 \\ 3 & 1 \end{pmatrix} \qquad 2.137$$

Our claim is that X is another generalized inverse of A. We verify this claim by checking the Penrose conditions as follows:

$$AXA = \begin{pmatrix} 1 & -1 & 1 \\ -1 & 1 & -1 \end{pmatrix} \begin{pmatrix} 0 & 0 \\ 3 & 2 \\ 3 & 1 \end{pmatrix} \begin{pmatrix} 1 & -1 & 1 \\ -1 & 1 & -1 \end{pmatrix}$$
$$= \begin{pmatrix} 1 & -1 & 1 \\ -1 & 1 & -1 \end{pmatrix} = A$$

$$XAX = \begin{pmatrix} 0 & 0 \\ 3 & 2 \\ 3 & 1 \end{pmatrix} \begin{pmatrix} 1 & -1 & 1 \\ -1 & 1 & -1 \end{pmatrix} \begin{pmatrix} 0 & 0 \\ 3 & 2 \\ 3 & 1 \end{pmatrix}$$
$$= \begin{pmatrix} 0 & 0 \\ 0 & -1 \\ 0 & -2 \end{pmatrix} \neq X$$

$$(AX)^T = \begin{pmatrix} 0 & -1 \\ 0 & 1 \end{pmatrix}^T \neq AX$$

$$(XA)^T = \begin{pmatrix} 0 & 0 & 0 \\ 1 & -1 & 1 \\ 2 & -2 & 2 \end{pmatrix}^T \neq XA \qquad 2.138$$

Since only Eq. 2.113 is satisfied among the four Penrose conditions of Eqs. 2.113 through 2.116, X is a generalized inverse, and is neither a reflexive generalized inverse nor a pseudoinverse. Note that, since a pseudoinverse is also a generalized inverse and is unique, once we get it, we can compute every generalized inverse using Eq. 2.134.

Pseudoinverse. [2.3]
(1) For a given $A \in R^{m \times n}$, the pseudoinverse $A^\# \in R^{n \times m}$ is unique, whereas A^- and A_r^- are not necessarily unique. Let the sets of A^-, A_r^-, and $A^\#$ be S^-, S_r^-, and $S^\#$, respectively; then, the following inclusion holds:

$$S^\# \subset S_r^- \subset S^- \qquad 2.139$$

(2) $(A^\#)^\# = A$.
(3) $(A^T)^\# = (A^\#)^T$.
(4) $A^\# = (A^T A)^\# A^T = A^T (AA^T)^\#$.

For $A \in R^{m \times n}$, if $m < n$ and rank $A = m$, then AA^T is nonsingular and

$$A^\# = A^T (AA^T)^{-1} \qquad 2.140$$

If $m > n$ and rank $A = n$, then $A^T A$ is nonsingular and

$$A^\# = (A^T A)^{-1} A^T \qquad 2.141$$

If $m = n$ and rank $A = m$, then

$$A^\# = A^{-1} \qquad 2.142$$

Example 2.22

$A \in R^{2 \times 3}$ is defined as follows:

$$A = \begin{pmatrix} 1 & -1 & 0 \\ 0 & 1 & 1 \end{pmatrix} \qquad 2.143$$

Since A is full rank and $m = 2 < 3 = n$, we can use Eq. 2.140 to compute $A^\#$; namely,

---

[2.3] Rao and Mitra 1971; Boullion and Odell 1971; Kodama and Suda 1978.

$$A^\# = A^T(AA^T)^{-1}$$

$$= \begin{pmatrix} 1 & 0 \\ -1 & 1 \\ 0 & 1 \end{pmatrix} \left\{ \begin{pmatrix} 1 & -1 & 0 \\ 0 & 1 & 1 \end{pmatrix} \begin{pmatrix} 1 & 0 \\ -1 & 1 \\ 0 & 1 \end{pmatrix} \right\}^{-1}$$

$$= \begin{pmatrix} 1 & 0 \\ -1 & 1 \\ 0 & 1 \end{pmatrix} \begin{pmatrix} 2 & -1 \\ -1 & 2 \end{pmatrix}^{-1} \qquad 2.144$$

$$= \frac{1}{3} \begin{pmatrix} 2 & 1 \\ -1 & 1 \\ 1 & 2 \end{pmatrix}$$

Equation 2.141 does not work for A because $m < n$ and

$$\det(A^T A) = \det \left\{ \begin{pmatrix} 1 & 0 \\ -1 & 1 \\ 0 & 1 \end{pmatrix} \begin{pmatrix} 1 & -1 & 0 \\ 0 & 1 & 1 \end{pmatrix} \right\}$$

$$= \det \begin{pmatrix} 1 & -1 & 0 \\ -1 & 2 & 1 \\ 0 & 1 & 1 \end{pmatrix} \qquad 2.145$$

$$= 0$$

and, therefore, $(A^T A)^{-1}$ is not defined. For any two matrices, M_1 and M_2, that can define a product $M_1 M_2$, the following equation holds:

$$\operatorname{rank}(M_1 M_2) \le \min(\operatorname{rank} M_1, \operatorname{rank} M_2) \qquad 2.146$$

Therefore, for $A \in R^{m \times n}$, $m < n$,

$$\operatorname{rank}(A^T A) \le \min(\operatorname{rank} A^T, \operatorname{rank} A)$$
$$= \operatorname{rank} A \le \min(m, n) = m \qquad 2.147$$

Since $A^T A \in R^{n \times n}$, $A^T A$ is not full rank and $\det(A^T A) = 0$.

(5) $A^\# A$, $AA^\#$, $E - A^\# A$ and $E - AA^\#$ are all symmetric and idempotent, where E represents an identity matrix of appropriate dimension.

Example 2.23

For matrix A as used in Example 2.22, we can compute $A^\# A$, $AA^\#$, $E - A^\# A$, and $E - AA^\#$ as follows:

$$A^\# A = \frac{1}{3} \begin{pmatrix} 2 & -1 & 1 \\ -1 & 2 & 1 \\ 1 & 1 & 2 \end{pmatrix}$$

$$AA^{\#} = \begin{pmatrix} 1 & 0 \\ 0 & 1 \end{pmatrix}$$

$$E - A^{\#}A = \frac{1}{3}\begin{pmatrix} 1 & 1 & -1 \\ 1 & 1 & -1 \\ -1 & -1 & 1 \end{pmatrix} \qquad 2.148$$

$$E - AA^{\#} = \begin{pmatrix} 0 & 0 \\ 0 & 0 \end{pmatrix}$$

Verify the idempotency of the four matrices.

(6) If $A \in R^{n \times n}$ is symmetric and idempotent, then, for any matrix $B \in R^{m \times n}$, the following equation holds (Maciejewski and Klein 1985):

$$A(BA)^{\#} = (BA)^{\#} \qquad 2.149$$

Example 2.24

We use $E - A^{\#}A$ as obtained in Eq. 2.148 as an example of a symmetric and idempotent matrix. Now, our $A \in R^{3 \times 3}$, and $B \in R^{2 \times 3}$ in Eq. 2.149 are as follows:

$$A = \frac{1}{3}\begin{pmatrix} 1 & 1 & -1 \\ 1 & 1 & -1 \\ -1 & -1 & 1 \end{pmatrix}$$

$$B = \begin{pmatrix} 1 & 2 & 3 \\ -1 & 0 & 1 \end{pmatrix} \qquad 2.150$$

$(BA)^{\#}$ is computed as

$$(BA)^{\#} = \left\{\frac{1}{3}\begin{pmatrix} -4 & -4 & 4 \\ 2 & 2 & -2 \end{pmatrix}\right\}^{\#}$$

$$= \begin{pmatrix} -0.2 & 0.1 \\ -0.2 & 0.1 \\ 0.2 & -0.1 \end{pmatrix} \qquad 2.151$$

$A(BA)^{\#}$ is, then, given by

$$A(BA)^{\#} = \frac{1}{3}\begin{pmatrix} 1 & 1 & -1 \\ 1 & 1 & -1 \\ -1 & -1 & 1 \end{pmatrix}\begin{pmatrix} -0.2 & 0.1 \\ -0.2 & 0.1 \\ 0.2 & -0.1 \end{pmatrix}$$

$$= \begin{pmatrix} -0.2 & 0.1 \\ -0.2 & 0.1 \\ 0.2 & -0.1 \end{pmatrix} \qquad 2.152$$

Equations 2.151 and 2.152 indicate that $A(BA)^{\#} = (BA)^{\#}$. The same result is obtained for any 2×3 matrices of B. This relationship will be used later to simplify the computation for utilizing kinematic redundancy.

Properties 2, 3, 4, and 6 are readily obtained by verifying Eqs. 2.113 through 2.116. Property 5 is obvious from Eqs. 2.115 and 2.116. For further properties, see Rao and Mitra (1971), Boullion and Odell (1971), and Kodama and Suda (1978).

2.4.3 Solving Linear Equations

The pseudoinverse has wide applications in solving various types of linear problems. The following theorem is particularly significant within the scope of this book.

Theorem 2.5 (Least-Squares Solutions)

For a linear equation of $\boldsymbol{x} \in R^n$

$$\boldsymbol{A}\boldsymbol{x} = \boldsymbol{y} \qquad 2.153$$

where $\boldsymbol{A} \in R^{m \times n}$ and $\boldsymbol{y} \in R^m$, the general form of the least-squares solutions is given by

$$\boldsymbol{x} = \boldsymbol{A}^{\#}\boldsymbol{y} + (\boldsymbol{E} - \boldsymbol{A}^{\#}\boldsymbol{A})\boldsymbol{z} \qquad 2.154$$

where $\boldsymbol{z} \in R^n$ is an arbitrary vector, and $\boldsymbol{E}$ is an identity matrix. The minimum norm solution among all the solutions provided by Eq. 2.154 is

$$\boldsymbol{x} = \boldsymbol{A}^{\#}\boldsymbol{y} \qquad 2.155$$

_____

For Eq. 2.153, the least-squares solution is the $\boldsymbol{x}$ that minimizes the error norm—namely,

$$\min \| \boldsymbol{y} - \boldsymbol{A}\boldsymbol{x} \| \qquad 2.156$$

where $\| * \|$ denotes the Euclidean norm of a vector $*$. The least-squares solution is not necessarily unique. Every solution is obtained by changing $\boldsymbol{z}$. Equation 2.155 implies the solution that also minimizes $\| \boldsymbol{x} \|$ among all the solutions given by Eq. 2.154.

When at least one exact solution exists for Eq. 2.153, Eq. 2.154 yields the general form of all the exact solutions. Note that the first and second terms of Eq. 2.154 are perpendicular to each other. Indeed,

$$\begin{aligned}(\boldsymbol{A}^{\#}\boldsymbol{y})^T(\boldsymbol{E} - \boldsymbol{A}^{\#}\boldsymbol{A})\boldsymbol{z} &= \boldsymbol{y}^T(\boldsymbol{A}^{\#})^T(\boldsymbol{E} - \boldsymbol{A}^{\#}\boldsymbol{A})\boldsymbol{z} \\ &= \boldsymbol{y}^T\{(\boldsymbol{E} - \boldsymbol{A}^{\#}\boldsymbol{A})\boldsymbol{A}^{\#}\}^T \boldsymbol{z} \\ &= 0 \qquad\qquad 2.157\end{aligned}$$

where Eq. 2.114 and the symmetry of $\boldsymbol{E} - \boldsymbol{A}^{\#}\boldsymbol{A}$ were used.

***Example* 2.25**

We revisit Example 2.18. The linear equation of Eq. 2.153 is now given with

$$A = \begin{pmatrix} 1 & -1 & 1 \\ -1 & 1 & -1 \end{pmatrix} \qquad 2.158$$

and y_1 and $y_2 \in R^2$ are the following vectors:

$$y_1 = \begin{pmatrix} -1 \\ 1 \end{pmatrix}$$
$$y_2 = \begin{pmatrix} 1 \\ 0 \end{pmatrix} \qquad 2.159$$

Recall that, in Example 2.18, Eq. 2.153 has an exact solution for y_1, but does not have one for y_2. The general solution for $y = y_1$ is computed by Eq. 2.154 and is investigated using

$$A^{\#} = \frac{1}{6}\begin{pmatrix} 1 & -1 \\ -1 & 1 \\ 1 & -1 \end{pmatrix}$$
$$E - A^{\#}A = \frac{1}{3}\begin{pmatrix} 2 & 1 & -1 \\ 1 & 2 & 1 \\ -1 & 1 & 2 \end{pmatrix} \qquad 2.160$$
$$E - AA^{\#} = \frac{1}{2}\begin{pmatrix} 1 & 1 \\ 1 & 1 \end{pmatrix}$$

For an arbitrary z, the error norm of Eq. 2.156 is identically zero. Indeed,

$$\begin{aligned} \| y_1 - Ax \| &= \| y_1 - \{AA^{\#}y_1 + A(E - A^{\#}A)z\} \| \\ &= \| (E - AA^{\#})y_1 \| \\ &= \| \frac{1}{2}\begin{pmatrix} 1 & 1 \\ 1 & 1 \end{pmatrix}\begin{pmatrix} -1 \\ 1 \end{pmatrix} \| \\ &= 0 \end{aligned} \qquad 2.161$$

Equation 2.161 verifies that x of Eq. 2.154 with an arbitrary z and $y = y_1$ is an exact solution of Eq. 2.153 with $y = y_1$. When $z = O$, we have

$$x = A^{\#}y_1 = \frac{1}{3}\begin{pmatrix} -1 \\ 1 \\ -1 \end{pmatrix} \stackrel{\triangle}{=} x_1 \qquad 2.162$$

When $z = \begin{pmatrix} 1 & 1 & 1 \end{pmatrix}^T$, for example, we have

$$x = A^{\#}y_1 + (E - A^{\#}A)z$$

$$= \frac{1}{3}\begin{pmatrix}-1\\1\\-1\end{pmatrix} + \frac{1}{3}\begin{pmatrix}2&1&-1\\1&2&1\\-1&1&2\end{pmatrix}\begin{pmatrix}1\\1\\1\end{pmatrix}$$

$$= \frac{1}{3}\begin{pmatrix}1\\5\\1\end{pmatrix} \triangleq x_2 \qquad 2.163$$

Thus x_1 and x_2 are two different solutions for Eq. 2.153; they have the following relationship:

$$\|x_1\| = \frac{1}{\sqrt{3}} < \sqrt{3} = \|x_2\| \qquad 2.164$$

In general, the solution with $z = O$ (namely, Eq. 2.155) provides the solution with minimum magnitude.

For $y = y_2$, the error norm of Eq. 2.156 becomes

$$\|y_2 - Ax\| = \|(E - AA^\#)y_2\|$$
$$= \|\frac{1}{2}\begin{pmatrix}1&1\\1&1\end{pmatrix}\begin{pmatrix}1\\0\end{pmatrix}\|$$
$$= \|\frac{1}{2}\begin{pmatrix}1\\1\end{pmatrix}\|$$
$$= \frac{1}{\sqrt{2}} \neq 0 \qquad 2.165$$

Equation 2.165 implies that, since Eq. 2.153 does not have an exact solution for $y = y_2$, x of Eq. 2.154 with an arbitrary z and $y = y_2$ is an approximate solution. Note that the norm of error is $1/\sqrt{2}$, regardless of z. When $z = O$, we have the following equation:

$$x = A^\# y_2 = \frac{1}{6}\begin{pmatrix}1\\-1\\1\end{pmatrix} \triangleq x_3 \qquad 2.166$$

For $z = (1\ 1\ 1)^T$, we obtain

$$x = A^\# y_2 + (E - A^\# A)z$$
$$= \frac{1}{6}(1\ -1\ 1) + \frac{1}{3}\begin{pmatrix}2&1&-1\\1&2&1\\-1&1&2\end{pmatrix}\begin{pmatrix}1\\1\\1\end{pmatrix}$$
$$= \frac{1}{6}\begin{pmatrix}5\\7\\5\end{pmatrix} \triangleq x_4 \qquad 2.167$$

Hence, we have a result similar to Eq. 2.164, as follows:

$$\| \boldsymbol{x}_3 \| = 0.2887 < 1.6583 = \| \boldsymbol{x}_4 \| \qquad 2.168$$

In general, $\boldsymbol{x}$ of Eq. 2.154 with $\boldsymbol{z} = \boldsymbol{O}$ and $\boldsymbol{y} = \boldsymbol{y}_2$ provides the approximate solution of Eq. 2.153 that has the minimum magnitude.

2.4.4 Weighted Pseudoinverse

Theorem 2.5 offers the general form of the least-squares solutions (Eq. 2.154), and the least-squares solution with the minimum norm based on the Euclidean norm (Eq. 2.155). In many physical problems, the components of $\boldsymbol{x}$ or $\boldsymbol{y}$ may have different physical dimensions. Even if the components are physically consistent, the significance of magnitude can be different. For example, if $\boldsymbol{y}$ is the joint torque vector of a robot manipulator, a moderate value of torque of a large motor would have critical meaning to a small motor. In these cases, it would be necessary to evaluate the magnitude of error vector and the magnitude of solution based on an appropriate weighting of the components. In this subsection, we derive a result similar to Theorem 2.5, but for the weighted norm.

Let $\| \boldsymbol{a} \|_W$ represent the weighted norm such that

$$\| \boldsymbol{a} \|_W = \sqrt{\boldsymbol{a}^T \boldsymbol{W} \boldsymbol{a}} \qquad 2.169$$

where $\boldsymbol{W} \in R^{n \times n}$ is a symmetric positive definite matrix. A symmetric positive definite matrix can be represented as follows:

$$\boldsymbol{W} = \boldsymbol{W}_0^T \boldsymbol{W}_0 \qquad 2.170$$

where $\boldsymbol{W}_0 \in R^{n \times n}$ is nonsingular. In Eq. 2.170, $\boldsymbol{W}_0$ is not unique. However, for any symmetric and positive definite matrix $\boldsymbol{W}$, there exists a unique symmetric and positive definite matrix $\boldsymbol{W}_0$ satisfying Eq. 2.170, which is called the *square root* of $\boldsymbol{W}$.

The following two equivalences can be readily shown:

$$\min \| \boldsymbol{y} - \boldsymbol{A}\boldsymbol{x} \|_P \iff \min \| \boldsymbol{P}_0 \boldsymbol{y} - \boldsymbol{P}_0 \boldsymbol{A} \boldsymbol{x} \| \qquad 2.171$$

$$\min \| \boldsymbol{x} \|_Q \iff \min \| \boldsymbol{Q}_0 \boldsymbol{x} \| \qquad 2.172$$

where $\boldsymbol{P}, \boldsymbol{P}_0 \in R^{m \times m}$ and $\boldsymbol{Q}, \boldsymbol{Q}_0 \in R^{n \times n}$, and $\boldsymbol{P}_0, \boldsymbol{Q}_0$ are any matrices that satisfy Eq. 2.170 for symmetric positive definite $\boldsymbol{P}$ and $\boldsymbol{Q}$, respectively. Since $\boldsymbol{P}_0$ is nonsingular, Eq. 2.153 can be equivalently transformed to

$$\boldsymbol{P}_0 \boldsymbol{A} \boldsymbol{x} = \boldsymbol{P}_0 \boldsymbol{y} \qquad 2.173$$

2.4 Generalized Inverse and Pseudoinverse

It is obvious that $x^* = Q_0 x$ computed from x that minimizes $\| P_0 y - P_0 A x \|$ minimizes $\| P_0 y - P_0 A Q_0^{-1} x^* \|$, and vice versa, because Q_0 is nonsingular. Therefore, the general form of the least-squares solutions of $P_0 A x = P_0 y$ can be represented by

$$x = Q_0^{-1} x^*$$
$$x^* = (P_0 A Q_0^{-1})^\# P_0 y + \{E - (P_0 A Q_0^{-1})^\# (P_0 A Q_0^{-1})\} z \qquad 2.174$$

We can see from Theorem 2.5 that $x^* = (P_0 A Q_0^{-1})^\# P_0 y$—namely $x = Q_0^{-1}(P_0 A Q_0^{-1})^\# P_0 y$—is the one that minimizes $\| x^* \| = \| Q_0 x \|$ among all the least-squares solutions. From Eqs. 2.171 and 2.172, we can conclude that Eq. 2.174 provides the general form of the least P-weighted-norm solutions, and that $x = Q_0^{-1}(P_0 A Q_0^{-1})^\# P_0 y$ is the minimum Q-weighted-norm solution among all the least P-weighted-norm solutions. We summarize this discussion in the following theorem.

Theorem 2.6 (Weighted-Norm Solutions)

For a linear equation of $x \in R^n$,

$$A x = y \qquad 2.175$$

where $A \in R^{m \times n}$ and $y \in R^m$, the general form of the least P-weighted-norm solutions is given by

$$x = Q_0^{-1} A^{*\#} P_0 y + Q_0^{-1}(E - A^{*\#} A^*) z \qquad 2.176$$

$$A^* \triangleq P_0 A Q_0^{-1} \qquad 2.177$$

where P_0 and Q_0 are defined for P and Q by Eq. 2.170, and $z \in R^n$ is an arbitrary vector. The minimum Q-weighted-norm solution among all the solutions provided by Eq. 2.176 is

$$x = Q_0^{-1} A^{*\#} P_0 y \qquad 2.178$$

where $Q_0^{-1} A^{*\#} P_0$ is called a *weighted pseudoinverse* of A.

Example 2.26

We again discuss a linear equation of Eq. 2.175 with

$$A = \begin{pmatrix} 1 & -1 & 1 \\ -1 & 1 & -1 \end{pmatrix}$$
$$y = y_2 = \begin{pmatrix} 1 \\ 0 \end{pmatrix} \qquad 2.179$$

Suppose $P \in R^{2\times 2}$ and $Q \in R^{3\times 3}$ are given as follows:

$$P = \begin{pmatrix} 1 & 0 \\ 0 & 4 \end{pmatrix}$$

$$Q = \begin{pmatrix} 9 & 0 & 0 \\ 0 & 4 & 0 \\ 0 & 0 & 1 \end{pmatrix} \qquad 2.180$$

We choose P_0 and Q_0 as the square roots of P and Q; namely,

$$P_0 = \begin{pmatrix} 1 & 0 \\ 0 & 2 \end{pmatrix}$$

$$Q_0 = \begin{pmatrix} 3 & 0 & 0 \\ 0 & 2 & 0 \\ 0 & 0 & 1 \end{pmatrix} \qquad 2.181$$

We have

$$A^* = P_0 A Q_0^{-1}$$

$$= \begin{pmatrix} 1 & 0 \\ 0 & 2 \end{pmatrix} \begin{pmatrix} 1 & -1 & 1 \\ -1 & 1 & -1 \end{pmatrix} \begin{pmatrix} 3 & 0 & 0 \\ 0 & 2 & 0 \\ 0 & 0 & 1 \end{pmatrix}^{-1}$$

$$= \frac{1}{6} \begin{pmatrix} 2 & -3 & 6 \\ -4 & 6 & -12 \end{pmatrix} \qquad 2.182$$

$$A^{*\#} = \begin{pmatrix} 0.0490 & -0.0980 \\ -0.0735 & 0.1469 \\ 0.1469 & -0.2939 \end{pmatrix}$$

$$E - A^{*\#} A^* = \begin{pmatrix} 0.9184 & 0.1224 & -0.2449 \\ 0.1224 & 0.8163 & 0.3673 \\ -0.2449 & 0.3673 & 0.2653 \end{pmatrix}$$

The weighted-norm solutions are compared with the solutions we obtained in Example 2.25. When $z = O$, we obtain the following solution using Eq. 2.178:

$$x = Q_0^{-1} A^{*\#} P_0 y_2$$

$$= \begin{pmatrix} 3 & 0 & 0 \\ 0 & 2 & 0 \\ 0 & 0 & 1 \end{pmatrix}^{-1} \begin{pmatrix} 0.0490 & -0.0980 \\ -0.0735 & 0.1469 \\ 0.1469 & -0.2939 \end{pmatrix} \begin{pmatrix} 1 & 0 \\ 0 & 2 \end{pmatrix} \begin{pmatrix} 1 \\ 0 \end{pmatrix}$$

$$= \begin{pmatrix} 0.0163 \\ -0.0367 \\ 0.1469 \end{pmatrix} \triangleq x_5 \qquad 2.183$$

2.4 Generalized Inverse and Pseudoinverse

When $z = (1 \ 1 \ 1)^T$, we get the following solution from Eq. 2.176:

$$x = x_5 + Q_0^{-1}(E - A^{*\#}A^*)z$$

$$= \begin{pmatrix} 0.0163 \\ -0.0367 \\ 0.1469 \end{pmatrix} + \begin{pmatrix} 3 & 0 & 0 \\ 0 & 2 & 0 \\ 0 & 0 & 1 \end{pmatrix}^{-1}$$

$$\cdot \begin{pmatrix} 0.9184 & 0.1224 & -0.2449 \\ 0.1224 & 0.8163 & 0.3673 \\ -0.2449 & 0.3673 & 0.2653 \end{pmatrix} \begin{pmatrix} 1 \\ 1 \\ 1 \end{pmatrix}$$

$$= \begin{pmatrix} 0.2816 \\ 0.6163 \\ 0.5347 \end{pmatrix} \triangleq x_6 \qquad 2.184$$

Note that, since Eq. 2.175 with A and y defined by Eq. 2.179 has no exact solution, as we demonstrated in Example 2.18, x_5 and x_6 are also approximations.

Let's compare x_5 and x_6 with x_3 and x_4 in Eqs. 2.166 and 2.167, respectively. The P-weighted norms of the error vectors are as follows:

$$\|y_2 - Ax_3\|_P = \|y_2 - Ax_4\|_P = 1.1180$$
$$> 0.8944 = \|y_2 - Ax_5\|_P = \|y_2 - Ax_6\|_P \qquad 2.185$$

On the other hand, the Euclidean norms of the error vectors are

$$\|y_2 - Ax_5\| = \|y_2 - Ax_6\| = 0.8246$$
$$> \frac{1}{\sqrt{2}} = \|y_2 - Ax_3\| = \|y_2 - Ax_4\| \qquad 2.186$$

Note that the Q-weighted norms of the solutions are

$$\|x_5\|_Q = 0.1714 < 1.5872 = \|x_6\|_Q \qquad 2.187$$

Equations 2.185 and 2.186 show that x_3 and x_4 are better approximations than x_5 and x_6 are, in the sense of the Euclidean norms, and that, on the contrary, x_5 and x_6 are better than x_3 and x_4 are, in the sense of P-weighed norms.

2.4.5 Mappings and Projections

Pseudoinverses possess significant geometric characteristics as projections. The linear mapping by a linear equation

$$y = Ax \qquad 2.188$$

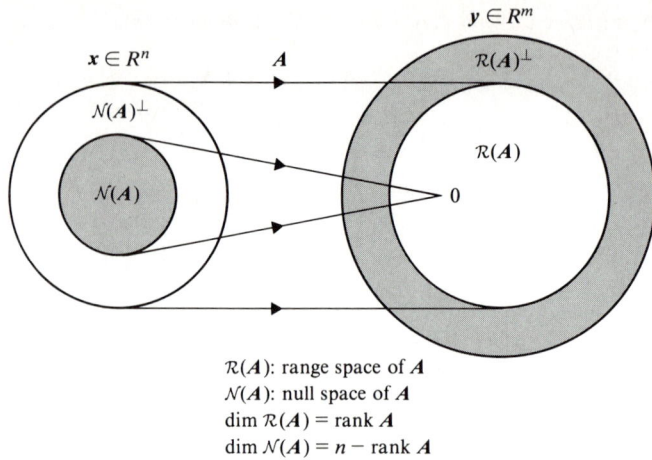

$\mathcal{R}(A)$: range space of A
$\mathcal{N}(A)$: null space of A
dim $\mathcal{R}(A)$ = rank A
dim $\mathcal{N}(A)$ = n − rank A

Figure 2.4 Linear mapping by Eq. 2.188.

where $A \in R^{m \times n}$, $y \in R^m$, and $x \in R^n$, is shown in Fig. 2.4. Let $\mathcal{R}(A) \subset R^m$ and $\mathcal{N}(A) \subset R^n$ be the range space[2.4] of A, and the null space[2.5] of A, respectively. We represent the orthogonal complements[2.6] of $\mathcal{R}(A)$ and $\mathcal{N}(A)$ by $\mathcal{R}(A)^\perp \subset R^m$ and $\mathcal{N}(A)^\perp \subset R^n$, respectively. Four subspaces of $\mathcal{R}(A)$, $\mathcal{R}(A)^\perp$, $\mathcal{N}(A)$, and $\mathcal{N}(A)^\perp$ have the following relationships with the pseudoinverse of A.

$$\mathcal{R}(A) = \mathcal{N}(A^{\#})^\perp = \mathcal{R}(AA^{\#}) = \mathcal{N}(E - AA^{\#}) \qquad 2.189$$

$$\mathcal{R}(A)^\perp = \mathcal{N}(A^{\#}) = \mathcal{N}(AA^{\#}) = \mathcal{R}(E - AA^{\#}) \qquad 2.190$$

$$\mathcal{N}(A) = \mathcal{R}(A^{\#})^\perp = \mathcal{N}(A^{\#}A) = \mathcal{R}(E - A^{\#}A) \qquad 2.191$$

$$\mathcal{N}(A)^\perp = \mathcal{R}(A^{\#}) = \mathcal{R}(A^{\#}A) = \mathcal{N}(E - A^{\#}A) \qquad 2.192$$

These four relationships are illustrated conceptually in Fig. 2.5. We can interpret $\mathcal{R}(A)^\perp$ and $\mathcal{N}(A)^\perp$ as follows: If y has a nonzero component in $\mathcal{R}(A)^\perp$, it does not have an exact solution for Eq. 2.188. If x has a nonzero component in $\mathcal{N}(A)^\perp$, its mapping by Eq. 2.88 is nonzero.

A square matrix $M \in R^{n \times n}$ is called an *orthogonal projection* if its mapping Mx of $x \in R^n$ is perpendicular to $x - Mx$ for any x. It is

---

[2.4] The set of all ys obtained by computing Eq. 2.188 for every $x \in R^n$ makes a linear subspace in R^m. This linear subspace is termed the *range space*.

[2.5] The set of all xs that provide $y = O$ in Eq. 2.188 makes a linear space in R^n. This linear subspace is termed the *null space* of A.

[2.6] The *orthogonal complement* of a linear subspace implies a set of all the vectors perpendicular to all the vectors in the subspace. The orthogonal complement is also a linear subspace of the whole space.

noteworthy that $AA^\#$, $A^\#A$, $E-AA^\#$, and $E-A^\#A$ are all orthogonal projections.

Example 2.27

Equation 2.188 with
$$A = \begin{pmatrix} 1 & -1 & 1 \\ -1 & 1 & -1 \end{pmatrix} \qquad 2.193$$

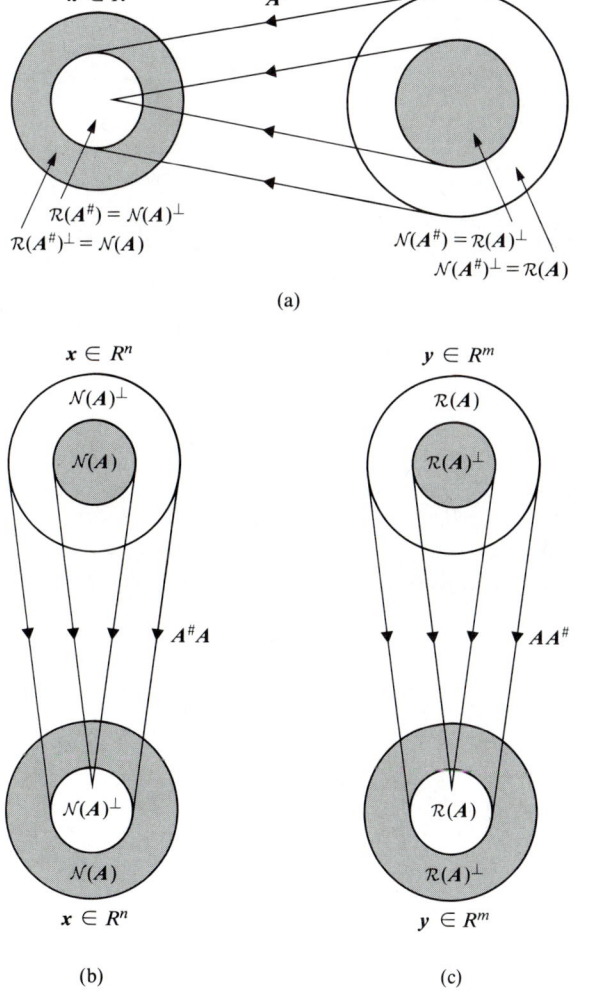

Figure 2.5 Relationships among the pseudoinverse, the range space, the null space, and their orthogonal complements. (a) Mapping by $A^\#$. (b) Projection by $A^\#A$. (c) Projection by $AA^\#$.

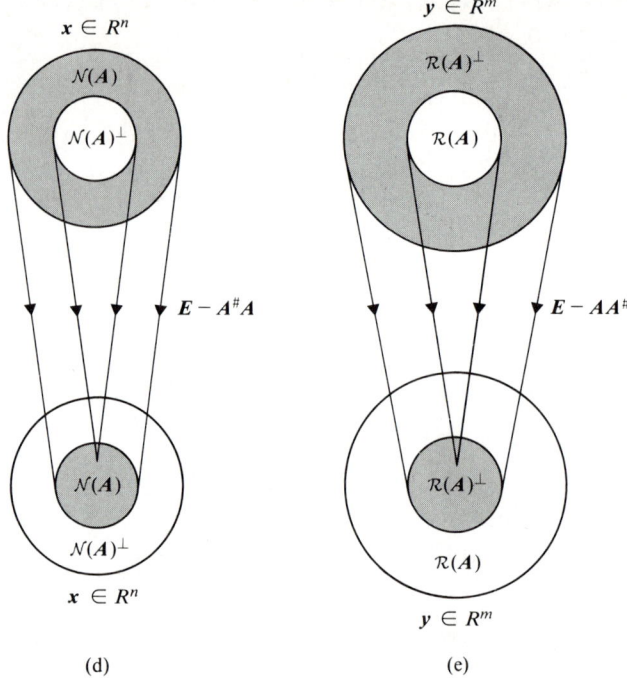

Figure 2.5 (continued) (d) Projection by $E - A^\# A$. (e) Projection by $E - AA^\#$.

is again used as an example. The orthogonal projections are computed as follows:

$$AA^\# = \frac{1}{2}\begin{pmatrix} 1 & -1 \\ -1 & 1 \end{pmatrix}$$

$$E - AA^\# = \frac{1}{2}\begin{pmatrix} 1 & 1 \\ 1 & 1 \end{pmatrix}$$

$$A^\# A = \frac{1}{3}\begin{pmatrix} 1 & -1 & 1 \\ -1 & 1 & -1 \\ 1 & -1 & 1 \end{pmatrix} \qquad 2.194$$

$$E - A^\# A = \frac{1}{3}\begin{pmatrix} 2 & 1 & -1 \\ 1 & 2 & 1 \\ -1 & 1 & 2 \end{pmatrix}$$

2.4 Generalized Inverse and Pseudoinverse

Now, we define the following vectors:

$$x_0 = \begin{pmatrix} 1 \\ 1 \\ 1 \end{pmatrix}$$
$$y_0 = \begin{pmatrix} 1 \\ 0 \end{pmatrix}$$

(2.195)

We can decompose y_0 into two orthogonal vectors, which are the members of $\mathcal{R}(A)$ and $\mathcal{R}(A)^\perp$ and are denoted by y_R and $y_{R\perp}$, respectively. Similarly, x_0 can be decomposed into two orthogonal vectors, which are the members of $\mathcal{N}(A)$ and $\mathcal{N}(A)^\perp$ and are denoted by x_N and $x_{N\perp}$, respectively. They are computed using Eq. 2.194 as follows:

$$\begin{aligned}
y_R &= AA^\# y_0 \\
&= \frac{1}{2} \begin{pmatrix} 1 & -1 \\ -1 & 1 \end{pmatrix} \begin{pmatrix} 1 \\ 0 \end{pmatrix} \\
&= \frac{1}{2} \begin{pmatrix} 1 \\ -1 \end{pmatrix} \\
y_{R\perp} &= (E - AA^\#) y_0 \\
&= \frac{1}{2} \begin{pmatrix} 1 & 1 \\ 1 & 1 \end{pmatrix} \begin{pmatrix} 1 \\ 0 \end{pmatrix} \\
&= \frac{1}{2} \begin{pmatrix} 1 \\ 1 \end{pmatrix}
\end{aligned}$$

(2.196)

$$\begin{aligned}
x_N &= (E - A^\# A) x_0 \\
&= \frac{1}{3} \begin{pmatrix} 2 & 1 & -1 \\ 1 & 2 & 1 \\ -1 & 1 & 2 \end{pmatrix} \begin{pmatrix} 1 \\ 1 \\ 1 \end{pmatrix} = \frac{2}{3} \begin{pmatrix} 1 \\ 2 \\ 1 \end{pmatrix} \\
x_{N\perp} &= A^\# A x_0 \\
&= \frac{1}{3} \begin{pmatrix} 1 & -1 & 1 \\ -1 & 1 & -1 \\ 1 & -1 & 1 \end{pmatrix} \begin{pmatrix} 1 \\ 1 \\ 1 \end{pmatrix} = \frac{1}{3} \begin{pmatrix} 1 \\ -1 \\ 1 \end{pmatrix}
\end{aligned}$$

(2.197)

We check the results of Eqs. 2.196 and 2.197 by the following equations:

$$y_R + y_{R\perp} = \frac{1}{2} \begin{pmatrix} 1 \\ -1 \end{pmatrix} + \frac{1}{2} \begin{pmatrix} 1 \\ 1 \end{pmatrix} = \begin{pmatrix} 1 \\ 0 \end{pmatrix} = y_0$$

$$x_N + x_{N\perp} = \frac{2}{3} \begin{pmatrix} 1 \\ 2 \\ 1 \end{pmatrix} + \frac{1}{3} \begin{pmatrix} 1 \\ -1 \\ 1 \end{pmatrix} = \begin{pmatrix} 1 \\ 1 \\ 1 \end{pmatrix} = x_0$$

$$y_R^T y_{R\perp} = \frac{1}{2}(1 \quad -1) \cdot \frac{1}{2}\begin{pmatrix} 1 \\ 1 \end{pmatrix} = 0$$

$$x_N^T x_{N\perp} = \frac{2}{3}(1 \quad 2 \quad 1) \cdot \frac{1}{3}\begin{pmatrix} 1 \\ -1 \\ 1 \end{pmatrix} = 0 \qquad 2.198$$

Note that

$$\text{rank}\,[A \quad y_R] = \text{rank}\begin{pmatrix} 1 & -1 & 1 & 0.5 \\ -1 & 1 & -1 & -0.5 \end{pmatrix}$$

$$= \text{rank}\,A \qquad 2.199$$

Therefore, Eq. 2.188 with $y = y_R$ has exact solutions. An exact solution is the *pseudoinverse solution*; namely,

$$x = A^\# y_R$$

$$= \frac{1}{6}\begin{pmatrix} 1 & -1 \\ -1 & 1 \\ 1 & -1 \end{pmatrix} \cdot \frac{1}{2}\begin{pmatrix} 1 \\ -1 \end{pmatrix} = \frac{1}{6}\begin{pmatrix} 1 \\ -1 \\ 1 \end{pmatrix} \qquad 2.200$$

This solution is equal to the pseudoinverse approximation of Eq. 2.188, with $y = y_0$. Indeed,

$$x = A^\# y_0$$

$$= \frac{1}{6}\begin{pmatrix} 1 & -1 \\ -1 & 1 \\ 1 & -1 \end{pmatrix}\begin{pmatrix} 1 \\ 0 \end{pmatrix} = \frac{1}{6}\begin{pmatrix} 1 \\ -1 \\ 1 \end{pmatrix} \qquad 2.201$$

Also, note that

$$y = A x_N$$

$$= \begin{pmatrix} 1 & -1 & 1 \\ -1 & 1 & -1 \end{pmatrix} \cdot \frac{2}{3}\begin{pmatrix} 1 \\ 2 \\ 1 \end{pmatrix} = \begin{pmatrix} 0 \\ 0 \end{pmatrix} \qquad 2.202$$

2.4.6 Computation of Pseudoinverse

In this subsection, three computational methods of pseudoinverse are summarized.

Computation by Singular Value Decomposition. The computational algorithms for SVD (see, for example, Golub and Van Loan 1983) are reliable schemes. By Eq. 2.78, the pseudoinverse of a matrix can be computed using SVD, as in Example 2.12.

Greville's Algorithm. [2.7]   Greville (1962) proposed a recursive algorithm for computing the pseudoinverse of $A \in R^{m \times n}$. Let a_i and A^i represent the ith column of A and the matrix consisting of the first i columns of A, respectively. The algorithm follows.

Algorithm 2.1   (Greville's Algorithm)

1. $A^{1\#} = O$ if $a_1 = O$.
   $A^{1\#} = (a_1^T a_1)^{-1} a_1^T$ if $a_1 \neq O$.
   $i = 2$.

2. $d_i = A^{i-1\#} a_i$.
   $c_i = a_i - A^{i-1} d_i$.

3. If $c_i \neq O$, then $b_i^T = c_i^\#$.
   If $c_i = O$, then $b_i^T = (1 + d_i^T d_i)^{-1} d_i^T A^{i-1\#}$.

4. Prepare $A^{i\#}$ for the next iteration:
$$A^{i\#} = \begin{pmatrix} A^{i-1\#} - d_i b_i^T \\ b_i^T \end{pmatrix}$$

5. If $i = n$, then $A^\# = A^{n\#}$ and end.
   If $i < n$, then $i = i + 1$ and go to step 2.

Since this algorithm is concise, it is easy to implement. Indeed, the author programmed this algorithm for a minicomputer, *NOVA 03* (Data General), using *assembly language*, for real-time control of a kinematically redundant manipulator, which will be discussed in Chapter 4. Based on the author's experience, however, this method seems computationally a little sensitive. When it was computed in 32-bit single-precision floating-point arithmetics, it had relatively large computational error for some matrices. The computation was satisfactory when computed in 64-bit double-precision floating-point arithmetic. Thus, computation in 64-bit double-precision floating-point arithmetic, or higher, is strongly recommended.

Example 2.28

We now compute the pseudoinverse of A given by Eq. 2.193 using Algorithm 2.1. In the following, the number indicates the step number of Algorithm 2.1.

1. $$a_1 = \begin{pmatrix} 1 \\ -1 \end{pmatrix} \neq O$$

[2.7] Boullion and Odell 1971.

$$A^{1\#} = (a_1^T a_1)^{-1} a_1^T$$
$$= \{(1 \quad -1)\begin{pmatrix}1\\-1\end{pmatrix}\}^{-1}\begin{pmatrix}1\\-1\end{pmatrix}^T$$
$$= \frac{1}{2}(1 \quad -1)$$
$$i = 2$$

2. $$d_2 = A^{1\#} a_2$$
$$= \frac{1}{2}(1 \quad -1)\begin{pmatrix}-1\\1\end{pmatrix}$$
$$= -1$$
$$c_2 = a_2 - A^1 d_2$$
$$= \begin{pmatrix}-1\\1\end{pmatrix} - \begin{pmatrix}1\\-1\end{pmatrix}(-1)$$
$$= \begin{pmatrix}0\\0\end{pmatrix} = O$$

3. $$b_2^T = (1 + d_2^T d_2)^{-1} d_2^T A^{1\#}$$
$$= \{1 + (-1)(-1)\}^{-1}(-1) \cdot \frac{1}{2}(1 \quad -1)$$
$$= \frac{1}{4}(-1 \quad 1)$$

4. $$A^{2\#} = \begin{pmatrix}A^{1\#} - d_2 b_2^T\\b_2^T\end{pmatrix}$$
$$= \begin{pmatrix}\frac{1}{2}(1 \quad -1) - (-1)\cdot\frac{1}{4}(-1 \quad 1)\\\frac{1}{4}(-1 \quad 1)\end{pmatrix}$$
$$= \frac{1}{4}\begin{pmatrix}1 & -1\\-1 & 1\end{pmatrix}$$

5. $$i = 2 < 3 = m$$
$$i = 3$$

2. $$d_3 = A^{2\#} a_3$$
$$= \frac{1}{4}\begin{pmatrix}1 & -1\\-1 & 1\end{pmatrix}\begin{pmatrix}1\\-1\end{pmatrix}$$
$$= \frac{1}{2}\begin{pmatrix}1\\-1\end{pmatrix}$$
$$c_3 = a_3 - A^2 d_3$$

2.4 Generalized Inverse and Pseudoinverse

$$= \begin{pmatrix} 1 \\ -1 \end{pmatrix} - \begin{pmatrix} 1 & -1 \\ -1 & 1 \end{pmatrix} \cdot \frac{1}{2} \begin{pmatrix} 1 \\ -1 \end{pmatrix}$$

$$= \begin{pmatrix} 0 \\ 0 \end{pmatrix} = O$$

3. $$\boldsymbol{b_3}^T = (1 + \boldsymbol{d_3}^T \boldsymbol{d_3})^{-1} \boldsymbol{d_3}^T \boldsymbol{A}^{2\#}$$

$$= \{1 + \frac{1}{2}(1 \quad -1) \cdot \frac{1}{2}\begin{pmatrix} 1 \\ -1 \end{pmatrix}\}^{-1}$$

$$\cdot \frac{1}{2}(1 \quad -1) \cdot \frac{1}{4}\begin{pmatrix} 1 & -1 \\ -1 & 1 \end{pmatrix}$$

$$= \frac{1}{6}(1 \quad -1)$$

4. $$\boldsymbol{A}^{3\#} = \begin{pmatrix} \boldsymbol{A}^{2\#} - \boldsymbol{d_3} \boldsymbol{b_3}^T \\ \boldsymbol{b_3}^T \end{pmatrix}$$

$$= \begin{pmatrix} \frac{1}{4}\begin{pmatrix} 1 & -1 \\ -1 & 1 \end{pmatrix} - \frac{1}{2}\begin{pmatrix} 1 \\ -1 \end{pmatrix} \cdot \frac{1}{6}(1 \quad -1) \\ \frac{1}{6}(1 \quad -1) \end{pmatrix}$$

$$= \frac{1}{6}\begin{pmatrix} 1 & -1 \\ -1 & 1 \\ 1 & -1 \end{pmatrix}$$

5. $$\boldsymbol{A}^{\#} = \boldsymbol{A}^{3\#}$$

Pseudoinverse of Full Rank Matrices. When a matrix $\boldsymbol{A} \in R^{m \times n}$ is full rank, the pseudoinverse of $\boldsymbol{A}$ is computed using the regular inverse of a nonsingular matrix. From Eqs. 2.140 and 2.141, the pseudoinverse is computed as follows.

Algorithm 2.2

1. If $m < n$ and rank $\boldsymbol{A} = m$, then

$$\boldsymbol{A}^{\#} = \boldsymbol{A}^T (\boldsymbol{A} \boldsymbol{A}^T)^{-1} \qquad 2.203$$

2. If $m > n$ and rank $\boldsymbol{A} = n$, then

$$\boldsymbol{A}^{\#} = (\boldsymbol{A}^T \boldsymbol{A})^{-1} \boldsymbol{A}^T \qquad 2.204$$

From Eq. 2.142, $A^\# = A^{-1}$ if $m = n$ and rank $A = m$. Note that Eq. 2.204 suggests that the pseudoinverse of a vector $a \in R^n$ can be computed by

$$a^\# = a^T / \| a \|^2 \qquad 2.205$$

Example 2.22 also works as an example of Algorithm 2.2.

Klein and Huang's Algorithm for Computing Eq. 2.154. In order to control kinematical redundancy as discussed in Chapters 3, 4, and 5, it is extremely important to compute the solution of Eq. 2.154 for a matrix $A \in R^{m \times n}$ ($m < n$) and a specified $z \in R^n$. Assuming that A is full rank, we can compute it efficiently using Eq. 2.203, without actually computing any inverse matrix. The following algorithm was proposed by Klein and Huang (1983).

Algorithm 2.3 (Klein and Huang 1983)

1. $p = y - Az$, $B = AA^T$.
2. Solve $p = Bq$ by the Gaussian elimination.
3. $x = A^T q + z$.

We can derive Algorithm 2.3 readily by substituting Eq. 2.203 into Eq. 2.154 and setting $q = (AA^T)^{-1}(y - Az)$.

Example 2.29

We compute the following equation:

$$x = A^\# y + (E - A^\# A) z \qquad 2.206$$

where A, y, and z are given as

$$A = \begin{pmatrix} 1 & -1 & 0 \\ 0 & 1 & 1 \end{pmatrix}$$

$$y = \begin{pmatrix} 1 \\ 0 \end{pmatrix} \qquad 2.207$$

$$z = \begin{pmatrix} 1 \\ 1 \\ 1 \end{pmatrix}$$

$A^\#$ is computed as

$$A^\# = A^T (AA^T)^{-1} = \frac{1}{3} \begin{pmatrix} 2 & 1 \\ -1 & 1 \\ 1 & 2 \end{pmatrix} \qquad 2.208$$

2.4 Generalized Inverse and Pseudoinverse

Substituting Eqs. 2.207 and 2.208 into Eq. 2.206 results in

$$x = \frac{1}{3}\begin{pmatrix} 2 & 1 \\ -1 & 1 \\ 1 & 2 \end{pmatrix}\begin{pmatrix} 1 \\ 0 \end{pmatrix} + \{\begin{pmatrix} 1 & 0 & 0 \\ 0 & 1 & 0 \\ 0 & 0 & 1 \end{pmatrix}$$

$$-\frac{1}{3}\begin{pmatrix} 2 & 1 \\ -1 & 1 \\ 1 & 2 \end{pmatrix}$$

$$\begin{pmatrix} 1 & -1 & 0 \\ 0 & 1 & 1 \end{pmatrix}\}\begin{pmatrix} 1 \\ 1 \\ 1 \end{pmatrix}$$

$$= \begin{pmatrix} 1 \\ 0 \\ 0 \end{pmatrix}$$

2.209

The total computation involved in the procedure is 53 multiplications, 31 additions, and 1 division.

Now, we apply Algorithm 2.3 as follows:

1.
$$p = y - Az$$
$$= \begin{pmatrix} 1 \\ 0 \end{pmatrix} - \begin{pmatrix} 1 & -1 & 0 \\ 0 & 1 & 1 \end{pmatrix}\begin{pmatrix} 1 \\ 1 \\ 1 \end{pmatrix}$$
$$= \begin{pmatrix} 1 \\ -2 \end{pmatrix}$$
$$B = AA^T$$
$$= \begin{pmatrix} 2 & -1 \\ -1 & 2 \end{pmatrix}$$

2. Solve $\begin{pmatrix} 1 \\ -2 \end{pmatrix} = \begin{pmatrix} 2 & -1 \\ -1 & 2 \end{pmatrix} q$ by Gaussian elimination.

$$q = \begin{pmatrix} 0 \\ -1 \end{pmatrix}$$

3.
$$x = A^T q + z$$
$$= \begin{pmatrix} 1 & 0 \\ -1 & 1 \\ 0 & 1 \end{pmatrix}\begin{pmatrix} 0 \\ -1 \end{pmatrix} + \begin{pmatrix} 1 \\ 1 \\ 1 \end{pmatrix}$$
$$= \begin{pmatrix} 1 \\ 0 \\ 0 \end{pmatrix}$$

The total computation needed is 27 multiplications, 21 additions, and 2 divisions. Note that, in this example, the computation of the Gaussian elimination is as expensive as that for $(AA^T)^{-1} \in R^{2\times 2}$ matrix. For larger matrices, the Gaussian elimination is far more computationally efficient.

2.5 Variational Methods

2.5.1 Euler–Lagrange Equation

Many problems of optimization are described as follows. Find $\boldsymbol{x}(t) \in R^n$, $t_1 \leq t \leq t_2$, that minimizes

$$Q = \int_{t_1}^{t_2} f(\boldsymbol{x}(t), \dot{\boldsymbol{x}}(t), t)\, dt \qquad 2.210$$

where $f(*)$ is a scalar function. We assume that $f(*)$ is continuous with respect to $\boldsymbol{x}$, $\dot{\boldsymbol{x}}$, and t, and has continuous partial derivatives with respect to $\boldsymbol{x}$ and $\dot{\boldsymbol{x}}$.

We assume $\boldsymbol{x}(t_1)$ and $\boldsymbol{x}(t_2)$ are given. Now a variation $\boldsymbol{h}(t) \in R^n$ is a differentiable vector that satisfies $\boldsymbol{h}(t_1) = \boldsymbol{h}(t_2) = \boldsymbol{O}$. Figure 2.6 shows the conceptual image of variation. The following theorem shows the principle of variational methods (Luenberger 1969).

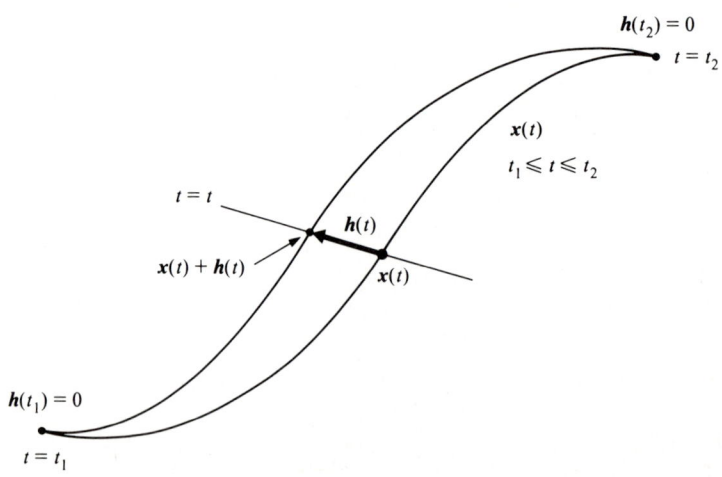

Figure 2.6 Conceptual image of variation.

2.5 Variational Methods

Theorem 2.7

A necessary condition for $x(t) \in R^n$ to give an extreme value of Eq. 2.210 is that the following equation be satisfied for all $h(t) \in R^n$:

$$\delta Q(x; h) \triangleq \frac{d}{ds} \int_{t_1}^{t_2} f(x^*, \dot{x}^*, t) \, dt \Big|_{s=0} = O \qquad 2.211$$

$$x^* = x + sh$$
$$\dot{x}^* = \dot{x} + s\dot{h}$$

For a given $h(t)$, as s changes, the shape of trajectory $x^*(t)$ ($t_1 \leq t \leq t_2$) changes continuously. When $s = 0$, $x^*(t)$ coincides with $x(t)$. When $s = \pm 1$, $x^*(t)$ coincides with $x(t) \pm h(t)$. Equation 2.211 implies that Q for $x^*(t)$ should take a stationary value at $s = 0$ in order that $x(t)$ provides an extreme value of Q. This condition must be fulfilled for an arbitrary $h(t)$ that satisfies $h(t_1) = h(t_2) = O$.

Equation 2.211 can be computed as follows:

$$\delta Q(x; h) = \int_{t_1}^{t_2} \frac{\partial f}{\partial x} h + \frac{\partial f}{\partial \dot{x}} \dot{h} \, dt = 0 \qquad 2.212$$

The second term of the integrand of Eq. 2.212 is computed by partial integration, as follows:

$$\int_{t_1}^{t_2} \frac{\partial f}{\partial \dot{x}} \dot{h} \, dt = \frac{\partial f}{\partial \dot{x}} h \Big|_{t_1}^{t_2} - \int_{t_1}^{t_2} \frac{d}{dt}\left(\frac{\partial f}{\partial \dot{x}}\right) h \, dt \qquad 2.213$$

Substituting Eq. 2.213 with $h(t_1) = h(t_2) = 0$, Eq. 2.212 becomes

$$\delta Q(x; h) = \int_{t_1}^{t_2} \left\{ \frac{\partial f}{\partial x} - \frac{d}{dt}\left(\frac{\partial f}{\partial \dot{x}}\right) \right\} h \, dt = 0 \qquad 2.214$$

In Eq. 2.214, the coefficient of h is continuous and h satisfies $h(t_1) = h(t_2) = O$. The $x(t)$ that satisfies Eq. 2.214 for all $h(t)$, must satisfy

$$\frac{d}{dt}\left(\frac{\partial f}{\partial \dot{x}}\right) - \frac{\partial f}{\partial x} = O \qquad 2.215$$

because, if the ith component of the left-hand side of Eq. 2.215 is positive (or negative) for $t_1 < t_3 \leq t \leq t_4 < t_2$, then Eq. 2.214 will be unsatisfied if we choose, for example,

$$h_i(t) = (t - t_3)^2 (t - t_4)^2 \quad \text{for } t_3 \leq t \leq t_4$$
$$= 0 \quad \text{for the other } t$$
$$h_j(t) = 0 \quad \text{for } j \neq i \text{ and } t_1 \leq t \leq t_2$$

Hence, the following corollary is derived from Theorem 2.7.

Corollary 2.1 (Euler–Lagrange Equation)

A necessary condition for $\boldsymbol{x}(t)$ to provide an extreme value of

$$Q = \int_{t_1}^{t_2} f(\boldsymbol{x}(t), \dot{\boldsymbol{x}}(t), t)\, dt \qquad 2.216$$

is that $\boldsymbol{x}(t)$ should satisfy the following equation:

$$\frac{d}{dt}\left(\frac{\partial f}{\partial \dot{\boldsymbol{x}}}\right) - \frac{\partial f}{\partial \boldsymbol{x}} = \boldsymbol{O} \qquad 2.217$$

Equation 2.217 is called the *Euler–Lagrange equation*; it is the fundamental equation of the variational methods.

Example 2.30

Let $\boldsymbol{x}(t) = (x_1(t)\ x_2(t))^T$. We prove that the shortest trajectory that connects $\boldsymbol{x}(t_1)$ and $\boldsymbol{x}(t_2)$ is the straight line passing through these two points. The cost function is given by

$$Q = \int_{t_1}^{t_2} f(\boldsymbol{x}, \dot{\boldsymbol{x}}, t)\, dt = \int_{t_1}^{t_2} \sqrt{\dot{x}_1^2 + \dot{x}_2^2}\, dt \qquad 2.218$$

which computes the total length of trajectory. The Euler–Lagrange equation is computed as follows:

$$\frac{\partial f}{\partial \dot{\boldsymbol{x}}} = \left(\frac{\dot{x}_1}{\sqrt{\dot{x}_1^2 + \dot{x}_2^2}}\quad \frac{\dot{x}_2}{\sqrt{\dot{x}_1^2 + \dot{x}_2^2}}\right) \qquad 2.219$$

$$\frac{d}{dt}\left(\frac{\partial f}{\partial \dot{\boldsymbol{x}}}\right) - \frac{\partial f}{\partial \boldsymbol{x}} = \frac{d}{dt}\left(\frac{\dot{x}_1}{\sqrt{\dot{x}_1^2 + \dot{x}_2^2}}\quad \frac{\dot{x}_2}{\sqrt{\dot{x}_1^2 + \dot{x}_2^2}}\right) = 0 \qquad 2.220$$

Equation 2.220 yields

$$\frac{\dot{x}_1}{\sqrt{\dot{x}_1^2 + \dot{x}_2^2}} = c_0 \qquad c_0 : \text{constant} \qquad 2.221$$

Hence, we have

$$\frac{\dot{x}_2}{\dot{x}_1} = \pm \sqrt{\frac{1 - c_0^2}{c_0^2}} \qquad (c_0 \neq 0) \qquad 2.222$$

Equation 2.222 implies that the shortest trajectory is a straight line.

2.5.2 Transversality Condition

In Section 2.5.1 $\boldsymbol{x}(t_1)$ and $\boldsymbol{x}(t_2)$ were assumed to be given. Now, let us relax this rather strict condition. We assume $\boldsymbol{x}(t_2)$ to be

$$\boldsymbol{x}(t_2) \in \{\boldsymbol{x} : g(\boldsymbol{x}) = 0\} \qquad 2.223$$

where $g(\boldsymbol{x})$ is a scalar function. Our problem is to find a necessary condition for $\boldsymbol{x}(t)$ that minimizes Eq. 2.210 among all the trajectories starting from $\boldsymbol{x}(t_1)$ and shooting $\boldsymbol{x}$ that satisfies $g(\boldsymbol{x}) = 0$ at $t = t_2$. The variation $\boldsymbol{h}(t)$ is now such that $\boldsymbol{h}(t_1) = \boldsymbol{O}$ and

$$g(\boldsymbol{x}(t_2) + \boldsymbol{h}(t_2)) = 0 \qquad 2.224$$

where $\boldsymbol{x}(t_2)$ is the value of the optimal trajectory $\boldsymbol{x}(t)$ at $t = t_2$ and $\boldsymbol{h}(t)$ is assumed to be infinitesimal. Equation 2.213 becomes

$$\int_{t_1}^{t_2} \frac{\partial f}{\partial \dot{\boldsymbol{x}}} \boldsymbol{h}\, dt = \frac{\partial f}{\partial \dot{\boldsymbol{x}}}\Big|_{t_2} \boldsymbol{h}(t_2) - \int_{t_1}^{t_2} \frac{d}{dt}\left(\frac{\partial f}{\partial \dot{\boldsymbol{x}}}\right) \boldsymbol{h}\, dt \qquad 2.225$$

Hence, Eq. 2.212 is now calculated as follows:

$$\delta Q(\boldsymbol{x}; \boldsymbol{h}) = \frac{\partial f}{\partial \dot{\boldsymbol{x}}}\Big|_{t_2} \boldsymbol{h}(t_2) + \int_{t_1}^{t_2} \left\{\frac{\partial f}{\partial \boldsymbol{x}} - \frac{d}{dt}\left(\frac{\partial f}{\partial \dot{\boldsymbol{x}}}\right)\right\} \boldsymbol{h}\, dt = 0 \qquad 2.226$$

Compare Eq. 2.226 with Eq. 2.214. For Eq. 2.226 to hold for all $\boldsymbol{h}(t)$ such that $\boldsymbol{h}(t_1) = \boldsymbol{O}$ and $g(\boldsymbol{x}(t_2) + \boldsymbol{h}(t_2)) = 0$, the following conditions must be satisfied:

$$\frac{d}{dt}\left(\frac{\partial f}{\partial \dot{\boldsymbol{x}}}\right) - \frac{\partial f}{\partial \boldsymbol{x}} = \boldsymbol{O} \qquad 2.227$$

$$\frac{\partial f}{\partial \dot{\boldsymbol{x}}}\Big|_{t_2} \boldsymbol{h}(t_2) = 0 \qquad 2.228$$

Equation 2.227 is derived from the fact that Eq. 2.226 should be satisfied even for $\boldsymbol{h}(t_1) = \boldsymbol{h}(t_2) = \boldsymbol{O}$. From Eqs. 2.226 and 2.227, Eq. 2.228 is yielded. Equation 2.227 is the same Euler–Lagrange equation as Eq. 2.217. Since $\boldsymbol{h}(t_2)$ is infinitesimal, subtracting $g(\boldsymbol{x}(t_2)) = 0$ from Eq. 2.224 results in

$$\frac{\partial g}{\partial \boldsymbol{x}} \boldsymbol{h}(t_2) = 0 \qquad 2.229$$

Equation 2.229 must hold for any infinitesimal $\boldsymbol{h}(t_2)$. Geometrically, Eq. 2.229 implies that $\boldsymbol{h}(t_2)$ must be a tangential vector of the manifold $g(\boldsymbol{x}) = 0$ at $\boldsymbol{x}(t_2)$, because $(\partial g/\partial \boldsymbol{x})^T$ is nothing but a normal vector. From Eq. 2.228, $\partial f/\partial \dot{\boldsymbol{x}}\big|_{t_2}$ must be perpendicular to all of the tangential vectors at $\boldsymbol{x}(t_2)$. Hence, we have

$$\frac{\partial f}{\partial \dot{\boldsymbol{x}}}\Big|_{t_2} = \alpha \frac{\partial g}{\partial \boldsymbol{x}} \qquad 2.230$$

where α is an appropriate scalar. Equation 2.230 implies that $\partial f / \partial \dot{\boldsymbol{x}}|_{t_2}$ of the optimal trajectory should be perpendicular to the manifold of $g(\boldsymbol{x}) = 0$ at $\boldsymbol{x}(t_2)$. Equation 2.230 is called a *transversality condition*. Figure 2.7 shows this geometric relationship. When $\boldsymbol{x}(t_1)$ is not given, but a manifold where $\boldsymbol{x}(t_1)$ should be is specified, we can obtain the equivalent condition by repeating the same procedure.

Example 2.31

Assume that $\boldsymbol{x} = (x_1\ x_2)^T$. $\boldsymbol{x}(t)$ starts from $\boldsymbol{x}(t_1) = (x_1(t_1)\ x_2(t_1))^T$ and arrives at a line $g(\boldsymbol{x}) = x_1 - k = 0$ at $t = t_2$. Find the shortest trajectory.

The cost function is provided by Eq. 2.218. In Example 2.30, it was shown that the shortest trajectory should satisfy Eq. 2.222, regardless of the boundary conditions. When we substitute Eq. 2.219 and $\partial g / \partial \boldsymbol{x} = (1\ 0)$ into Eq. 2.230, the transversality condition becomes

$$\left(\frac{\dot{x}_1}{\sqrt{\dot{x}_1^2 + \dot{x}_2^2}}\ \frac{\dot{x}_2}{\sqrt{\dot{x}_1^2 + \dot{x}_2^2}} \right) = \alpha(1\ 0) \qquad 2.231$$

Therefore, $\dot{x}_2 = 0$ is required from the transversality condition. Consequently, the optimal trajectory becomes a straight line parallel to the x_1 axis. Figure 2.8 shows the relationship. This example proves that the shortest trajectory from a point to a line is a straight line that passes through the point and is perpendicular to the line.

2.5.3 Lagrange Multiplier Theorem

Theorem 2.7 and Corollary 2.1 provide necessary conditions when $\boldsymbol{x}(t)$ is subject to no constraint other than $\boldsymbol{x}(t_1)$ and $\boldsymbol{x}(t_2)$ or the manifolds at $t = t_1$ and t_2. These problems are called *unconstrained optimization problems*. Lagrange multipliers are used to solve optimization problems with constraints.

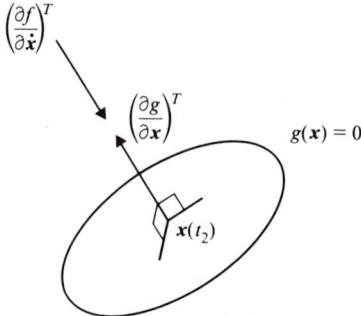

Figure 2.7 Transversality condition of variational methods.

2.5 Variational Methods

The theorem to be provided in this subsection offers the rigorous mathematical background for such optimization. We begin with the following problem: Find $\boldsymbol{x}$ that minimizes $f(\boldsymbol{x})$ under the constraints of

$$\boldsymbol{g}(\boldsymbol{x}) = \begin{pmatrix} g_1(\boldsymbol{x}) \\ g_2(\boldsymbol{x}) \\ \vdots \\ g_m(\boldsymbol{x}) \end{pmatrix} = \boldsymbol{O} \in R^m \qquad 2.232$$

We need the following definition.

Definition 2.5 (Regular Point)

An $\boldsymbol{x}_0$ that satisfies the constraints Eq. 2.232 is said to be a *regular point* if $(\partial g_i/\partial \boldsymbol{x})^T$ $(i = 1, \cdots, m)$ are linearly independent at $\boldsymbol{x} = \boldsymbol{x}_0$.

Example 2.32

Let $\boldsymbol{x}$ and its constraint equation $\boldsymbol{g}(\boldsymbol{x}) = \boldsymbol{O}$ be as follows:

$$\boldsymbol{x} = \begin{pmatrix} x_1 \\ x_2 \end{pmatrix} \qquad 0 \leq x_1 \leq 2\pi, \ 0 \leq x_2 \leq 2\pi \qquad 2.233$$

$$\boldsymbol{g}(\boldsymbol{x}) = \begin{pmatrix} \cos x_1 + \cos(x_1 + x_2) \\ \sin x_1 + \sin(x_1 + x_2) \end{pmatrix} = \boldsymbol{O} \qquad 2.234$$

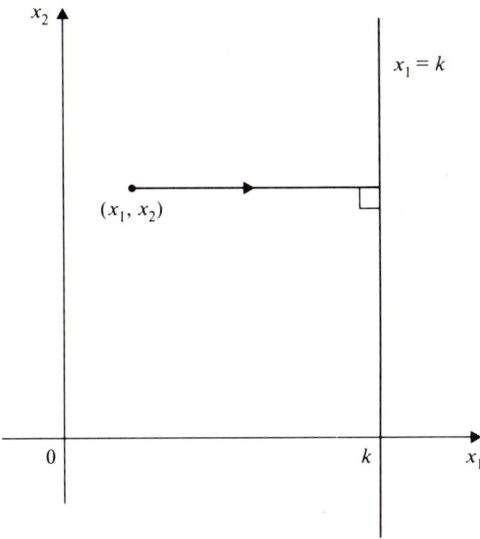

Figure 2.8 Optimal trajectory: Example 2.31.

Note that the solution of Eq. 2.234 is given by

$$\begin{cases} x_1 : \text{arbitrary} \\ x_2 = \pi \end{cases} \qquad 2.235$$

Now, we show that no x that satisfies Eq. 2.234 is a regular point. We compute

$$\frac{\partial g}{\partial x} = \begin{pmatrix} \partial g_1/\partial x \\ \partial g_2/\partial x \end{pmatrix}$$

$$= \begin{pmatrix} -\sin x_1 - \sin(x_1 + x_2) & -\sin(x_1 + x_2) \\ \cos x_1 + \cos(x_1 + x_2) & \cos(x_1 + x_2) \end{pmatrix} \qquad 2.236$$

Therefore,

$$\frac{\partial g}{\partial x}\bigg|_{x_2 = \pi} = \begin{pmatrix} 0 & \sin x_1 \\ 0 & -\cos x_1 \end{pmatrix} \qquad 2.237$$

Equation 2.237 implies that $(\partial g_1/\partial x)^T$ and $(\partial g_2/\partial x)^T$ are linearly dependent. Since every x that satisfies $g(x) = O$ should have $x_2 = \pi$, from Eq. 2.235, we can conclude that no x satisfying Eq. 2.234 is a regular point.

Definition 2.5 means that an x_0 satisfying Eq. 2.232 is a regular point if rank $\{\partial g/\partial x\} = m$ at $x = x_0$. The following important theorem is derived using the concept of a regular point.

Theorem 2.8 (Lagrange Multiplier Theorem)

Let x_0 provide an extreme value of $f(x)$ and satisfy the constraints of

$$g(x) = \begin{pmatrix} g_1(x) \\ g_2(x) \\ \vdots \\ g_m(x) \end{pmatrix} = O \in R^m \qquad 2.238$$

If x_0 is a regular point of the constraints, then there exist scalars λ_i ($i = 1, \cdots, m$) that provide the stationary value[2.8] of the following function at x_0:

$$f(x) + \lambda^T g(x) \qquad 2.239$$
$$\lambda = \text{col}(\lambda_1, \lambda_2, \cdots, \lambda_m) \in R^m$$

where the λ_is are called the *Lagrange multipliers*.

[2.8] Equation 2.239 is said to have a *stationary value* at $x = x_0$ when the partial derivative of Eq. 2.239 with respect to x becomes equal to zero at $x = x_0$.

2.5 Variational Methods

Example 2.33 (Gradient Method)

Let $p(\boldsymbol{x})$ be a potential function of $\boldsymbol{x} \in R^n$. Find the direction of infinitesimal change $\Delta \boldsymbol{x}$ that provides the largest change in $p(\boldsymbol{x})$. Assume that $\Delta \boldsymbol{x}$ has a constant magnitude, δ. Hence, we have the following constraint:

$$g(\Delta \boldsymbol{x}) = \Delta \boldsymbol{x}^T \Delta \boldsymbol{x} - \delta^2 = 0 \qquad \delta > 0 \qquad 2.240$$

The change of $p(\boldsymbol{x})$ is represented by

$$\Delta p = \frac{\partial p}{\partial \boldsymbol{x}} \Delta \boldsymbol{x} \qquad 2.241$$

Using the Lagrange multiplier λ, the following augmented function is prepared:

$$p^* = \Delta p + \lambda \ g(\Delta \boldsymbol{x}) \qquad 2.242$$

Since $\partial g / \partial \Delta \boldsymbol{x} = 2 \Delta \boldsymbol{x}^T \neq \boldsymbol{O}$, $\Delta \boldsymbol{x}$ is always a regular point on the constraint manifold $g(\Delta \boldsymbol{x}) = 0$. From the Lagrange multiplier theorem, λ must give a stationary value of p^*. Therefore,

$$\frac{\partial p^*}{\partial \Delta \boldsymbol{x}} = \frac{\partial p}{\partial \boldsymbol{x}} + 2 \lambda \Delta \boldsymbol{x}^T = \boldsymbol{O} \qquad 2.243$$

Equation 2.243 implies that $\Delta \boldsymbol{x}$ must be parallel to $(\partial p / \partial \boldsymbol{x})^T$. Therefore, the direction of $\Delta \boldsymbol{x}$ that maximizes Δp of Eq. 2.241 is that of the gradient of $p(\boldsymbol{x})$. Namely,

$$\Delta \boldsymbol{x} = \delta \ (\frac{\partial p}{\partial \boldsymbol{x}})^T / \| \frac{\partial p}{\partial \boldsymbol{x}} \| \qquad 2.244$$

Substituting Eq. 2.244 into Eq. 2.241, the change of $p(\boldsymbol{x})$ obtained by the optimal $\Delta \boldsymbol{x}$ becomes

$$\Delta p = \delta \ \| \frac{\partial p}{\partial \boldsymbol{x}} \| \qquad 2.245$$

Consequently, the direction of the gradient indicates the best direction of $\Delta \boldsymbol{x}$, and the Euclidean norm of the gradient represents the ratio of the change of $p(\boldsymbol{x})$ and the magnitude of $\Delta \boldsymbol{x}$ in the best direction.

We now use the lagrange multiplier theorem to obtain a necessary condition for $\boldsymbol{x}(t)$ to minimize Eq. 2.210 under the constraints of Eq. 2.232. The following theorem is the extension of Corollary 2.1 to constrained problems (Luenberger 1969).

Theorem 2.9

Let $\boldsymbol{x}(t) \in R^n$ satisfy the following constraints:

$$g(x) = \begin{pmatrix} g_1(x) \\ g_2(x) \\ \vdots \\ g_m(x) \end{pmatrix} = O \in R^m \qquad 2.246$$

A necessary condition for $x(t)$ to provide an extreme value of

$$Q = \int_{t_1}^{t_2} f(x(t), \dot{x}(t), t)\, dt \qquad 2.247$$

is that the following equation is satisfied:

$$\frac{d}{dt}\left(\frac{\partial f}{\partial \dot{x}}\right) - \frac{\partial f}{\partial x} - \lambda^T(t)\frac{\partial g}{\partial x} = O \qquad 2.248$$

where $\lambda(t) = \mathrm{col}(\lambda_1, \lambda_2, \cdots, \lambda_m)$, and λ_i ($i = 1, \cdots, m$) are Lagrange multipliers. Note that $f(*)$ and $g_i(*)$ are assumed to be real functions with continuous second partial derivatives.

Example 2.34

Let two points, $x_1 = (x_1\ y_1\ z_1)^T$ and $x_2 = (x_2\ y_2\ z_2)^T$, be on the following sphere:

$$g(x, y, z) = x^2 + y^2 + z^2 - r^2 = 0 \qquad 2.249$$

Find the trajectory on the sphere that goes from x_1 to x_2 with the minimum length.

The cost function is given by

$$Q = \int_{t_1}^{t_2} f(x, \dot{x}, t)\, dt$$
$$= \int_{t_1}^{t_2} \sqrt{\dot{x}^2 + \dot{y}^2 + \dot{z}^2}\, dt \qquad 2.250$$

From Theorem 2.9, the trajectory should satisfy

$$\frac{d}{dt}\left(\frac{\partial f}{\partial \dot{x}}\right) - \frac{\partial f}{\partial x} - \lambda \frac{\partial g}{\partial x} = O \qquad 2.251$$

where

$$\frac{\partial f}{\partial \dot{x}} = \left(\frac{\dot{x}}{\sqrt{\dot{x}^2 + \dot{y}^2 + \dot{z}^2}}\ \frac{\dot{y}}{\sqrt{\dot{x}^2 + \dot{y}^2 + \dot{z}^2}}\ \frac{\dot{z}}{\sqrt{\dot{x}^2 + \dot{y}^2 + \dot{z}^2}}\right)$$

$$\frac{\partial f}{\partial x} = O$$

$$\frac{\partial g}{\partial x} = (2x\ 2y\ 2z)$$

Consequently, we have

$$\frac{d}{dt}\left(\frac{\dot{x}}{\sqrt{\dot{x}^2+\dot{y}^2+\dot{z}^2}}\right) = 2\lambda x$$

$$\frac{d}{dt}\left(\frac{\dot{y}}{\sqrt{\dot{x}^2+\dot{y}^2+\dot{z}^2}}\right) = 2\lambda y \qquad 2.252$$

$$\frac{d}{dt}\left(\frac{\dot{z}}{\sqrt{\dot{x}^2+\dot{y}^2+\dot{z}^2}}\right) = 2\lambda z$$

Eliminating λ in Eq. 2.252, we can get two differential equations, which sufficiently describe a curve in the xyz space. However, this computation is very complex. To simplify the computation, we prepare the following assumptions.

1. $\boldsymbol{x}(t)$ moves with constant velocity; namely,

$$\dot{x}^2 + \dot{y}^2 + \dot{z}^2 = v^2 \qquad v > 0. \qquad 2.253$$

2. $\boldsymbol{x}_1 = (r\ 0\ 0)^T$.

Note that the first assumption imposes nothing on the shape of the trajectory. The second assumption also does not lose generality because, by changing the direction of the coordinate frame, we can transform any $\boldsymbol{x}_1$ into the second assumption.

Using the first assumption, Eq. 2.252 becomes

$$\ddot{x} = 2\lambda v x$$
$$\ddot{y} = 2\lambda v y \qquad 2.254$$
$$\ddot{z} = 2\lambda v z$$

Differentiating Eq. 2.249 twice, we have

$$x\ddot{x} + y\ddot{y} + z\ddot{z} + \dot{x}^2 + \dot{y}^2 + \dot{z}^2 = 0 \qquad 2.255$$

Substituting Eqs. 2.253 and 2.254 into Eq. 2.255 and using Eq. 2.249, we calculate λ as follows:

$$\lambda = -\frac{1}{2}\frac{v}{r^2} \qquad 2.256$$

Substituting Eq. 2.256, Eq. 2.254 forms three independent differential equations. Their solutions are obtained as follows:

$$x = a_1 \sin(\omega t + \theta_1)$$
$$y = a_2 \sin(\omega t + \theta_2) \qquad 2.257$$
$$z = a_3 \sin(\omega t + \theta_3)$$

where $\omega \triangleq v/r$, and a_i and θ_i ($i = 1, 2, 3$) are constants. Equation 2.257

includes six constants. Four of them can be eliminated using Eq. 2.249 and the second assumption. Accordingly, we have

$$\begin{aligned} x &= r\ \cos \omega t \\ y &= a_2 \sin \omega t \\ z &= a_3 \sin \omega t \end{aligned} \qquad 2.258$$

The cross-product of $\boldsymbol{x}_1 = (r\ 0\ 0)^T$ and $\boldsymbol{x}(t) = (x\ y\ z)^T$ becomes

$$\boldsymbol{x}_1 \times \boldsymbol{x}(t) = \begin{pmatrix} 0 \\ -a_3\, r\ \sin\ \omega t \\ a_2\, r\ \sin\ \omega t \end{pmatrix} \qquad 2.259$$

Equation 2.259 shows that the direction of $\boldsymbol{x}_1 \times \boldsymbol{x}(t)$ is invariant, which implies that the trajectory must be on a great circle passing through $\boldsymbol{x}_1$ and $\boldsymbol{x}_2$. Figure 2.9 shows the trajectory. The trajectory on a given surface connecting two points with minimum length, is generally called a *geodesic line*. Geodesic lines of a sphere are great circles.

2.5.4 Kuhn–Tucker Theorem

The Lagrange multiplier theorem reduces a minimization (maximization) problem subject to equality constraints to a constraint-free stationary value problem. This classical result of variational methods was extended to minimization (maximization) under inequality constraints by Kuhn and Tucker (1951). The main result to be obtained in this subsection is called the *Kuhn–Tucker theorem*, after these researchers. This theorem has served as the basis of *nonlinear programming*.

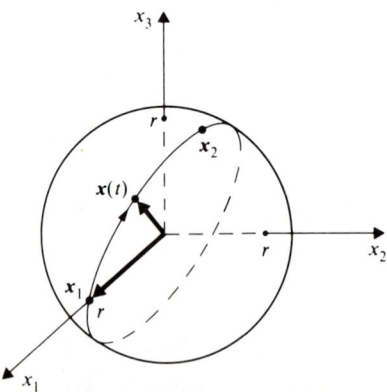

Figure 2.9 Optimal trajectory: Example 2.34.

Now, our problem is to minimize a scalar function $f(\boldsymbol{x})$, $\boldsymbol{x} \in R^n$, under the following constraints:

$$\boldsymbol{g}(\boldsymbol{x}) = \begin{pmatrix} g_1(\boldsymbol{x}) \\ g_2(\boldsymbol{x}) \\ \vdots \\ g_m(\boldsymbol{x}) \end{pmatrix} = \boldsymbol{O} \in R^m \qquad 2.260$$

$$\boldsymbol{r}(\boldsymbol{x}) = \begin{pmatrix} r_1(\boldsymbol{x}) \\ r_2(\boldsymbol{x}) \\ \vdots \\ r_l(\boldsymbol{x}) \end{pmatrix} \leq \boldsymbol{O} \in R^l \qquad 2.261$$

where Eq. 2.261 implies $r_i(\boldsymbol{x}) \leq 0$, for $i = 1, \cdots, l$. First, we have to extend the definition of a regular point (Definition 2.5) as follows.

Definition 2.6 (Regular Point Extended)

Let J be the set of indices of active inequality conditions; namely, $r_j(\boldsymbol{x}) = 0$, for $j \in J$. An $\boldsymbol{x}_0$ that satisfies the constraints of Eqs. 2.260 and 2.261 is said to be a *regular point* if $(\partial g_i/\partial \boldsymbol{x})^T$, $(i = 1, \cdots, m)$, and $(\partial r_j/\partial \boldsymbol{x})^T$, $(j \in J)$, at $x = x_0$ are linearly independent.

Note that, Definition 2.6, the set J comprises the indices of the active constraints among all the inequality constraints. Hence, considering the equality constraints and the active inequality constraints as the total equality constraints, Definition 2.6 is nothing but Definition 2.5. Using the concept of extended regular point, the Kuhn–Tucker theorem (Kuhn and Tucker 1951) is now given.

Theorem 2.10 (Kuhn-Tucker Theorem)

Let $\boldsymbol{x}_0$ provide a local minimum of $f(\boldsymbol{x})$ and satisfy the following equality and inequality constraints:

$$\boldsymbol{g}(\boldsymbol{x}) = \begin{pmatrix} g_1(\boldsymbol{x}) \\ g_2(\boldsymbol{x}) \\ \vdots \\ g_m(\boldsymbol{x}) \end{pmatrix} = \boldsymbol{O} \in R^m \qquad 2.262$$

$$\boldsymbol{r}(\boldsymbol{x}) = \begin{pmatrix} r_1(\boldsymbol{x}) \\ r_2(\boldsymbol{x}) \\ \vdots \\ r_l(\boldsymbol{x}) \end{pmatrix} \leq \boldsymbol{O} \in R^l \qquad 2.263$$

If x_0 is a regular point for Eqs. 2.262 and 2.263, then there exist vectors $\boldsymbol{\lambda} = \text{col}\,(\lambda_1, \lambda_2, \cdots, \lambda_m) \in R^m$ and $\boldsymbol{\mu} = \text{col}\,(\mu_1, \mu_2, \cdots, \mu_l) \geq \boldsymbol{O} \in R^l$ that provide the stationary value of the following function at x_0:

$$f(x) + \boldsymbol{\lambda}^T g(x) + \boldsymbol{\mu}^T r(x) \qquad 2.264$$

and that satisfy

$$\boldsymbol{\mu}^T r(x_0) = 0 \qquad 2.265$$

Note that, since $\boldsymbol{\mu} \geq \boldsymbol{O}$ and $r(x) \leq \boldsymbol{O}$, Eq. 2.265 implies that $\mu_i = 0$ for $r_i(x) < 0$, and that $\mu_i \geq 0$ for $r_i(x) = 0$.

Example 2.35

Minimize the function

$$f(x) = \frac{1}{2} x^T A x + b^T x \qquad 2.266$$

$$x = \begin{pmatrix} x_1 \\ x_2 \end{pmatrix}$$

$$A = \begin{pmatrix} 2 & -1 \\ -1 & 2 \end{pmatrix}$$

$$b = \begin{pmatrix} 1 \\ 0 \end{pmatrix}$$

under the following constraints:

$$g(x) = c^T x - c_0 = 0 \qquad 2.267$$

$$c = \begin{pmatrix} 1 \\ -1 \end{pmatrix}$$

$$c_0 = 2$$

$$r(x) = d^T x \leq 0 \qquad 2.268$$

$$d = \begin{pmatrix} 1 \\ 1 \end{pmatrix}$$

Computing the partial derivative of Eq. 2.264 with respect to x and combining Eq. 2.265, we obtain the following equations:

$$\frac{\partial}{\partial x}\{f(x) + \boldsymbol{\lambda}^T g(x) + \boldsymbol{\mu}^T r(x)\}$$
$$= x^T A + b^T + \lambda c^T + \mu d^T$$
$$= \boldsymbol{O} \qquad 2.269$$

$$\mu d^T x = 0 \qquad 2.270$$

Equations 2.267, 2.269, and 2.270 include four equations and four unknowns. That is,

$$2x_1 - x_2 + 1 + \lambda + \mu = 0 \qquad 2.271$$

$$-x_1 + 2x_2 - \lambda + \mu = 0 \qquad 2.272$$

$$\mu(x_1 + x_2) = 0 \qquad 2.273$$

$$x_1 - x_2 = 2 \qquad 2.274$$

From Eq. 2.273, there are two cases: $\mu = 0$ or $x_1 + x_2 = 0$. For the former, the solution becomes

$$\begin{cases} x_1 = 1/2 \\ x_2 = -3/2 \\ \lambda = -7/2 \\ \mu = 0 \end{cases} \qquad 2.275$$

For the latter, the solution is

$$\begin{cases} x_1 = 1 \\ x_2 = -1 \\ \lambda = -7/2 \\ \mu = -1/2 \end{cases} \qquad 2.276$$

From the Kuhn–Tucker theorem, μ must be nonnegative. Consequently, the optimal solution is given by Eq. 2.275.

In Example 2.35, the cost function is given as a quadratic function and the equality and inequality constraints are linear. This kind of problem is called a *quadratic programming problem*.

2.6 Pontryagin's Maximum Principle

2.6.1 Classical Variational Methods and the Maximum Principle

Optimal control problems in this section are concerned with the dynamics described by the following ordinary differential equation:

$$\dot{x} = f(x, u) \qquad 2.277$$

where $x \in R^n$ is the state variable and $u \in R^m$ is the input variable, and where $\dot{x}$ denotes the derivative of x with respect to time. Our problem is to find $u(t)$, $t_1 \le t \le t_2$, that transfers the state variable from a given initial value $x(t_1)$ to a given final value $x(t_2)$ under the condition that the whole trajectory minimizes the following cost function:

$$Q = \int_{t_1}^{t_2} f_0(x, u)\, dt \qquad 2.278$$

In most engineering problems, u is subject to practical constraints. In other words, u must be an entry of a given subset U of an m-dimensional Euclidean space. Namely,

$$u \in U \subset R^m \qquad 2.279$$

Basically, this problem is a variational problem, and *Pontryagin's maximum principle* (MP) is an extension of the classical variational method discussed in Section 2.5. The difference is that MP can deal with closed sets of U, whereas the classical variational methods work only for open sets of U. Needless to say, problems with closed sets of U are extremely significant from the engineering point of view.

The joint torque vector $\tau \in R^m$ of a robot manipulator usually has the following constraints:

$$|\tau_i| \le \tau_{i\max} \qquad \tau_{i\max} > 0 \text{ and } i = 1, \cdots, m \qquad 2.280$$

Many important problems would not have solutions if they had to be solved using the classical variational methods with the following open-set constraints:

$$|\tau_i| < \tau_{i\max} \qquad \tau_{i\max} > 0 \text{ and } i = 1, \cdots, m \qquad 2.281$$

In MP, an input variable must be an entry of a mathematical class. The input variable satisfying all the imposed conditions is called an *admissible control*. The conditions for admissible control are summarized as follows.

Definition 2.7 (Admissible Control)

An input variable $u \in R^m$ is called an *admissible control* if it satisfies the following four conditions:

1. u is an entry of a given subset $U \subset R^m$.
2. $u(t)$ is defined over $t_1 \le t \le t_2$.
3. $u(t)$ is piecewise continuous and is defined at a discontinuous point t_d by the left limit; that is,

$$u(t_d) = u(t_d - 0) \qquad 2.282$$

4. $u(t)$ is continuous at $t = t_1$ and $t = t_2$.

2.6 Pontryagin's Maximum Principle

The mathematical conditions on $f(x, u)$ in Eq. 2.277 and on $f_0(x, u)$ in Eq. 2.278 are summarized as follows:

1. $f(x, u)$ is defined over arbitrary $x \in R^n$ and $u \in U$.
2. $f(x, u)$ is continuous with respect to x and u.
3. $f(x, u)$ is continuously differentiable with respect to x and u.
4. $f_0(x, u)$ and $\partial f_0 / \partial x$ are defined over arbitrary $x \in R^n$ and $u \in U$.
5. $f_0(x, u)$ and $\partial f_0 / \partial x$ are continuous with respect to x and u.

2.6.2 The Maximum Principle

Equation 2.277 is represented component-wise as follows:

$$\dot{x}_i = f_i(x, u) \quad (i = 1, \cdots, n) \qquad 2.283$$

$$x = \text{col}(x_1, x_2, \cdots, x_n) \in R^n$$

Now, using $f_0(x, u)$ in Eq. 2.278, we define an additional state variable:

$$\dot{x}_0 \stackrel{\Delta}{=} f_0(x, u), \qquad x_0(t_1) = 0 \qquad 2.284$$

It is obvious that

$$Q = x_0(t_2) \qquad 2.285$$

Combining Eqs. 2.283 and 2.284, we obtain the augmented dynamics equation:

$$\dot{x}_0 = f_0(x_0, u)$$
$$x_0 = \text{col}(x_0, x_1, \cdots, x_n) \in R^{n+1}$$
$$f_0 = \text{col}(f_0, f_1, \cdots, f_n) \in R^{n+1} \qquad 2.286$$

We introduce a vector $\psi_0 \in R^{n+1}$ that has the same dimension as x_0, and call it the adjoint vector. The *Hamiltonian* is defined using the adjoint vector as follows:

$$H_0(\psi_0, x_0, u) = \psi_0^T f_0(x_0, u) \qquad 2.287$$

Using the Hamiltonian, the behaviors of $x_0(t)$ and $\psi_0(t)$ are determined by the following Hamilton form:

$$\dot{x}_0 = \left(\frac{\partial H_0}{\partial \psi_0}\right)^T \qquad 2.288$$

$$\dot{\psi}_0 = -\left(\frac{\partial H_0}{\partial x_0}\right)^T \qquad 2.289$$

Note that Eq. 2.288 is equivalent to Eq. 2.286 and, therefore, Eqs. 2.288 and 2.289 determine only the behavior of the adjoint vector. The following theorem is a necessary condition for optimality (Pontryagin, Boltyanskii, Gamkrelidze, and Mishchenko 1962).

Theorem 2.11 (Pontryagin's Maximum Principle)

Let $\boldsymbol{u}(t)$, $t_1 \leq t \leq t_2$, be an admissible control that transfers the state variable from $\boldsymbol{x}(t_1)$ to $\boldsymbol{x}(t_2)$. It is a necessary condition for $\boldsymbol{x}(t)$ and $\boldsymbol{u}(t)$ to be optimal in the sense of Eq. 2.278 that there exists a nonzero continuous vector $\boldsymbol{\psi}_0(t)$, $t_1 \leq t \leq t_2$, such that

$$H_0(\boldsymbol{\psi}_0(t), \boldsymbol{x}_0(t), \boldsymbol{u}(t)) = \sup_{u \in U} H_0(\boldsymbol{\psi}_0(t), \boldsymbol{x}_0(t), \boldsymbol{u})^{\dagger 2.9} \qquad 2.290$$

$$\psi_0(t_2) \leq 0, \quad \sup_{u \in U} H_0(\boldsymbol{\psi}_0(t_2), \boldsymbol{x}_0(t_2), \boldsymbol{u}) = 0 \qquad 2.291$$

where $\boldsymbol{x}_0 \in R^{n+1}$ is defined by Eqs. 2.284 and 2.286, and H_0 and $\boldsymbol{\psi}_0 \in R^{n+1}$ are defined by Eqs. 2.287 and 2.289. If $\boldsymbol{\psi}_0(t)$, $\boldsymbol{x}_0(t)$, and $\boldsymbol{u}(t)$ satisfy Eqs. 2.288, 2.289, and 2.290, then $\psi_0(t)$ and $\sup_{u \in U} H_0(\boldsymbol{\psi}_0(t), \boldsymbol{x}_0(t), \boldsymbol{u})$ become constant. Therefore, Eq. 2.291 is satisfied at an arbitrary time t, $t_1 \leq t \leq t_2$.

It is not easy to interpret the physical meaning of the Hamiltonian and the adjoint vector in Eqs. 2.287 through 2.289. Equations 2.288 and 2.289 are in a format known as the *canonical equations of Hamilton* (Goldstein 1980). The Hamilton's equations are often used to formulate dynamic behaviors of mechanical systems, and are equivalent to the Lagrange formulation of dynamics for rigid body systems. In a typical case, the Hamiltonian represents the total energy of the system, and ψ_0 indicates the momentum. We might say that the Hamiltonian in Eq. 2.287 denotes the generalized total energy, and the adjoint vector represents the generalized momentum of the system.

Example 2.36

Suppose that the system equation and the cost function are given as

$$\dot{\boldsymbol{x}} = \boldsymbol{A}\boldsymbol{x} + \boldsymbol{B}\boldsymbol{u} \qquad \boldsymbol{x} \in R^n \qquad 2.292$$

$$Q = \int_{t_1}^{t_2} \boldsymbol{u}^T \boldsymbol{u} \, dt \qquad 2.293$$

[†2.9] Let S be an upper-bounded set of real numbers. There exists the smallest real number y that satisfies $x \leq y$ for all $x \in S$. We say that y is the *supremum* of x and defined by $y = \sup_{x \in S} x$.

2.6 Pontryagin's Maximum Principle

The goal is to find a control $u(t)$ that transfers the state variable $x(t)$ from a given initial state, $x(t_1)$, to a given final state, $x(t_2)$, where the final time t_2 could be either specified or unspecified.

The Hamiltonian becomes

$$H_0 = \psi_0 \, u^T u + \psi^T (Ax + Bu) \qquad 2.294$$

Hence, Eqs. 2.288 and 2.289 are computed as follows:

$$\begin{pmatrix} \dot{x}_0 \\ \dot{x} \end{pmatrix} = \begin{pmatrix} \partial H_0 / \partial \psi_0 \\ (\partial H_0 / \partial \psi)^T \end{pmatrix}$$

$$= \begin{pmatrix} u^T u \\ Ax + Bu \end{pmatrix} \qquad 2.295$$

$$\begin{pmatrix} \dot{\psi}_0 \\ \dot{\psi} \end{pmatrix} = \begin{pmatrix} -\partial H_0 / \partial x_0 \\ -(\partial H_0 / \partial x)^T \end{pmatrix}$$

$$= \begin{pmatrix} 0 \\ -A^T \psi \end{pmatrix} \qquad 2.296$$

We assume that u has no restriction. Then, H_0 has its maximum value only when $\psi_0 < 0$. The u that maximizes H_0 is obtained from

$$\frac{\partial H_0}{\partial u} = \psi_0 \, u^T + \psi^T B = O \qquad 2.297$$

Hence,

$$u = -\frac{1}{\psi_0} B^T \psi \qquad 2.298$$

Note that ψ_0 is constant from Eq. 2.296. Substituting Eq. 2.298 into the lower parts of Eqs. 2.295 and 2.296, we have

$$\dot{x} = Ax - BB^T \frac{\psi}{\psi_0}$$
$$\dot{\psi} = -A^T \psi \qquad 2.299$$

If we define $\bar{\psi} \triangleq \psi/\psi_0$, then Eq. 2.299 becomes

$$\dot{x} = Ax - BB^T \frac{\psi}{\psi_0}$$
$$\dot{\bar{\psi}} = -A^T \bar{\psi} \qquad 2.300$$

Consequently, Eq. 2.300 represents the differential equation that the optimal trajectory should satisfy. Note that Eq. 2.300 includes $2n$ differential equations, where n is the dimension of x. Since both $x(t_1)$ and $x(t_2)$ are given, we have $2n$ boundary conditions. The relationship between the number of differential equations and the number of boundary conditions will be discussed in detail in Section 2.6.7.

2.6.3 Minimum Time Control

The goal of minimum time control is to find the trajectory $\boldsymbol{x}(t)$ and the input $\boldsymbol{u}(t)$ starting from the initial state $\boldsymbol{x}(t_1)$ and arriving at the final state $\boldsymbol{x}(t_2)$ with minimum time. This problem is important not only from a theoretical standpoint, but also from a practical one. We can formulate problems of minimum time control by setting $f_0(\boldsymbol{x}, \boldsymbol{u}) = 1$ in Eq. 2.278.

Taking account of $f_0(\boldsymbol{x}, \boldsymbol{u}) = 1$, Theorem 2.11 is simplified. We redefine the Hamiltonian as follows:

$$H(\boldsymbol{\psi}, \boldsymbol{x}, \boldsymbol{u}) = \boldsymbol{\psi}^T \boldsymbol{f}(\boldsymbol{x}, \boldsymbol{u}), \qquad \boldsymbol{\psi} \in R^n \qquad 2.301$$

The dynamics equations are represented by the following Hamilton form:

$$\dot{\boldsymbol{x}} = \left(\frac{\partial H}{\partial \boldsymbol{\psi}}\right)^T \qquad 2.302$$

$$\dot{\boldsymbol{\psi}} = -\left(\frac{\partial H}{\partial \boldsymbol{x}}\right)^T \qquad 2.303$$

Theorem 2.12 (Minimum Time Control)

Let $\boldsymbol{u}(t)$, $t_1 \leq t \leq t_2$, be an admissible control that transfers the state variable from $\boldsymbol{x}(t_1)$ to $\boldsymbol{x}(t_2)$. It is a necessary condition for $\boldsymbol{x}(t)$ and $\boldsymbol{u}(t)$ to be minimum time control that there exists a nonzero continuous vector $\boldsymbol{\psi}(t)$, $t_1 \leq t \leq t_2$, such that

$$H(\boldsymbol{\psi}(t), \boldsymbol{x}(t), \boldsymbol{u}(t)) = \sup_{\boldsymbol{u} \in U} H(\boldsymbol{\psi}(t), \boldsymbol{x}(t), \boldsymbol{u}) \qquad 2.304$$

$$\sup_{\boldsymbol{u} \in U} H(\boldsymbol{\psi}(t_1), \boldsymbol{x}(t_1), \boldsymbol{u}) \geq 0 \qquad 2.305$$

where H and $\psi \in R^n$ are defined by Eqs. 2.301 and 2.303. If $\boldsymbol{\psi}(t)$, $\boldsymbol{x}(t)$, and $\boldsymbol{u}(t)$ satisfy Eqs. 2.302, 2.303, and 2.304, then $\sup_{\boldsymbol{u} \in U} H(\boldsymbol{\psi}(t), \boldsymbol{x}(t), \boldsymbol{u})$ in Eq. 2.305 becomes constant.

Example 2.37

A linear dynamical system is given by

$$\dot{\boldsymbol{x}} = \boldsymbol{A}\boldsymbol{x} + \boldsymbol{b}\, u \qquad 2.306$$

$$\boldsymbol{x} = \begin{pmatrix} x_1 \\ x_2 \end{pmatrix}, \quad \boldsymbol{A} = \begin{pmatrix} 0 & 1 \\ 0 & 0 \end{pmatrix}, \quad \boldsymbol{b} = \begin{pmatrix} 0 \\ 1 \end{pmatrix}$$

2.6 Pontryagin's Maximum Principle

with the constraint on the input variable

$$|u| \leq 1 \qquad 2.307$$

Obtain the minimum time trajectory $\boldsymbol{x}(t)$ and control $\boldsymbol{u}(t)$ that transfer the state variable from a given $\boldsymbol{x}(t_1)$ to the origin, $\boldsymbol{x}(t_2) = (0\ 0)^T$. The Hamiltonian and the dynamic equations become

$$H(\boldsymbol{\psi}, \boldsymbol{x}, \boldsymbol{u}) = \boldsymbol{\psi}^T(\boldsymbol{A}\boldsymbol{x} + \boldsymbol{b}\,u) \qquad 2.308$$

$$\dot{\boldsymbol{x}} = \left(\frac{\partial H}{\partial \boldsymbol{\psi}}\right)^T = \boldsymbol{A}\boldsymbol{x} + \boldsymbol{b}\,u \qquad 2.309$$

$$\dot{\boldsymbol{\psi}} = -\left(\frac{\partial H}{\partial \boldsymbol{x}}\right)^T = -\boldsymbol{A}^T\boldsymbol{\psi} = \begin{pmatrix} 0 \\ -\psi_1 \end{pmatrix} \qquad 2.310$$

From Eq. 2.310, $\boldsymbol{\psi}(t)$ is integrated as follows:

$$\boldsymbol{\psi}(t) = \begin{pmatrix} \psi_1 \\ \psi_2 \end{pmatrix} = \begin{pmatrix} c_1 \\ -c_1\,t + c_2 \end{pmatrix} \qquad 2.311$$

From Theorem 2.12, the optimal $u(t)$ is one that maximizes Eq. 2.308. Therefore, from Eqs. 2.308 and 2.311, $u(t)$ maximizes

$$\boldsymbol{\psi}^T \boldsymbol{b}\, u = (-c_1\,t + c_2)\,u \qquad 2.312$$

Accordingly, $u(t)$ is given by

$$u(t) = \begin{cases} 1, & \text{if } -c_1\,t + c_2 \geq 0 \\ -1, & \text{if } -c_1\,t + c_2 < 0 \end{cases} \qquad 2.313$$

Equation 2.313 implies that $u(t)$ is a step input and has at most one switching, as shown in Fig. 2.10. For each of $u(t) = 1$ and -1, the behavior of the state variable is computed from Eq. 2.306 as follows:

$$u(t) = 1: \quad \begin{aligned} x_1 &= \frac{1}{2}(t + c_3)^2 + c_4 \\ x_2 &= t + c_3 \end{aligned} \qquad 2.314$$

$$u(t) = -1: \quad \begin{aligned} x_1 &= -\frac{1}{2}(t - c_3)^2 + c_4 \\ x_2 &= -t + c_3 \end{aligned} \qquad 2.315$$

We eliminate t, so that Eqs. 2.314 and 2.315 yield the following equations:

$$x_1 = \frac{1}{2}x_2^2 + c_4;\ u(t) = 1 \qquad 2.316$$

$$x_1 = -\frac{1}{2}x_2^2 + c_4;\ u(t) = -1 \qquad 2.317$$

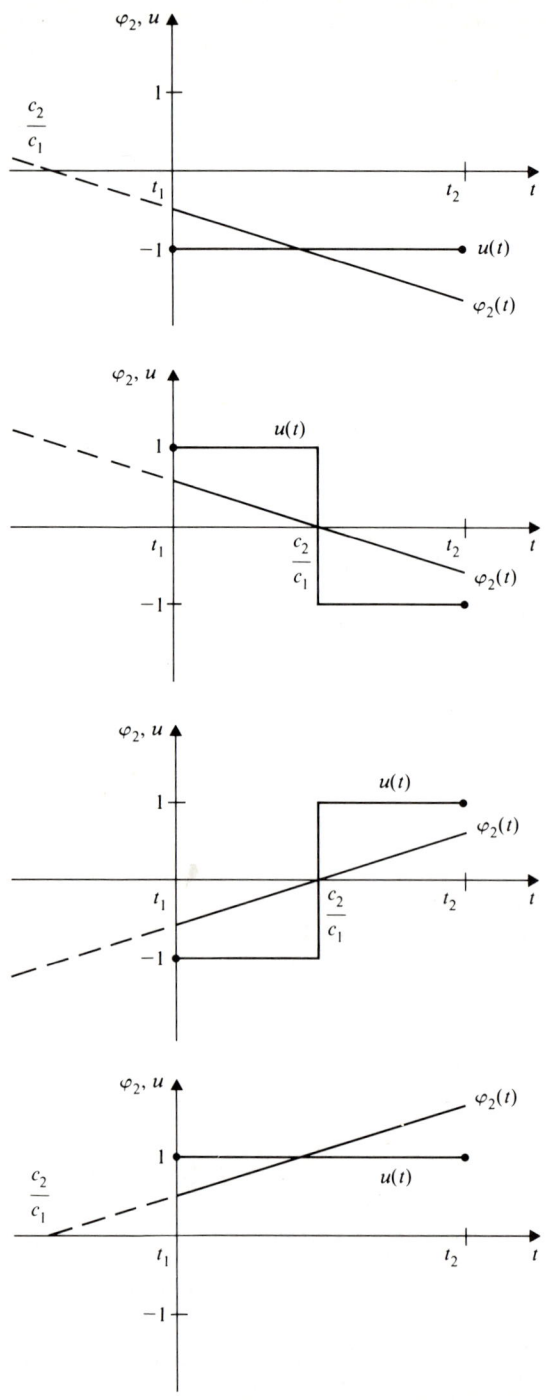

Figure 2.10 Switching of input variable: Eq. 2.313.

2.6 Pontryagin's Maximum Principle 89

Note that, if $u(t) = 1$ $(u(t) = -1))$, the state variable moves in the direction of positive (negative) x_2, because $\dot{x}_2 = u$ from Eq. 2.306. The behaviors in the state space are shown for $u(t) = 1$ and $u(t) = -1$ in Fig. 2.11. Considering that the number of switching is at most once, the behavior of the state variable $x(t)$ is represented in the state space as shown in Fig. 2.12.

For linear time-invariant dynamical systems, the following theorem is provided (Pontryagin, Boltyanskii, Gamkrelidze, and Mishchenko 1962).

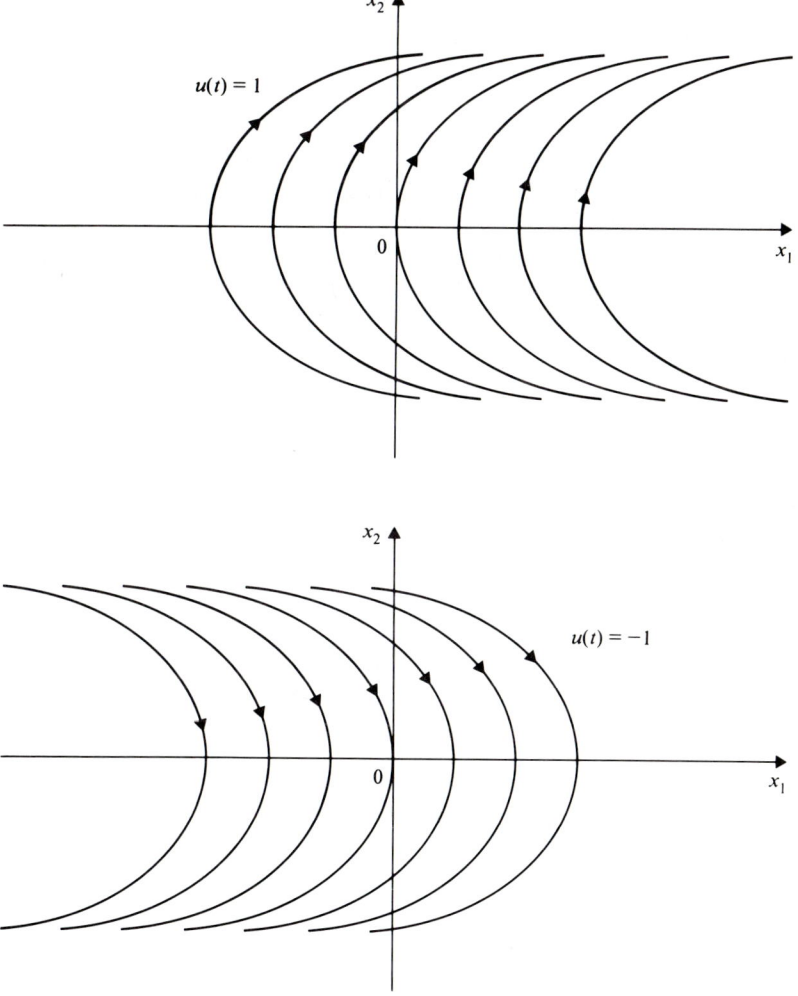

Figure 2.11 Behavior of state variable for $u(t) = 1$ and $u(t) = -1$.

90   Chapter 2   Mathematical Toolbox

Theorem 2.13   (Linear Time-Invariant Systems)

Let U be a rectangular parallel-piped domain

$$u_{i\min} \leq u_i \leq u_{i\max} \qquad (i = 1, \cdots, m) \qquad 2.318$$

allowed for the input variable of a linear time-invariant dynamical system

$$\dot{\boldsymbol{x}} = \boldsymbol{A}\boldsymbol{x} + \boldsymbol{B}\boldsymbol{u}, \qquad 2.319$$

where $\boldsymbol{x} \in R^n$, $\boldsymbol{u} \in R^m$, $\boldsymbol{A} \in R^{n \times n}$, and $\boldsymbol{B} \in R^{n \times m}$. The input variable of time optimal control is determined for each nontrivial solution of

$$\dot{\boldsymbol{\psi}} = -\boldsymbol{A}^T \boldsymbol{\psi} \qquad 2.320$$

as the one that maximizes

$$\boldsymbol{\psi}^T(t) \boldsymbol{B} \boldsymbol{u}(t) \qquad 2.321$$

where $u_i(t)$, $(i = 1, \cdots, m)$ become step functions and take either $u_{i\min}$ or $u_{i\max}$. If all the eigenvalues of $\boldsymbol{A}$ are real, the number of switchings is not more than $n - 1$.

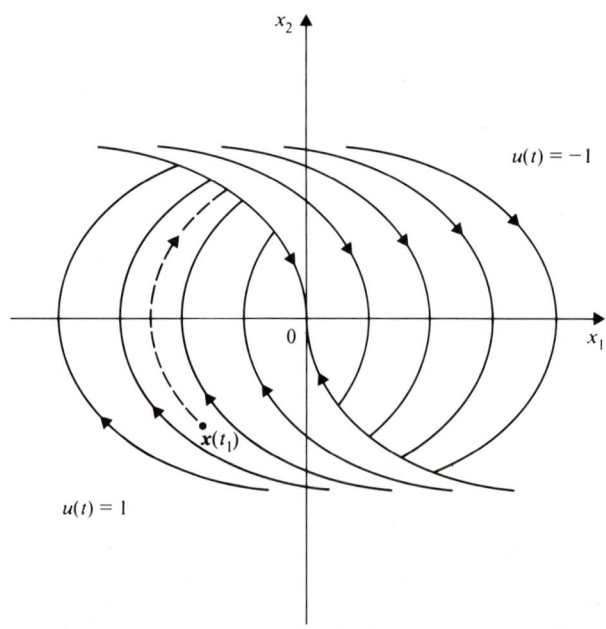

Figure 2.12 Behavior of optimal trajectory in state space (a fixed-point problem).

The control in which the input variable takes either of the maximum or minimum value is called *bang-bang control*.

Minimum time control of robot manipulators has been studied by many researchers since Kahn and Roth (1971). For general cases, the application of MP results in a *two-point boundary value problem* that is, as we shall explain later in the section, computationally very expensive and difficult. A successful approach is to assume that the geometric trajectory of the manipulator is preplanned, and to reduce the problem to the one-dimensional problem of determining the minimum time motion along the trajectory (Shin and McKay 1985; Bobrow, Dubowsky, and Gibson 1985; Dubowsky, Norris, and Shiller 1986).

2.6.4 Transversality Conditions

We have so far discussed optimal control problems for which $\boldsymbol{x}(t_1)$ and $\boldsymbol{x}(t_2)$ are given. Such problems are called *fixed-point problems*. In this subsection, we deal with the cases where the domains X_1 and X_2 on which $\boldsymbol{x}(t_1)$ and $\boldsymbol{x}(t_2)$ should lie are given. It is necessary to determine the $\boldsymbol{x}(t_1)$ and $\boldsymbol{x}(t_2)$ that optimize the motion, in the sense of Eq.2.278. These problems are called *fixed-domain problems*.

We assume that the domains X_1 and X_2 are k_1- and k_2-dimensional smooth manifolds,[†2.10] respectively. The transversality condition of MP is defined as follows.

Definition 2.8 (Transversality Condition)

We say that $\boldsymbol{\psi}(t)$ satisfies the *transversality condition* at $t = t_1$ (or $t = t_2$) if $\boldsymbol{\psi}(t_1)$ (or $\boldsymbol{\psi}(t_2)$) is perpendicular to the tangent plane of X_1 (or X_2) at $\boldsymbol{x}(t_1)$ (or $\boldsymbol{x}(t_2)$).

When the hypersurfaces of X_1 and X_2 are respectively represented by

$$\boldsymbol{g}_i(\boldsymbol{x}) = \begin{pmatrix} g_1^i(\boldsymbol{x}) \\ g_2^i(\boldsymbol{x}) \\ \vdots \\ g_{n-k_i}^i(\boldsymbol{x}) \end{pmatrix} = \boldsymbol{O} \in R^{n-k_i} \qquad i = 1, 2 \qquad 2.322$$

$\boldsymbol{\psi}(t)$ satisfies the transversality condition at $t = t_1$ (or $t = t_2$) if and only if $\boldsymbol{\psi}(t_1)$ (or $\boldsymbol{\psi}(t_2)$) is included in the range space of $(\partial \boldsymbol{g}_1/\partial \boldsymbol{x}|_{x=x_1})^T$ (or $(\partial \boldsymbol{g}_2/\partial \boldsymbol{x}|_{x=x_2})^T$) because the column vectors of the matrix are the set of

---

[†2.10] Let the intersection of hypersurfaces $g_i(\boldsymbol{x}) = 0$, $i = 1, \cdots, m$ with $\boldsymbol{x} \in R^n$, be represented by S. If all the points in S are regular points (Definition 2.5), S is said to be an $(n-m)$-dimensional smooth manifold in R^n.

all the normal vectors of the manifold at $\boldsymbol{x}_1$ (or $\boldsymbol{x}_2$) and, therefore, all the tangential vectors of the manifold at $\boldsymbol{x}_1$ (or $\boldsymbol{x}_2$) are perpendicular to any linear combination of them. Hence, from Eq. 2.189, we can use the following equations as the transversality condition:

$$[\boldsymbol{E} - (\frac{\partial \boldsymbol{g}_1}{\partial \boldsymbol{x}}|_{x=x_1})^T \{(\frac{\partial \boldsymbol{g}_1}{\partial \boldsymbol{x}}|_{x=x_1})^T\}^{\#}]\boldsymbol{\psi}(t_1)$$

$$= \{\boldsymbol{E} - (\frac{\partial \boldsymbol{g}_1}{\partial \boldsymbol{x}}|_{x=x_1})^{\#}(\frac{\partial \boldsymbol{g}_1}{\partial \boldsymbol{x}}|_{x=x_1})\}\boldsymbol{\psi}(t_1) = \boldsymbol{O} \qquad 2.323$$

and

$$\{\boldsymbol{E} - (\frac{\partial \boldsymbol{g}_2}{\partial \boldsymbol{x}}|_{x=x_2})^{\#}(\frac{\partial \boldsymbol{g}_2}{\partial \boldsymbol{x}}|_{x=x_2})\}\boldsymbol{\psi}(t_2) = \boldsymbol{O} \qquad 2.324$$

where Eq. 2.116 was used to derive Eq. 2.323. $\boldsymbol{E}$ is an identity matrix with appropriate dimension. Note that since rank $\partial \boldsymbol{g}_i / \partial \boldsymbol{x} = n - k_i$, the ranks of the coefficient matrices of Eqs. 2.323 and 2.324 are k_1 and k_2, respectively, and, therefore, the numbers of independent conditions included in the equations are k_1 and k_2, respectively. The following theorem is MP for fixed-domain problems.

Theorem 2.14 (Fixed-Domain Problems)

Let $\boldsymbol{u}(t)$, $t_1 \leq t \leq t_2$, be an admissible control that transfers the state variable from a domain X_1 to a domain X_2. A necessary condition for $\boldsymbol{x}(t)$ and $\boldsymbol{u}(t)$ being the optimal solution of the fixed-domain problem in the sense of Eq. 2.278 is that Theorem 2.11 be satisfied and there be a nonzero continuous vector $\boldsymbol{\psi}(t)$ that satisfies the transversality condition at $t = t_1$ and $t = t_2$.

Example 2.38

A dynamical system is represented by Eqs. 2.306 and 2.307 (Example 2.37). Consider the case where the state variable is transferred from a given $\boldsymbol{x}(t_1)$ to the x_2 axis with the minimum time. Vector $\boldsymbol{\psi}(t)$ is given by Eq. 2.311. Since the tangential vector of the manifold X_2 (x_2 axis) is $\boldsymbol{a} = (0\ 1)^T$, $\boldsymbol{\psi}(t_2)$ must satisfy

$$\boldsymbol{\psi}^T(t_2)\,\boldsymbol{a} = -c_1 t_2 + c_2 = 0 \qquad 2.325$$

which means that the signum of the coefficient of Eq. 2.312 does not change for $t_1 \leq t \leq t_2$. This implies that the only switching point is $t = t_2$, because we know from Theorem 2.13 that this system has no more than one switching point. Therefore, $u(t)$, $t_1 \leq t < t_2$, is equal to either 1 or -1 depending on $\boldsymbol{x}(t_1)$. Consequently, the optimal trajectory $\boldsymbol{x}(t)$ behaves as shown in Fig. 2.13.

2.6.5 *Maximum Principle for Nonautonomous Systems*

Equations 2.277 and 2.278 do not include time in $f(x, u)$, and $f_0(x, u)$, and, therefore, are called an *autonomous system*. The optimal control problem for nonautonomous systems is described by

$$\dot{x} = f(x, u, t) \qquad 2.326$$

$$Q = \int_{t_1}^{t_2} f_0(x, u, t)\, dt \qquad 2.327$$

Now, we derive MP for nonautonomous systems from Theorem 2.11 straightforwardly by defining time as the $(n+2)$th state variable; namely,

$$\dot{x}^* = f^*(x^*, u) \qquad 2.328$$

$$x_{n+1} = t$$

$$x^* = \text{col}(x_0, x_1, \cdots, x_n, x_{n+1}) \in R^{n+2}$$

$$f^* = \text{col}(f_0, f_1, \cdots, f_n, 1) \in R^{n+2}$$

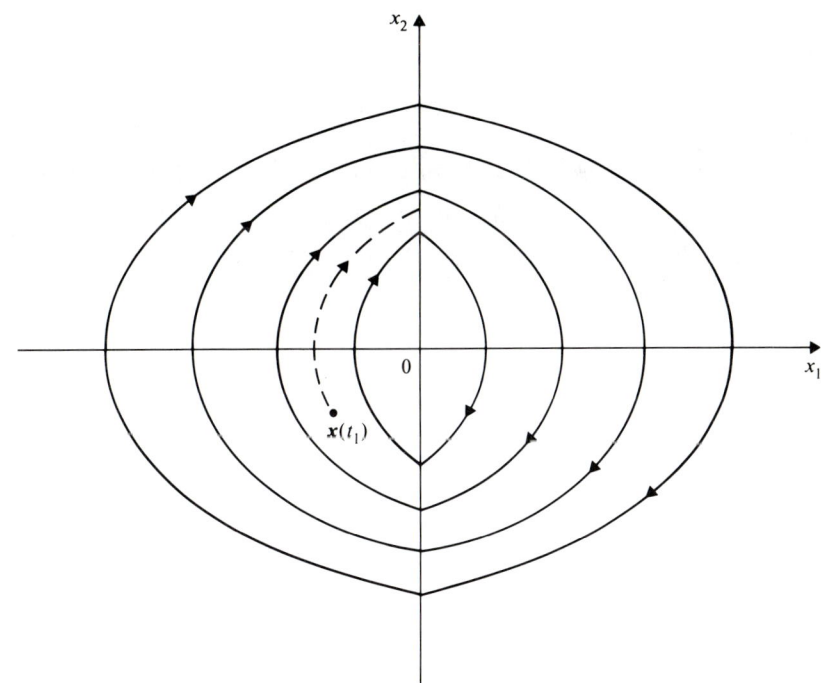

Figure 2.13 Behavior of optimal trajectory in state space (a fixed-domain problem).

$$Q = \int_{t_1}^{t_2} f_0(\boldsymbol{x}^*, \boldsymbol{u}) \, dt \qquad 2.329$$

where $\dot{x}_{n+1} = 1 = f_{n+1}$. Now, Eqs. 2.326 and 2.327 have been reduced to an autonomous system. Note that, when t_2 is not specified, Eqs. 2.328 and 2.329 form a fixed-domain problem. The Hamiltonian becomes

$$\begin{aligned}
H^*(\boldsymbol{\psi}^*, \boldsymbol{x}^*, \boldsymbol{u}) &= \boldsymbol{\psi}^{*T} \boldsymbol{f}^* = \boldsymbol{\psi}_0^T \boldsymbol{f}_0 + \psi_{n+1} \\
&= H_0(\boldsymbol{\psi}_0, \boldsymbol{x}_0, \boldsymbol{u}) + \psi_{n+1} \qquad 2.330 \\
\boldsymbol{\psi}^* &= \mathrm{col}\,(\psi_0, \psi_1, \cdots, \psi_n, \psi_{n+1}) \in R^{n+2} \\
\boldsymbol{\psi}_0 &= \mathrm{col}\,(\psi_0, \psi_1, \cdots, \psi_n) \in R^{n+1} \\
\boldsymbol{x}_0 &= \mathrm{col}\,(x_0, x_1, \cdots, x_n) \in R^{n+1} \\
\boldsymbol{f}_0 &= \mathrm{col}\,(f_0, f_1, \cdots, f_n) \in R^{n+1} \\
H_0(\boldsymbol{\psi}_0, \boldsymbol{x}_0, \boldsymbol{u}) &= \boldsymbol{\psi}_0^T \boldsymbol{f}_0
\end{aligned}$$

The Hamilton form of the dynamic equations is as follows:

$$\dot{\boldsymbol{x}}^* = \left(\frac{\partial H^*}{\partial \boldsymbol{\psi}^*}\right)^T \qquad 2.331$$

$$\dot{\boldsymbol{\psi}}^* = -\left(\frac{\partial H^*}{\partial \boldsymbol{x}^*}\right)^T \qquad 2.332$$

From Eq. 2.330, maximizing H^* is equivalent to maximizing H_0. On the other hand, when t_2 is not specified, the transversality condition concerning the x_{n+1} axis is that $\boldsymbol{\psi}^*(t_2)$ is perpendicular to $\boldsymbol{e} = (0\ 0\ \cdots\ 0\ 1)^T \in R^{n+2}$; namely,

$$\boldsymbol{\psi}^{*T}(t_2) \boldsymbol{e} = \psi_{n+1}(t_2) = 0. \qquad 2.333$$

The $(n+2)$th component of Eq. 2.332 is calculated as follows:

$$\dot{\psi}_{n+1} = -\boldsymbol{\psi}_0^T \frac{\partial \boldsymbol{f}_0}{\partial t} \qquad 2.334$$

From Theorem 2.11, we have

$$\sup_{u \in U} H^*(\boldsymbol{\psi}^*, \boldsymbol{x}^*, \boldsymbol{u}) = 0 \qquad 2.335$$

From Eqs. 2.330, 2.333, and 2.334, Eq. 2.335 is transformed into the following form:

$$\begin{aligned}
\sup_{u \in U} H_0(\boldsymbol{\psi}_0, \boldsymbol{x}_0, t, \boldsymbol{u}) &= -\psi_{n+1}(t) = \int_t^{t_2} \dot{\psi}_{n+1} \, dt \\
&= \int_{t_2}^{t} \boldsymbol{\psi}_0^T \frac{\partial \boldsymbol{f}_0}{\partial t} \, dt \qquad 2.336
\end{aligned}$$

2.6 Pontryagin's Maximum Principle

Note that Eq. 2.336 includes the final condition on ψ_{n+1} given by Eq. 2.333. We summarize this discussion by the following theorem.

Theorem 2.15 (Nonautonomous Systems)

Let $\boldsymbol{u}(t)$, $t_1 \leq t \leq t_2$, be an admissible control of Eq. 2.326, and transfer the state variable from $\boldsymbol{x}(t_1)$ to $\boldsymbol{x}(t_2)$. A necessary condition for $\boldsymbol{u}(t)$ and $\boldsymbol{x}(t)$ being optimal in the sense of Eq. 2.327 is that there exists a nonzero continuous vector $\boldsymbol{\psi}_0(t)$, $t_1 \leq t \leq t_2$, such that

$$H_0(\boldsymbol{\psi}_0(t), \boldsymbol{x}_0(t), t, \boldsymbol{u}(t)) = \sup_{u \in U} H_0(\boldsymbol{\psi}_0(t), \boldsymbol{x}_0(t), t, \boldsymbol{u}) \qquad 2.337$$

$$\psi_0(t) = \text{constant} \leq 0 \qquad 2.338$$

$$\sup_{u \in U} H_0(\boldsymbol{\psi}_0(t), \boldsymbol{x}_0(t), t, \boldsymbol{u}) = \int_{t_2}^{t} \boldsymbol{\psi}_0^T(t) \frac{\partial \boldsymbol{f}_0(\boldsymbol{x}_0(t), \boldsymbol{u}(t), t)}{\partial t} \, dt \qquad 2.339$$

where $\boldsymbol{x}_0$, H_0, and $\boldsymbol{\psi}_0 \in R^{n+1}$ are defined by Eq. 2.330. It is sufficient if Eqs. 2.338 and 2.339 are satisfied at an arbitrary time t, $t_1 \leq t \leq t_2$. Hence, Eqs. 2.338 and 2.339 can be replaced with

$$\psi_0(t_2) \leq 0 \qquad 2.340$$

$$\sup_{u \in U} H_0(\boldsymbol{\psi}_0(t_2), \boldsymbol{x}_0(t_2), t_2, \boldsymbol{u}) = 0 \qquad 2.341$$

Example 2.39

A system equation and a cost function are given as follows:

$$\begin{cases} \dot{x}_1 = x_2 \\ \dot{x}_2 = tx_2 + u \end{cases} \qquad 2.342$$

$$Q = \int_{t_1}^{t_2} u^2 \, dt \qquad 2.343$$

The goal is to find $u(t)$, $t_1 \leq t \leq t_2$, that brings the state variables x_1 and x_2 from given initial values to given final values, where t_2 is not specified. The Hamiltonian is computed by

$$H_0 = \psi_0 u^2 + \psi_1 x_2 + \psi_2(tx_2 + u) \qquad 2.344$$

and x_0 is defined by

$$\dot{x}_0 = \frac{\partial H_0}{\partial x_0} = u^2 \qquad 2.345$$

Differential equations of the adjoint variables become

$$\dot{\psi}_0 = -\left(\frac{\partial H_0}{\partial x_0}\right) = 0$$

$$\dot{\psi}_1 = -\left(\frac{\partial H_0}{\partial x_1}\right) = 0 \qquad 2.346$$

$$\dot{\psi}_2 = -\left(\frac{\partial H_0}{\partial x_2}\right) = -\psi_1 - \psi_2 t$$

The first two equations of Eq. 2.346 identify that $\psi_0(t) = c_0$ and $\psi_1(t) = c_1$, where c_0 and c_1 are constant.

When there is no constraint on $u(t)$, from Eq. 2.337 of Theorem 2.15, $u(t)$ should satisfy

$$\frac{\partial H_0}{\partial u} = 2\psi_0 u + \psi_2 = 0 \qquad 2.347$$

Note that we are assuming that $\psi_0 = c_0 < 0$. Otherwise, Eq. 2.344 does not have the maximum value in terms of u. Equation 2.347 provides the following optimal input:

$$u = -\frac{\psi_2}{2c_0} \qquad 2.348$$

Substituting Eq. 2.348 into Eqs. 2.342 and 2.346 and defining $\bar{\psi}_i = \psi_i/c_0$ ($i = 1, 2$), we obtain

$$\begin{aligned}\dot{x}_1 &= x_2 \\ \dot{x}_2 &= t x_2 - \frac{\bar{\psi}_2}{2} \\ \dot{\bar{\psi}}_1 &= 0 \\ \dot{\bar{\psi}}_2 &= -\bar{\psi}_1 - \bar{\psi}_2 t\end{aligned} \qquad 2.349$$

These are the differential equations that the optimal trajectory should satisfy. The four equations of Eq. 2.349 must be solved such that four boundary values of $x_1(t_1)$, $x_2(t_1)$, $x_1(t_2)$, and $x_2(t_2)$ are satisfied. We then obtain the optimal input by substituting $\bar{\psi}_2$ into $u = -\bar{\psi}_2/2$.

We now give the final condition of Theorem 2.15 by substituting Eq. 2.348 into Eq. 2.344 with $\bar{\psi}_i = \psi_i/c_0$ as follows:

$$\sup_{u \in U} H_0(t_2)$$

$$= c_0\left\{-\frac{\bar{\psi}_2(t_2)}{2}\right\}^2 + c_0 \bar{\psi}_1 x_2(t_2) + c_0 \bar{\psi}_2(t_2)\left\{t_2 x_2(t_2) - \frac{\bar{\psi}_2(t_2)}{2}\right\}$$

$$= c_0\left\{-\frac{1}{4}\bar{\psi}_2^2(t_2) + t_2 x_2(t_2)\bar{\psi}_2(t_2) + \bar{c}_1 x_2(t_2)\right\}$$

$$= 0 \qquad 2.350$$

where $\bar{c}_1 \triangleq c_1/c_0$. Therefore, we have

$$\bar{\psi}_2^2(t_2) - 4t_2 x_2(t_2)\bar{\psi}_2(t_2) - 4\bar{c}_1 x_2(t_2) = 0 \qquad 2.351$$

Equation 2.351 can be considered as the fifth condition for the five variables—x_1, x_2, ψ_1, ψ_2, and t—involved in Eq. 2.349.

2.6.6 Fixed-Time Problems

When the final time t_2 is specified explicitly for nonautonomous systems, we can derive a necessary condition for optimal control by choosing t as a state variable, similar to our approach in the previous subsection. In this case, the transversality condition for the new state variable is unnecessary, since it has a fixed final value. By eliminating the corresponding equations—Eqs. 2.339 and 2.341 in Theorem 2.15—we obtain the following theorem.

Theorem 2.16 (Fixed-Time Problems)

Let $\boldsymbol{u}(t)$, $t_1 \leq t \leq t_2$, be an admissible control of Eq. 2.326 that transfers the state variable from $\boldsymbol{x}(t_1)$ to $\boldsymbol{x}(t_2)$, where t_1 and t_2 are given. A necessary condition for $\boldsymbol{u}(t)$ and $\boldsymbol{x}(t)$ to be optimal in the sense of Eq. 2.327 is that there exists a nonzero continuous vector $\boldsymbol{\psi}_0(t)$, $t_1 \leq t \leq t_2$, such that

$$H_0(\boldsymbol{\psi}_0(t), \boldsymbol{x}_0(t), t, \boldsymbol{u}(t)) = \sup_{u \in U} H_0(\boldsymbol{\psi}_0(t), \boldsymbol{x}_0(t), t, \boldsymbol{u}) \qquad 2.352$$

$$\psi_0(t) = \text{constant} \leq 0 \qquad 2.353$$

where $\boldsymbol{x}_0 = \text{col}(x_0, x_1, \cdots, x_n) \in R^{n+1}$, and H_0 and $\boldsymbol{\psi}_0 \in R^{n+1}$ are defined by Eq. 2.330. Equation 2.353 is sufficient if it is satisfied at an arbitrary time t, $t_1 \leq t \leq t_2$.

Example 2.40

When t_2 is specified in Example 2.39, it forms a fixed-time problem. The approach to this problem using Theorem 2.16 is the same as what we saw in Example 2.39, except that we do not need Eq. 2.351. The fifth condition is now replaced with the specified t_2.

Finally, let us consider the case where $\boldsymbol{x}(t_2)$ (t_2 is given) is totally free. This will be used in Chapter 5 for global optimization of kinematic redundancy. We can consider it as the case where the domain X_2 spans the whole $\boldsymbol{x}$ space. Hence, the transversality condition is as follows:

$$\boldsymbol{\psi}(t_2) = \text{col}(\psi_1(t_2), \psi_2(t_2), \cdots, \psi_n(t_2)) = \boldsymbol{O} \qquad 2.354$$

Note that ψ_0 is a nonpositive constant from Theorem 2.16. From Eqs. 2.330 and 2.332, if $\boldsymbol{\psi}^*$ is a solution, $\alpha\boldsymbol{\psi}^*$ is also a solution, where α is a positive constant. Therefore, without loss of generality, we can conclude that $\psi_0 = -1$. Consequently, the following theorem is obtained.

Theorem 2.17 (Fixed-Time and Free End-State Problems)

Let $\boldsymbol{u}(t)$, $t_1 \leq t \leq t_2$, be an admissible control of Eq. 2.326. When t_1, t_2, and $\boldsymbol{x}(t_1)$ are given and $\boldsymbol{x}(t_2)$ is free, a necessary condition for $\boldsymbol{u}(t)$ and $\boldsymbol{x}(t)$ to be optimal in the sense of Eq. 2.327 is that there exists a nonzero continuous $\boldsymbol{\psi}_0(t)$, $t_1 \leq t \leq t_2$, such that

$$H_0(\boldsymbol{\psi}_0(t), \boldsymbol{x}_0(t), t, \boldsymbol{u}(t)) = \sup_{u \in U} H_0(\boldsymbol{\psi}_0(t), \boldsymbol{x}_0(t), t, \boldsymbol{u}) \qquad 2.355$$

$$\boldsymbol{\psi}_0(t_2) = \mathrm{col}(-1, 0, \cdots, 0) \qquad 2.356$$

where $\psi_0(t)$ is constantly equal to -1.

Example 2.41

In addition to the specified t_2, now we assume free end-states for Example 2.39; namely, $x_1(t_1)$ and $x_2(t_1)$ are given, and $x_1(t_2)$ and $x_2(t_2)$ are free. Equation 2.356 in Theorem 2.17 is equivalent to

$$\begin{aligned} c_0 &= -1 \\ \bar{\psi}_1(t_2) &= 0 \\ \bar{\psi}_2(t_2) &= 0 \end{aligned} \qquad 2.357$$

Therefore, we obtain the solution by solving the four differential equations of Eq. 2.349 with the boundary conditions on $x_1(t_1)$, $x_2(t_1)$, $\bar{\psi}_1(t_2)$, and $\bar{\psi}_2(t_2)$. From the second equation of Eq. 2.357 and the third equation of Eq. 2.349, the fourth equation of Eq. 2.349 becomes

$$\dot{\bar{\psi}}_2 = -\bar{\psi}_2 t \qquad 2.358$$

which has the following solution:

$$\bar{\psi}_2(t) = \bar{c}_2 e^{-t^2/2} \qquad 2.359$$

From the third condition of Eq. 2.357, we conclude that

$$\bar{\psi}_2(t) = 0 \qquad 2.360$$

Therefore, from $u = -\bar{\psi}_2/2$ and Eq. 2.360, the optimal input is

$$u(t) = 0 \qquad t_1 \leq t \leq t_2 \qquad 2.361$$

The optimal trajectories of the state variables are obtained from the first two equations of Eq. 2.349 as follows:

$$x_1(t) = x_1(t_1) + x_2(t_1) \int_{t_1}^{t} e^{(t^2 - t_1^2)/2} \, dt$$
$$x_2(t) = x_2(t_1) e^{(t^2 - t_1^2)/2}$$
2.362

The result of Eq. 2.361 is not surprising. It is physically obvious that the input of a free end-point problem with the cost function of Eq. 2.343 becomes identically zero.

The assumption of free end-point is trivial for Eqs. 2.342 and 2.343. However, free end-point problems find important applications when an appropriate system equation and an adequate cost function are provided. In Chapter 5, we apply Theorem 2.17 to the global trajectory planning of kinematically redundant manipulators.

2.6.7 Two-Point Boundary Value Problems

The computation of MP results in integration of a pair of Hamilton form differential equations, such as Eqs. 2.288 and 2.289, or Eqs. 2.302 and 2.303. The former pair has $2n + 2$ variables; the latter has $2n$ variables. This computation is not easy because all the boundary conditions are not simply given at the initial time.

Figure 2.14 shows how the boundary conditions are given for each theorem. In Theorem 2.11, as shown in Fig. 2.14(a), $n+1$ conditions of $x_0(t_1) = 0$ and $\pmb{x}(t_1)$ are given at $t = t_1$ and n conditions of $\pmb{x}(t_2)$ are given at $t = t_2$. The second equation of Eq. 2.291 is the final condition that can be evaluated at any t. Theorem 2.12, as shown in Fig. 2.14(b), has n conditions at each of $t = t_1$ and $t = t_2$ as $\pmb{x}(t_1)$ and $\pmb{x}(t_2)$, respectively. For fixed-domain problems the boundary conditions are a little complex. In Theorem 2.14, as shown in Fig. 2.14(c), $n + 1$ conditions are provided at $t = t_1$ as $x_0(t_1) = 0$, $\pmb{g}_1(\pmb{x}) = \pmb{O}$, which has $n - k_1$ conditions, and k_1 independent conditions of the transversality condition of Eq. 2.323. At $t = t_2$, n conditions are given as $\pmb{g}_2(\pmb{x}) = \pmb{O}$, which has $n - k_2$ conditions, and k_2 independent conditions of the transversality condition of Eq. 2.324. The second equation of Eq. 2.291 is the final condition. On the other hand, for fixed-time problems, Theorem 2.16 has $n + 1$ conditions of $x_0(t_1) = 0$ and $\pmb{x}(t_1)$ at the initial endpoint, whereas $t = t_2$ and $\pmb{x}(t_2)$ are $n + 1$ conditions at the final endpoint. Finally, Theorem 2.17 is slightly different from the others. As shown in Fig. 2.14(e), it has $n + 1$ conditions at each endpoint, where $x_0(t_1) = 0$ and $\pmb{x}(t_1)$ are the $n + 1$ conditions at $t = t_1$, and the conditions at $t = t_2$ are given as the adjoint vector of Eq. 2.356.

The boundary conditions mentioned here are almost equally split into both endpoints. The problems of integration for which boundary conditions are split

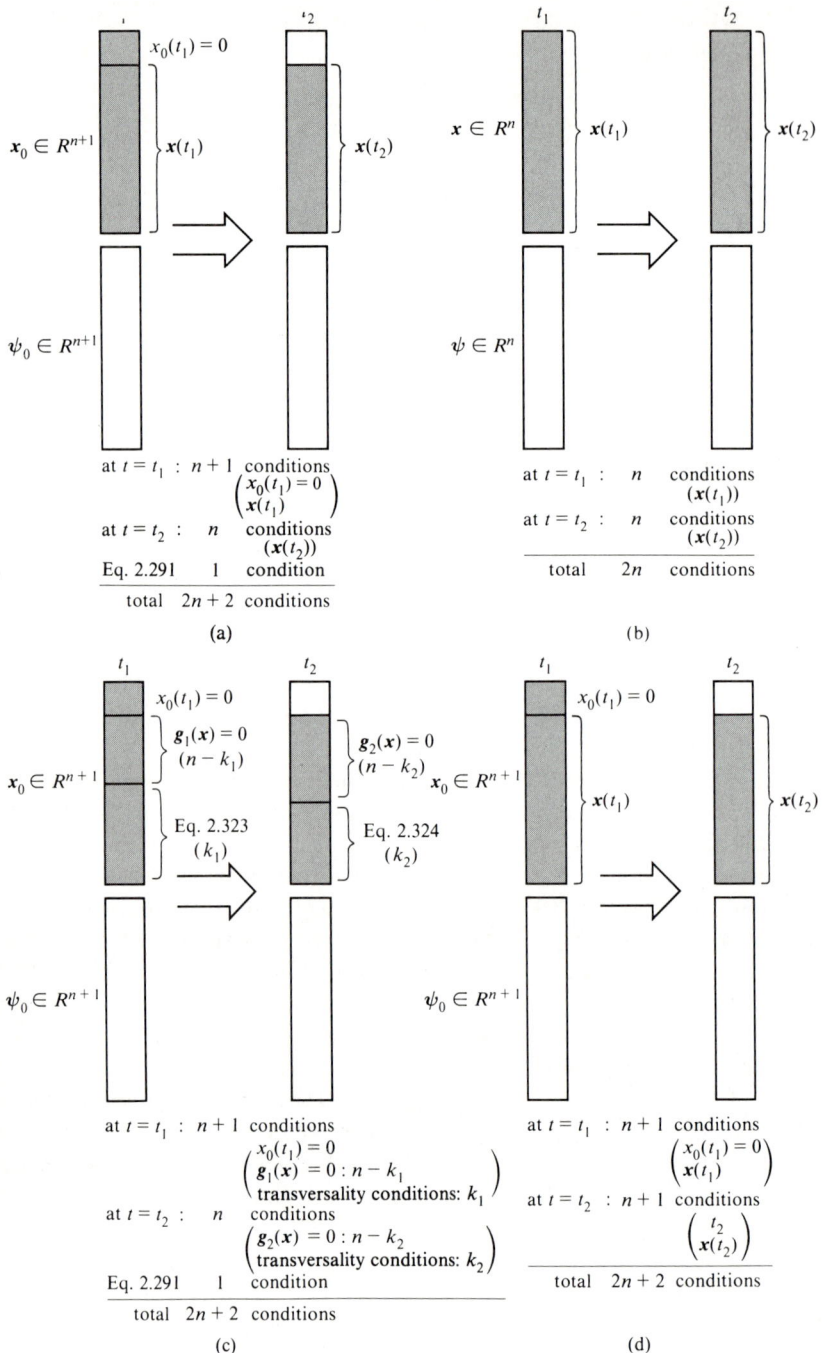

Figure 2.14 Boundary conditions for the maximum principle. (a) Theorem 2.11. (b) Theorem 2.12. (c) Theorem 2.14. (d) Theorem 2.16.

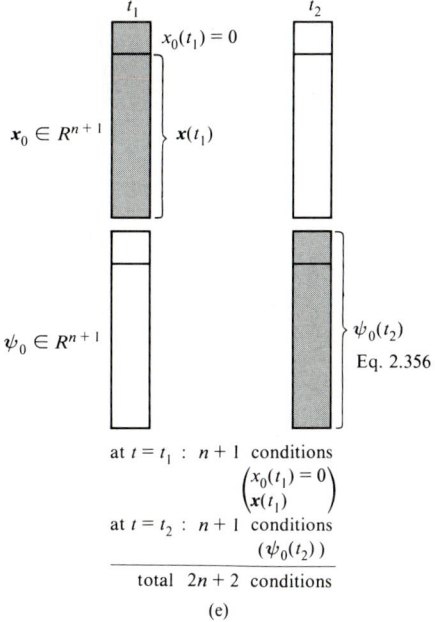

Figure 2.14 (continued) (e) Theorem 2.17.

into two places are called *two-point boundary value problems*. Generally, the computation of MP results in a two-point boundary value problem.

Basically, the computation of two-point boundary value problems is executed as follows. Make a guess for all of the unspecified conditions at $t = t_1$. Integrate the differential equations with the specified and guessed boundary conditions. Compare the result at $t = t_2$ with the boundary conditions specified at $t = t_2$. If they meet, then the result of computation is the solution; otherwise, modify the guess at $t = t_1$ based on the difference at $t = t_2$ and repeat this process until the result and the specified boundary condition at $t = t_2$ coincide with each other.

As you might imagine, the computation can be very time consuming and expensive. The systematic approaches to two-point boundary value problems are named the *shooting method* and the *relaxation method*. Their algorithms and programs are found in Press, Flannery, Teukolsky, and Vetterling (1986).

Summary

In this chapter, we have overviewed the mathematical tools that will be used in the following chapters, or that are closely related to the contents of those chapters. These mathematical techniques will be useful in dealing with many kinds of optimization problems of robotic systems. The other important optimization techniques that are not covered in this chapter include *linear programming* (LP) and *dynamic programming* (DP). Although they are not used in this book, these mathematical programming methods are also powerful tools and, indeed, have been used in robotics for minimum time control, collision-free path planning, and so on. The reader is encouraged to refer to Luenberger (1984) for LP and to Bellman (1961); and Bellman and Dreyfus (1962) for DP.

In preparing this chapter, the author found the following books informative and useful:

- *Singular value decomposition*: Golub and Van Loan (1983), Kodama and Suda (1978)

- *Pseudoinverse*: Boullion and Odell (1971), Rao and Mitra (1971)

- *Variational methods*: Luenberger (1969,1984)

- *Pontryagin's maximum principle*: Pontryagin, Boltyanskii, Gamkrelidze, and Mishchenko (1962)

References

Asada, H. 1983. A geometrical representation of manipulator dynamics and its application to arm design. *Trans. ASME J. Dyn. Sys., Meas. Contr.* 105 (3):131–136.

Bellman, R. 1961. *Dynamic Programming.* Princeton, NJ: Princeton University Press.

Bellman, R., and Dreyfus, S. 1962. *Applied Dynamic Programming.* Princeton, NJ: Princeton University Press.

Bobrow, J. E., Dubowsky, S., and Gibson, J. S. 1985. Time-optimal control of robotic manipulators along specified paths. *Int. J. Robotics Res.* 4 (3):3–17.

Boullion, T. L., and Odell, P. L. 1971. *Generalized inverse matrices.* New York: Wiley-Interscience.

Craig, J. J. 1989. *Introduction to robotics: Mechanics & control*, second edition. Reading, MA: Addison-Wesley.

Dubowsky, S., Norris, M. A., and Shiller, Z. 1986 (San Fransisco). Time optimal trajectory planning for robotic manipulators with obstacle avoidance: A CAD approach. *Proc. IEEE Int. Conf. Robotics and Automation*, pp. 1906–1912.

Goldstein, H. 1980 *Classical mechanics*, second edition. Reading, MA: Addison-Wesley.

Golub, G., and Van Loan, C. 1983. *Matrix computations.* Baltimore: Johns Hopkins University Press.

Iri, M., Kodama, S., and Suda, N. 1982. Singular value decomposition and its application to system control. *J. Society of Instrument and Control Engineers* 21(8): 763–772. *(in Japanese).*

Kahn, M. E., and Roth, B. E. 1971. The near minimum-time control of open-loop articulated kinematic chains. *Trans. ASME J. Dyn. Sys., Meas. Contr.* 93(3): 164–172.

Klein, C. A., and Huang, C. H. 1983. Review of pseudoinverse control for use with kinematically redundant manipulators. *IEEE Trans. Sys., Man, Cyber.* SMC-13 (3): 245–250.

Kodama, S., and Suda, N. 1978. *Matrix theory for systems control.* Tokyo: Society of Instruments and Control Engineers *(in Japanese).*

Kuhn, H. W., and Tucker, A. W. 1951. Nonlinear Programming. *Proc. Second Berkeley Symposium on Mathematical Statistics and Probability*, ed. J. Neyman, pp. 481–492, Berkeley: University of California Press.

Lawson, C., and Hanson, R. J. 1974. *Solving least squares problems.* Englewood Cliffs, NJ: Prentice-Hall.

Luenberger, D. G. 1969. *Optimization by vector space methods.* New York: John Wiley.

Luenberger, D. G. 1984. *Linear and non-linear programming*, second edition. Reading, MA: Addison-Wesley.

Maciejewski, A. A., and Klein, C. A. 1985. Obstacle avoidance for kinematically redundant manipulators in dynamically varying environments. *Int. J. Robotics Res.* 4 (3): 109–117.

Mine, H. 1966. *Operations Research: Vol. 1.* Tokyo: Asakura.

Moore, E. H. 1920. On the reciprocal of the general algebraic matrix. *Bulletin Amer. Math. Soc.* 26: 394–395.

Moore, E. H. 1935. Generalized analysis, part I. *Mem. Amer. Philos. Soc.* 1: 1–231.

Penrose, R. 1955. A generalized inverse for matrices. *Proc. Cambridge Phil. Soc.*, 51: 406–413.

Pontryagin, L. S., Boltyanskii, V. G., Gamkrelidze, R. V., and Mishchenko, E. F. 1962. *The mathematical theory of optimal processes.* (trans. from Russian by K. N. Trirogoff; ed. L. W. Neustadt). New York: John Wiley.

Press, W. H., Flannery, B. P., Teukolsky, S. A., and Vetterling, W. T. 1986. *Numerical recipes: The art of scientific computing.* Cambridge, England: Cambridge University Press.

Rao, S. S. 1984. *Optimization: theory and applications*, second edition. New Delhi: Wiley Eastern.

Rao, C. R., and Mitra, S. K. 1971. *Generalized inverse of matrices and its applications.* New York: Wiley.

Scales, L. E. 1985. *Introduction to non-linear optimization.* New York: Springer-Verlag.

Shin, K. G., and Mckay, N. D. 1985. Minimum-time control of robot manipulators with geometric path constraints. *IEEE Trans. Automatic Control.* 30 (6): 531–541.

Yoshikawa, T. 1984. Analysis and control of robot manipulators with redundancy. In *Robotics Research*, eds. M. Brady and R. Paul, pp. 735–747. Cambridge, MA: MIT Press.

CHAPTER 3

Differential Kinematics and Redundancy

3.1 Introduction

Inverse kinematics comprise the computation needed to solve joint angles from a given Cartesian position and orientation of the end-effector. This computation is fundamental in the control of robot manipulators. It is, in general, nonlinear algebraic computation, and there is no analytical closed-form solution for a general robot structure, even for 6-DOF robots. Pieper (1968) studied the solutions of 6-DOF manipulators and showed that a 6-DOF manipulator possibly has at most 16 joint-angle solutions for a given Cartesian position and orientation of the end-effector. He also showed that a 6-DOF manipulator with three succeeding joint axes intersecting at a point always has closed-form solutions.

When the end-effector moves along a continuous trajectory, we cannot arbitrarily switch from one kind of solution to another. Uchiyama (1979) pointed out that the trajectories of multiple solutions intersect at singular points and can be switched only there. The singular points of a robotic mechanism are defined as the singular points of the implicit function—namely, the joint angles that make the Jacobian matrix of the nonlinear mapping not full rank.

Differential kinematics of robot manipulators were introduced by Whitney (1969). He proposed that we use differential relationships to solve for the joint motion from a given Cartesian motion of the end-effector. The relationship between the end-effector velocity and the joint velocity is represented by a linear algebraic equation. The coefficient of the linear equation is the Jacobian matrix and a nonlinear function of joint angles. Whitney named this method *resolved motion rate control*.

Although the computation of closed-form nonlinear solutions, if available, is generally less than that for resolved motion rate control, it is an important advantage of resolved motion rate control that the linearity of the equation allows the development of general discussion, whereas the computation of closed-form solutions depends on the specific kinematic design of a robot manipulator. Particularly, the differential kinematics are natural and unique in Cartesian trajectory control when the manipulator dynamics are considered (Takase 1976; Luh, Walker, and Paul 1981) because rigid-body dynamics and kinematics are naturally connected at the acceleration level. The relationship between the end-effector acceleration and the joint acceleration becomes a similar linear one, with the Jacobian matrix as a coefficient. This scheme is named *resolved acceleration control* (Luh, Walker and Paul 1981) as an extension of *resolved motion rate control*. It must be noted that differential kinematics are also closely related to end-effector force control.

The fundamental kinematic properties, including kinematic redundancy, are analyzed in this chapter in the framework of differential kinematics. Various concepts are established to analyze and evaluate robotic mechanisms. Some of the concepts are to be used as the basis of discussions in Chapters 4 and 5.

In Section 3.2, we provide an overview of kinematics computation associated with the Jacobian matrix. Section 3.3 describes manipulability and redundancy in light of matrix theory. Manipulability and redundancy are instantaneous or local properties derived directly from the Jacobian matrix. The quantitative measure of manipulability developed by Yoshikawa (1984, 1985) is introduced in Section 3.4.

3.2 Jacobian Matrix

3.2.1 Nonlinear Kinematics

Figure 3.1 shows the base coordinates, $O_0\text{-}x_0y_0z_0$, and the end-effector coordinates, $O_1\text{-}x_1y_1z_1$. The former is fixed to the robot base, and the latter is fixed to the end-effector. The position and orientation of the end-effector would be our most important concern when we use a robot manipulator. Let $\boldsymbol{h}_e \in R^3$ be a vector from O_0 to O_1 represented in the base coordinates. Also let $\boldsymbol{e}_1, \boldsymbol{e}_2$, and $\boldsymbol{e}_3 \in R^3$ be unit vectors in the directions of x_1, y_1, and z_1 axes, respectively, represented in the base coordinates. The orientation of the end-effector is described by the following orthogonal matrix:

$$A_e = (\,\boldsymbol{e}_1 \quad \boldsymbol{e}_2 \quad \boldsymbol{e}_3\,) \in R^{3\times 3} \qquad\qquad 3.1$$

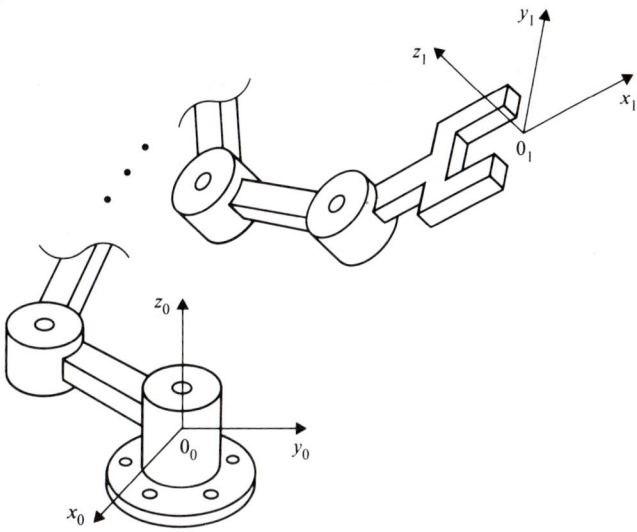

Figure 3.1 Base coordinates and end-effector coordinates.

Although Eq. 3.1 includes nine components, only three are independent, because e_1, e_2, and e_3 have to satisfy the following six relationships:

$$\begin{aligned} e_i^T e_i &= 1 & i &= 1, 2, 3 \\ e_i^T e_j &= 0 & i &\neq j \end{aligned} \qquad 3.2$$

It is well known, however, that there exists no perfect way to choose a set of three independent parameters in representing the orientation of a rigid body in three-dimensional space. A representative choice is *Z-Y-Z* Euler angles, which are defined as follows: (1) *Rotate O_0-$x_0y_0z_0$ coordinates about its z_0 axis by α*. (2) *Again, rotate the new O_0-$x_0y_0z_0$ coordinates about its y_0 axis by β*. (3) *Finally, rotate the newest O_0-$x_0y_0z_0$ coordinates about its z_0 axis by γ*. The α, β, and γ that coincide the coordinate frame obtained after three rotations with O_1-$x_1y_1z_1$ coordinates are called *Z-Y-Z* Euler angles. The sequence of rotations is shown in Fig. 3.2.

The orientation matrix $\boldsymbol{A}_e$ of the coordinates obtained after three rotations is represented by a function of α, β, and γ as follows:

$$\boldsymbol{A}_e = \begin{pmatrix} c_\alpha c_\beta c_\gamma - s_\alpha s_\gamma & -c_\alpha c_\beta s_\gamma - s_\alpha c_\gamma & c_\alpha s_\beta \\ s_\alpha c_\beta c_\gamma + c_\alpha s_\gamma & -s_\alpha c_\beta s_\gamma + c_\alpha c_\gamma & s_\alpha s_\beta \\ -s_\beta c_\gamma & s_\beta s_\gamma & c_\beta \end{pmatrix} \qquad 3.3$$

where $c_\alpha \triangleq \cos\alpha$, $s_\alpha \triangleq \sin\alpha$, and so on.

108 Chapter 3 Differential Kinematics and Redundancy

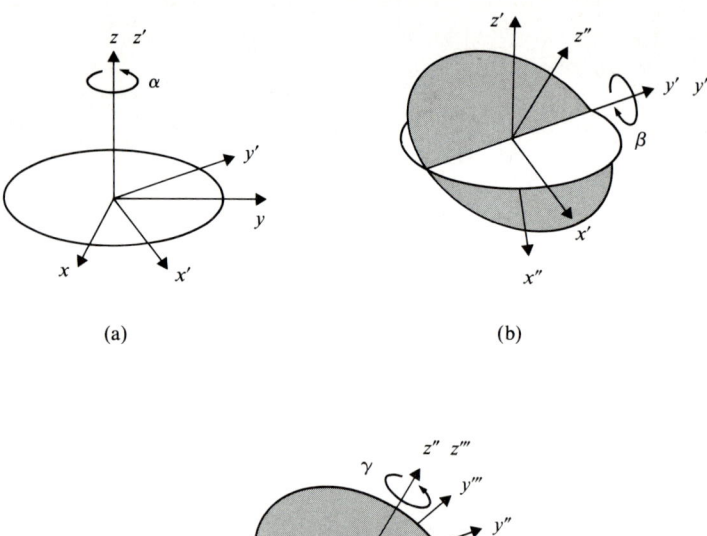

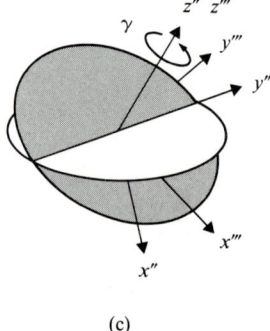

(c)

Figure 3.2 Sequence of Z-Y-Z Euler angles.

In contrast, for a given orientation matrix $\boldsymbol{A}_e$, the Z-Y-Z Euler angles are computed by (Paul 1983):

$$\alpha = \begin{cases} \phi \triangleq \text{ATAN2}(a_{23}, a_{13}) & \text{and} \\ \phi - \pi\,\text{sgn}(\phi) & \end{cases}$$
$$\beta = ATAN2(a_{13}\,c_\alpha + a_{23}\,s_\alpha, a_{33})$$
$$\gamma = ATAN2(-a_{11}\,s_\alpha + a_{21}\,c_\alpha, -a_{12}\,s_\alpha + a_{22}\,c_\alpha)$$

3.4

where a_{ij} is the (i,j) entry of $\boldsymbol{A}_e$. ATAN2($*,*$) is a FORTRAN built-in function that computes an arc-tangent from two arguments and provides a value from $-\pi$ to π. Note that there are two sets of solutions that equivalently represent the same orientation.

A problem in computing Z-Y-Z Euler angles using Eq. 3.4 is that ϕ cannot be obtained if $a_{23} = a_{13} = 0$, which implies that the direction of the z_1 axis is exactly the same as that of the z_0 axis. The problem is not a result of the algorithm being incomplete; rather, it is due to the intrinsic singularity of Z-Y-Z Euler angles. Physically, this corresponds to the fact that α and γ are not uniquely defined if $\beta = 0$. Note that this singularity is caused by representation only and has nothing to do with the mechanical structure of robots. This kind of singularity always appears in any other choice of three independent parameters, including X-Y-Z Euler angles.

In summary, the position and orientation of the end-effector in Fig. 3.1 are typically represented by six independent parameters as follows:

$$r = \begin{pmatrix} h_e \\ \alpha \\ \beta \\ \gamma \end{pmatrix} \in R^6 \qquad 3.5$$

The relationship between r and the joint angles of a robot manipulator is generally expressed by the following nonlinear equation:

$$r = f(\theta) \in R^6, \qquad \theta \in R^n \qquad 3.6$$

where n is the number of joints.

3.2.2 Resolved Motion Rate Control

Differentiating Eq. 3.5 with respect to time, the relationship between $\dot{r}$ and $\dot{\theta}$ is given by

$$\dot{r} = J(\theta)\dot{\theta} \qquad 3.7$$

$$J(\theta) \triangleq \frac{\partial f}{\partial \theta} \in R^{6 \times n} \qquad 3.8$$

where $J(\theta)$ is the Jacobian matrix and Eq. 3.8 is its formal definition. In *resolved motion rate control*, proposed by Whitney (1969), we compute $\dot{\theta}$ by solving the linear equation of Eq. 3.6 for given θ and $\dot{r}$.

It would be computationally inefficient to try to evaluate the Jacobian matrix using the closed analytical form of $\partial f/\partial \theta$. Whitney (1972) studied efficient computation of the Jacobian matrix, where the Jacobian matrix is redefined by the following equation:

$$\dot{r}_\omega = J_\omega(\theta)\dot{\theta} \qquad 3.9$$

$$\dot{r}_\omega \triangleq \begin{pmatrix} \dot{h}_e \\ \omega \end{pmatrix} \in R^6 \qquad 3.10$$

where $\omega \in R^3$ is the angular velocity of the end-effector coordinates represented in terms of the base coordinates. Note that $\dot{r}$ and $\dot{r}_\omega$ are obtained from each other as shown by the following equation:

$$\omega = K \begin{pmatrix} \dot{\alpha} \\ \dot{\beta} \\ \dot{\gamma} \end{pmatrix} \qquad 3.11$$

$$\boldsymbol{K} \triangleq \begin{pmatrix} 0 & -\sin\alpha & \cos\alpha\sin\beta \\ 0 & \cos\alpha & \sin\alpha\sin\beta \\ 1 & 0 & \cos\beta \end{pmatrix}$$

The three column vectors of $\boldsymbol{K}$ are unit vectors in the directions of rotational axes about which α, β, and γ are measured, respectively. Therefore $\boldsymbol{J}(\boldsymbol{\theta})$ and $\boldsymbol{J}_\omega(\boldsymbol{\theta})$ are related to each other by

$$\boldsymbol{J}_\omega(\boldsymbol{\theta}) = \begin{pmatrix} \boldsymbol{E}_3 & \boldsymbol{O} \\ \boldsymbol{O} & \boldsymbol{K} \end{pmatrix} \boldsymbol{J}(\boldsymbol{\theta}) \qquad 3.12$$

where $\boldsymbol{E}_3 \in R^{3\times 3}$ is an identity matrix.

The intrinsic singularity of Z-Y-Z Euler angles again appears when we compute $(\dot\alpha\ \dot\beta\ \dot\gamma)^T$ (or $\dot{\boldsymbol{J}}(\boldsymbol{\theta})$) from $\boldsymbol{\omega}$ (or $\boldsymbol{J}_\omega(\boldsymbol{\theta})$) when $\beta = 0$. The singularity is not a problem in the reverse computation. The expression of Eq. 3.9 is free from this kind of singularity. Therefore, it would be a good idea to use Eq. 3.9 as a differential relationship, rather than Eq. 3.7. We call $\boldsymbol{J}_\omega(\boldsymbol{\theta})$ the *basic Jacobian matrix* (Khatib 1987).

A drawback of the basic Jacobian matrix is that the integration of $\boldsymbol{\omega}$ has no clear physical meaning. We overcame this drawback by considering that $\boldsymbol{\omega}$ and $\boldsymbol{A}_e$ are related by the following differential equation:

$$\dot{\boldsymbol{A}}_e = \boldsymbol{\Omega}\,\boldsymbol{A}_e \qquad 3.13$$

$$\boldsymbol{\Omega} \triangleq \begin{pmatrix} 0 & -\omega_z & \omega_y \\ \omega_z & 0 & -\omega_x \\ -\omega_y & \omega_x & 0 \end{pmatrix} \qquad 3.14$$

where the skew symmetric matrix of Eq. 3.14 physically means the cross-product by $\boldsymbol{\omega}$. Eq. 3.13 is nothing but a collection of $\dot{\boldsymbol{e}}_i = \boldsymbol{\omega} \times \boldsymbol{e}_i$ ($i = 1, 2, 3$), where $\boldsymbol{e}_i$ is the ith column vector of $\boldsymbol{A}_e$. Accordingly, if the end-effector motion is described by $\boldsymbol{h}_e$ and $\boldsymbol{A}_e$ as a function of time, then we compute $\dot{\boldsymbol{r}}_\omega$ by taking their derivatives $\dot{\boldsymbol{h}}_e$ and $\dot{\boldsymbol{A}}_e$ and using Eq. 3.13 to obtain $\boldsymbol{\omega}$. The joint motion $\dot{\boldsymbol{\theta}}$ can now be computed from Eq. 3.9. On the other hand, when $\boldsymbol{\omega}$ and $\boldsymbol{A}_e$ are given, $\dot{\boldsymbol{A}}_e$ is obtained by Eq. 3.13. The change in orientation is computed by integrating $\dot{\boldsymbol{A}}_e$. Note that the integration should be done carefully such that $\boldsymbol{A}_e$ remains an orthogonal matrix even if numerical errors are involved.

3.2.3 Resolved Acceleration Control

Differentiating Eq. 3.9 with respect to time again, we have

$$\ddot{\boldsymbol{r}}_\omega = \boldsymbol{J}_\omega(\boldsymbol{\theta})\ddot{\boldsymbol{\theta}} + \dot{\boldsymbol{J}}_\omega(\boldsymbol{\theta})\dot{\boldsymbol{\theta}} \qquad 3.15$$

In *resolved acceleration control* (Takase 1976; Luh, Walker, and Paul 1981) we determine $\ddot{\boldsymbol{\theta}}$ for given $\ddot{\boldsymbol{r}}_w$, $\boldsymbol{\theta}$, and $\dot{\boldsymbol{\theta}}$. Note that $\boldsymbol{J}_\omega(\boldsymbol{\theta})$ is a function of $\boldsymbol{\theta}$ and $\dot{\boldsymbol{\theta}}$ and the second term of Eq. 3.15 is a quadratic function of $\dot{\boldsymbol{\theta}}$.

We compute $\dot{\boldsymbol{\omega}}$ in $\ddot{\boldsymbol{r}}_w$ from $\boldsymbol{A}_e$, $\dot{\boldsymbol{A}}_e$, and $\ddot{\boldsymbol{A}}_e$ by using the following equation:

$$\begin{aligned} \ddot{\boldsymbol{A}}_e &= \dot{\boldsymbol{\Omega}}\,\boldsymbol{A}_e + \boldsymbol{\Omega}^2\,\boldsymbol{A}_e \\ &= \dot{\boldsymbol{\Omega}}\,\boldsymbol{A}_e + \dot{\boldsymbol{A}}_e\boldsymbol{A}_e^T\dot{\boldsymbol{A}}_e \end{aligned} \qquad 3.16$$

We obtain the first half of Eq. 3.16 by collecting $\ddot{\boldsymbol{e}}_i = \dot{\boldsymbol{\omega}} \times \boldsymbol{e}_i + \boldsymbol{\omega} \times (\boldsymbol{\omega} \times \boldsymbol{e}_i)$ ($i = 1, 2, 3$). We derive the second half by substituting Eq. 3.13 into the first half with $\boldsymbol{A}_e^T \boldsymbol{A}_e = \boldsymbol{E}_3$.

The basic scenario of *resolved acceleration control* is as follows. *Given the desired trajectory of $\boldsymbol{h}_e$ and $\boldsymbol{A}_e$ as a function of time, compute $\dot{\boldsymbol{h}}_e$, $\ddot{\boldsymbol{h}}_e$, $\dot{\boldsymbol{A}}_e$, and $\ddot{\boldsymbol{A}}_e$. Obtain $\dot{\boldsymbol{\omega}}$ using Eq. 3.16. Solve $\ddot{\boldsymbol{\theta}}$ using Eq. 3.15 with known $\boldsymbol{\theta}$ and $\dot{\boldsymbol{\theta}}$, and with computed $\ddot{\boldsymbol{r}}_w$, $\boldsymbol{J}_\omega$, and $\dot{\boldsymbol{J}}_\omega$.* The necessary joint torque $\boldsymbol{\tau} \in R^n$ to realize the desired acceleration $\ddot{\boldsymbol{\theta}}$ is now calculated using the following dynamic model:

$$\boldsymbol{\tau} = \boldsymbol{A}(\boldsymbol{\theta})\ddot{\boldsymbol{\theta}} + \boldsymbol{B}(\boldsymbol{\theta},\dot{\boldsymbol{\theta}}) + \boldsymbol{C}(\boldsymbol{\theta}) \qquad 3.17$$

In Eq. 3.17, $\boldsymbol{A}(\boldsymbol{\theta}) \in R^{n \times n}$, $\boldsymbol{B}(\boldsymbol{\theta},\dot{\boldsymbol{\theta}}) \in R^n$, and $\boldsymbol{C}(\boldsymbol{\theta}) \in R^n$ are the inertia matrix, the torque due to the Coriolis and centrifugal forces, and the torque due to the gravitation, respectively. If the preceding dynamic model is exact, the obtained joint torque precisely creates the desired motion. Note that, in the preceding scenario, the joint acceleration plays a key role between the kinematic and dynamic equations of Eqs. 3.15 and 3.17, and the joint velocity and angle are treated as state variables, which can be measured by joint sensors or obtained by integration of the acceleration.

3.2.4 Closing the Feedback Loop

In real situations, both *resolved motion rate control* (RMRC) and *resolved acceleration control* (RAC) need feedback loops to work in the presence of various uncertainties, including modeling error of the dynamics. Let the desired and actual trajectories be represented by $\boldsymbol{r}_{wd}$ and $\boldsymbol{r}_w$, respectively. We synthesize feedback control such that the error vector $\Delta \boldsymbol{r}_w \stackrel{\triangle}{=} \boldsymbol{r}_{wd} - \boldsymbol{r}_w$ behaves as follows:

$$\Delta \dot{\boldsymbol{r}}_w + \boldsymbol{G}\,\Delta \boldsymbol{r}_w = \boldsymbol{O} \qquad \text{for RMRC} \qquad 3.18$$

$$\Delta \ddot{\boldsymbol{r}}_w + \boldsymbol{G}_1\,\Delta \dot{\boldsymbol{r}}_w + \boldsymbol{G}_2\,\Delta \boldsymbol{r}_w = \boldsymbol{O} \qquad \text{for RAC} \qquad 3.19$$

where $\boldsymbol{G}$, $\boldsymbol{G}_1$, and $\boldsymbol{G}_2$ are constant matrices that guarantee stability and must be chosen such that the poles of the linear systems of Eqs. 3.18 and 3.19

have negative real parts. Usually, they are chosen from diagonal matrices. From Eqs. 3.18 and 3.19, $\dot{r}_\omega$ of RMRC and $\ddot{r}_\omega$ of RAC are computed by

$$\dot{r}_\omega = \dot{r}_{\omega d} + G \Delta r_\omega \qquad \text{for RMRC} \qquad 3.20$$

$$\ddot{r}_\omega = \ddot{r}_{\omega d} + G_1 \Delta \dot{r}_\omega + G_2 \Delta r_\omega \qquad \text{for RAC} \qquad 3.21$$

Consequently, we solve $\dot{\theta}$ of RMRC and $\ddot{\theta}$ of RAC using Eqs. 3.9 and 3.15 by substituting Eqs. 3.20 and 3.21, respectively. Note that, to evaluate Eqs. 3.20 and 3.21, we must prepare $\ddot{r}_{\omega d}$, $\Delta \dot{r}_\omega$, and Δr_ω from the desired trajectories of h_e and A_e and the measured values of θ and $\dot{\theta}$. Now, our difficulty lies in the fact that the orientation part of Δr_ω cannot be computed straightforwardly from the desired and actual values of the orientation matrix—namely, A_{ed} and A_e—since the integrated value of ω has no direct relationship to A_e.

When two coordinate frames with different orientations are given, there exist a fixed unit vector and a rotational angle such that the first frame becomes parallel to the second one after the first one is rotated about the unit vector by the angle. This pair can be used to represent the orientation of the second frame relative to the first one. This is called the *equivalent angle-axis representation* (Craig 1989). The difference between A_{ed} and A_e can be expressed by the following vector:

$$\xi = \psi e_e \qquad \in R^3 \qquad 3.22$$

where e_e and ψ are the unit vector and angle of the *equivalent angle-axis representation*. The relationship is shown in Fig. 3.3.

We can compute e_e and ψ from A_e and A_{ed} as follows:

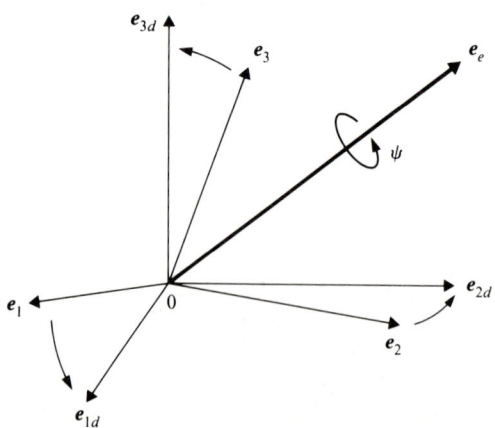

Figure 3.3 Equivalent angle-axis representation.

$$\psi = \text{ACOS}\{\frac{1}{2}(\boldsymbol{e}_1^T \boldsymbol{e}_{1d} + \boldsymbol{e}_2^T \boldsymbol{e}_{2d} + \boldsymbol{e}_3^T \boldsymbol{e}_{3d} - 1)\}$$

$$\boldsymbol{e}_e = \frac{1}{2\sin\psi}(\boldsymbol{e}_1 \times \boldsymbol{e}_{1d} + \boldsymbol{e}_2 \times \boldsymbol{e}_{2d} + \boldsymbol{e}_3 \times \boldsymbol{e}_{3d})$$

3.23

where $\boldsymbol{e}_i$ and $\boldsymbol{e}_{id}$ are the ith column vectors of $\boldsymbol{A}_e$ and $\boldsymbol{A}_{ed}$, respectively. ACOS(∗) is a FORTRAN built-in function that computes the arc-cosine and provides values between 0 and π. Note that $\boldsymbol{\xi}$ does not have a direct relationship to the error of the integrated $\boldsymbol{\omega}$. We anticipate that $\boldsymbol{\xi}$ will approximately suggest the velocity or acceleration to be generated to reduce the orientation error; namely, $\Delta\boldsymbol{r}_\omega$ is approximately represented by

$$\Delta\boldsymbol{r}_\omega = \begin{pmatrix} \boldsymbol{h}_{ed} - \boldsymbol{h}_e \\ \boldsymbol{\xi} \end{pmatrix}$$

3.24

To ease the computation of Eqs. 3.22 and 3.23, Luh, Walker, and Paul (1981) proposed that we use the following $\boldsymbol{\xi}'$ in place of $\boldsymbol{\xi}$:

$$\boldsymbol{\xi}' \overset{\Delta}{=} \sin\psi \; \boldsymbol{e}_e = \frac{1}{2}(\boldsymbol{e}_1 \times \boldsymbol{e}_{1d} + \boldsymbol{e}_2 \times \boldsymbol{e}_{2d} + \boldsymbol{e}_3 \times \boldsymbol{e}_{3d})$$

3.25

When $\boldsymbol{A}_e$ is close to $\boldsymbol{A}_{ed}$, ψ becomes small and $\boldsymbol{\xi}'$ closely approximates $\boldsymbol{\xi}$.

3.2.5 Computation of the Basic Jacobian Matrix

The basic Jacobian matrix is computed efficiently as follows (Whitney 1972):

$$\boldsymbol{J}_\omega = (\; \boldsymbol{J}_{\omega 1} \quad \boldsymbol{J}_{\omega 2} \quad \cdots \quad \boldsymbol{J}_{\omega n}\;)$$

3.26

where $\boldsymbol{J}_{\omega i} \in R^6$ implies the ith column vector of the basic Jacobian matrix and is computed as follows:

$$\boldsymbol{J}_{\omega i} = \begin{cases} \begin{pmatrix} \hat{\boldsymbol{z}}_{i-1} \times \boldsymbol{p}_n^{i-1} \\ \hat{\boldsymbol{z}}_{i-1} \end{pmatrix} & \text{(revolutive joint)} \\ \\ \begin{pmatrix} \hat{\boldsymbol{z}}_{i-1} \\ \boldsymbol{O} \end{pmatrix} & \text{(prismatic joint)} \end{cases}$$

3.27

where $\hat{\boldsymbol{z}}_{i-1} \in R^3$ is a unit vector in the direction of z axis of $(i-1)$th link frame and $\boldsymbol{p}_n^{i-1} \in R^3$ is a vector from the origin of the $(i-1)$th link frame to that of the nth one; see Fig. 3.4.

Both vectors are represented with respect to the base frame. We assume that the link frames are assigned according to the Denavit–Hartenberg convention (Denavit and Hartenberg 1955; Paul 1981). Namely, the $(i-1)$th frame is fixed to the $(i-1)$th link with the origin located on the ith joint axis. The direction of z axis of the $(i-1)$th frame is that of the ith joint axis. The origin of the nth frame is fixed at the effective point of the end-effector. The location of the origin of the $(i-1)$th frame on the ith joint is also determined

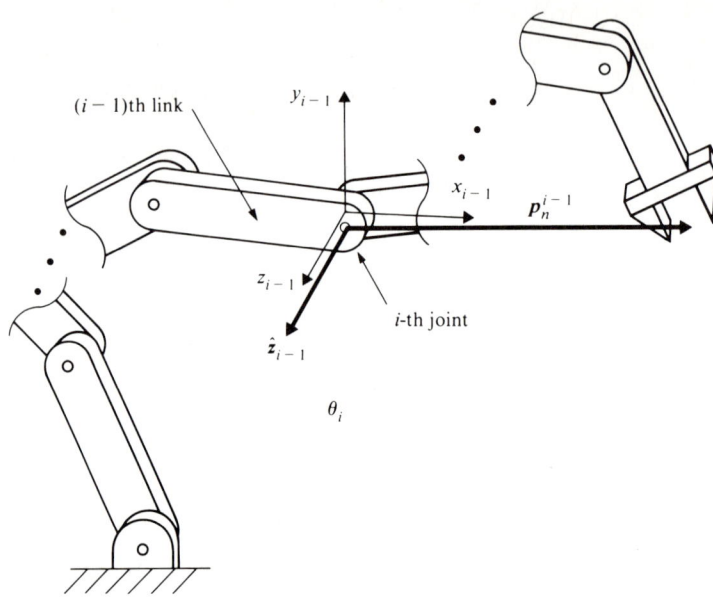

Figure 3.4 Computation of the basic Jacobian matrix.

by the Denavit–Hartenberg convention. Note that the computation of $J_{\omega i}$ in Eq. 3.27 is invariant to the location.

Based on the Whitney's formulation, Orin and Schrader (1984) provided the most efficient recursive computational algorithm when only the Jacobian matrix is computed. The computation of the basic Jacobian matrix and the other kinematics including $J_\omega \dot{\theta}$ involves a lot of computation overlapping with that of the inverse dynamics. Nakamura, Yokokohji, Hanafusa, and Yoshikawa (1986, 1987) proposed an efficient algorithm for simultaneous computations of kinematics and dynamics by eliminating the overlapping computations as much as possible. The algorithm is called *unified computation of kinematics and dynamics*. The algorithm is suitable for resolved acceleration control where all of the inverse dynamics and forward and inverse kinematics must be computed in real time.

3.3 Manipulability and Redundancy

3.3.1 *Manipulability*

Resolved motion rate control and resolved acceleration control result in the same type of computation—namely, solving Eqs. 3.9 and 3.15 for $\dot{\theta}$ and $\ddot{\theta}$, respectively. Both are linear equations, with the Jacobian matrix as a coefficient. In this section, the kinematic properties are analyzed on the basis of the Jacobian matrix in light of matrix theory (Nakamura 1979; Hanafusa, Yoshikawa, and Nakamura 1981, 1983).

In the rest of this chapter, r and $J(\theta)$ are not limited to the three-dimensional position and orientation with either Z-Y-Z Euler angles or ω. Now, $r \in R^m$ is used to generally represent the kinematic output of a robot manipulator. We call r the *manipulation variable*, and $J(\theta) \in R^{m \times n}$ is the corresponding Jacobian matrix. As a generalized expression of Eqs. 3.9 and 3.15, we use the following equation:

$$\delta r = J(\theta) \delta \theta \qquad 3.28$$

Equation 3.28 shows that δr is a linear mapping of $\delta \theta$ by the Jacobian matrix. Figure 3.5 illustrates the linear mapping by Eq. 3.28. Here, $\mathcal{R}(J)$ denotes the range space of $J(\theta)$, which is a subspace in the m-dimensional space of δr. The dimension of $\mathcal{R}(J)$ is rank $J(\theta) \leq \min(m, n)$. The range space is a space spanned by the column vectors of $J(\theta)$; it consists of all the mappings of every $\delta \theta$. Note that δr is kinematically realizable through an appropriate choice of $\delta \theta$ if and only if $\delta r \in \mathcal{R}(J)$. Also note that the orthogonal complement of the range space $\mathcal{R}(J)^{\perp}$ indicates the subspace made up of all of the kinematically unrealizable motions δr. Based on this consideration, the manipulable space and the degree of manipulability are defined as follows.

Definition 3.1 (Manipulable Space and Degree of Manipulability)

The range space of the Jacobian matrix and its dimension are called the *manipulable space* and the *degree of manipulability (DOM)*.

The motion δr is kinematically realizable if and only if it is a member of the manipulable space. Note that the manipulable space changes as the configuration of the manipulator changes. Since rank $J(\theta)$ degenerates at singular points, so does the manipulable space. At singular points, a manipulator cannot move in degenerated directions.

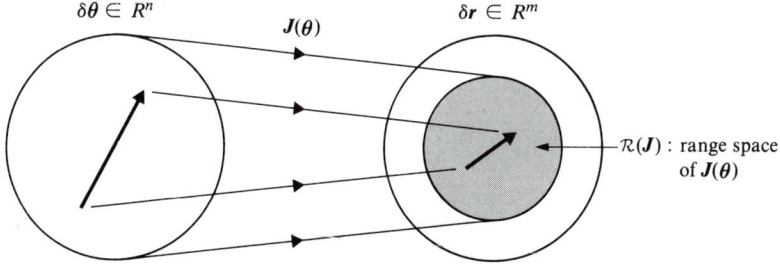

Figure 3.5 Linear mapping by Eq. 3.28.

3.3.2 Redundancy

If the number of joints is greater than the dimension of the manipulation variable—namely, $n > m$—a manipulator is said to be a *redundant manipulator*. This manipulator is characterized by the fact that there exist infinite solutions of inverse kinematics. We discuss redundancy in the framework of differential kinematics.

Let $\delta\boldsymbol{\theta}^1$ and $\delta\boldsymbol{\theta}^2$ be two distinctive solutions of Eq. 3.28. Then,

$$\boldsymbol{J}(\boldsymbol{\theta})\,\delta\boldsymbol{\theta}_e = \boldsymbol{J}(\boldsymbol{\theta})\,\delta\boldsymbol{\theta}^1 - \boldsymbol{J}(\boldsymbol{\theta})\,\delta\boldsymbol{\theta}^2 = \boldsymbol{O} \qquad 3.29$$

$$\delta\boldsymbol{\theta}_e \triangleq \delta\boldsymbol{\theta}^1 - \delta\boldsymbol{\theta}^2 \neq \boldsymbol{O} \qquad 3.30$$

Equations 3.29 and 3.30 imply that the difference of two solutions is always mapped to the zero vector in $\delta\boldsymbol{r}$ space. The variety of solutions is represented by that of the vectors to be mapped to the zero vector by the Jacobian matrix. It is known that such vectors span a linear subspace in n-dimensional space of $\delta\boldsymbol{\theta}$, which is called the *null space*. The null space of the Jacobian matrix, $\mathcal{N}(\boldsymbol{\theta})$, and its mapping are shown in Fig. 3.6.

The null space of the Jacobian matrix represents the variety of solutions of Eq. 3.28. Hence, we make the following definition.

Definition 3.2 (Redundant Space and Degree of Redundancy)

The null space of the Jacobian matrix and its dimension are called the *redundant space* and the *degree of redundancy* (DOR).

Note that, if $n \leq m$ and rank $\boldsymbol{J}(\boldsymbol{\theta}) = n$ (namely, DOM $= n$), there is no nonzero entry in the null space of $\boldsymbol{J}(\boldsymbol{\theta})$, since the column vectors of $\boldsymbol{J}(\boldsymbol{\theta})$ are all independent. Accordingly, DOR becomes zero and the solution of Eq. 3.28 becomes unique. A more general relationship between DOM and DOR is provided by using a well-known result from matrix theory that the range space and the null space of a matrix $\boldsymbol{M} \in R^{m \times n}$ should satisfy

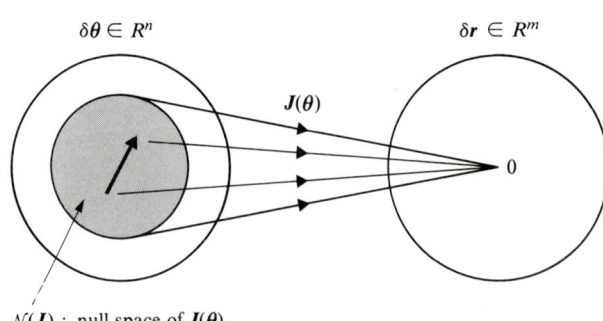

Figure 3.6 Null space of the Jacobian matrix.

3.3 Manipulability and Redundancy

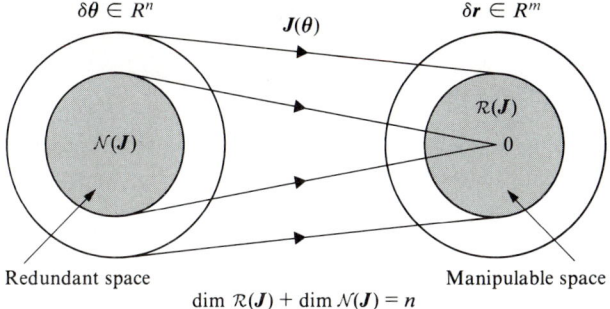

Figure 3.7 Relationship between the manipulable space and the redundant space.

$$\dim \mathcal{R}(M) + \dim \mathcal{N}(M) = n \qquad 3.31$$

The following theorem is obtained.

Theorem 3.1 (Degree of Manipulability and Degree of Redundancy)

The degree of manipulability (DOM) and the degree of redundancy (DOR) satisfy the following relationship regardless of $\boldsymbol{\theta}$:

$$\text{DOM} + \text{DOR} = n \qquad 3.32$$

Theorem 3.1 implies that DOR increases at singular points by the same number by which DOM decreases. The supplemented DOR at singular points, however, is momentary and would not be very useful since it arises just there. Figure 3.7 shows the relationship between the manipulable space and the redundant space.

3.3.3 Orthogonal Projections

There are two subspaces in each of the δr and $\delta \boldsymbol{\theta}$ spaces; the manipulable space (MS) and the orthogonal complement[3.1] of the manipulable space (OCMS) in the δr space, and the redundant space (RS) and the orthogonal complement of the redundant space (OCRS) in the $\delta \boldsymbol{\theta}$ space. The Jacobian matrix $\boldsymbol{J}(\boldsymbol{\theta})$ is a mapping from the whole $\delta \boldsymbol{\theta}$ space to MS. We shall use the mappings from one of these spaces to another space in Chapters 4 and 5 to control kinematic redundancy. In this subsection, we provide a theorem that summarizes four orthogonal projections: (1) from the whole δr space to MS, (2) from the whole δr space to OCMS, (3) from the whole $\delta \boldsymbol{\theta}$ space to RS,

---

[3.1] For the definition of orthogonal complements, see footnote 2.6 in Section 2.4.5.

and (4) from the whole $\delta\boldsymbol{\theta}$ space to OCRS. The following theorem is obtained straightforwardly from the result of Section 2.4.5.

Theorem 3.2 (Orthogonal Projections)

The orthogonal projections from the $\delta\boldsymbol{r}$ space to MS, from the $\delta\boldsymbol{r}$ space to OCMS, from the $\delta\boldsymbol{\theta}$ space to RS, and from the $\delta\boldsymbol{\theta}$ space to OCRS are:

$$\text{MS}: \quad \boldsymbol{J}(\boldsymbol{\theta})\boldsymbol{J}^{\#}(\boldsymbol{\theta}) \in R^{m \times m} \quad\quad 3.33$$

$$\text{OCMS}: \quad \boldsymbol{E}_m - \boldsymbol{J}(\boldsymbol{\theta})\boldsymbol{J}^{\#}(\boldsymbol{\theta}) \in R^{m \times m} \quad\quad 3.34$$

$$\text{RS}: \quad \boldsymbol{E}_n - \boldsymbol{J}^{\#}(\boldsymbol{\theta})\boldsymbol{J}(\boldsymbol{\theta}) \in R^{n \times n} \quad\quad 3.35$$

$$\text{OCRS}: \quad \boldsymbol{J}^{\#}(\boldsymbol{\theta})\boldsymbol{J}(\boldsymbol{\theta}) \in R^{n \times n} \quad\quad 3.36$$

where $\boldsymbol{J}^{\#}(\boldsymbol{\theta})$ is the pseudoinverse of the Jacobian matrix, and $\boldsymbol{E}_i$ is an $i \times i$ identity matrix.

We can physically interpret MS, OCMS, RS, and OCRS as follows: MS is a subspace of kinematically realizable $\delta\boldsymbol{r}$; OCMS is a subspace of kinematically unrealizable $\delta\boldsymbol{r}$; RS is a subspace of $\delta\boldsymbol{\theta}$ that causes zero motion of $\delta\boldsymbol{r}$; and OCRS is a subspace of $\delta\boldsymbol{\theta}$ that realizes $\delta\boldsymbol{r}$ motion with the minimum magnitude of $\delta\boldsymbol{\theta}$. When $\delta\boldsymbol{r}$ includes kinematically unrealizable components, multiplying $\delta\boldsymbol{r}$ by Eq. 3.33 filters out the components and the result becomes kinematically realizable. When you need to see the kinematically unrealizable components, they are given by multiplying $\delta\boldsymbol{r}$ by Eq. 3.34. On the other hand, when $\delta\boldsymbol{\theta}$ includes components in RS, multiplying the $\delta\boldsymbol{\theta}$ by Eq. 3.36 will filter out the components and provide the joint motion that creates the same end-effector motion with the minimum magnitude. The components in RS involved in the $\delta\boldsymbol{\theta}$ can be found by multiplying the $\delta\boldsymbol{\theta}$ by Eq. 3.35.

The range space, null space, and their orthogonal complements of $\boldsymbol{J}(\boldsymbol{\theta})$ have interesting relationships with those of $\boldsymbol{J}^T(\boldsymbol{\theta})$ and $\boldsymbol{J}^{\#}(\boldsymbol{\theta})$. These relationships are useful when we discuss the problems of statics and force control. The following four equations summarize the relationships:

$$\mathcal{R}(\boldsymbol{J}) = \mathcal{N}(\boldsymbol{J}^T)^{\perp} = \mathcal{N}(\boldsymbol{J}^{\#})^{\perp} = \text{MS} \quad\quad 3.37$$

$$\mathcal{N}(\boldsymbol{J}) = \mathcal{R}(\boldsymbol{J}^T)^{\perp} = \mathcal{R}(\boldsymbol{J}^{\#})^{\perp} = \text{RS} \quad\quad 3.38$$

$$\mathcal{R}(\boldsymbol{J}^T) = \mathcal{R}(\boldsymbol{J}^{\#}) = \mathcal{N}(\boldsymbol{J})^{\perp} = \text{OCRS} \quad\quad 3.39$$

$$\mathcal{N}(\boldsymbol{J}^T) = \mathcal{N}(\boldsymbol{J}^{\#}) = \mathcal{R}(\boldsymbol{J})^{\perp} = \text{OCMS} \quad\quad 3.40$$

3.4 Measure of Manipulability

3.4.1 Continuous Measure for Kinematic Quality

The degree of manipulability provides information about the quality of kinematic structure in executing tasks described by the manipulation variable. If DOM is equal to m, the manipulator can make any kinematic motion described by the manipulation variable. If DOM degenerates and becomes less than m, then there exist some directions in which the manipulator cannot move. In common situations, DOM is usually equal to m, and becomes less than m only at singular points. This change is discontinuous and, therefore, it is impossible to predict it by observing DOM. However, since this is the physical change of kinematic quality, it is continuous and should be able to be represented continuously by choosing an appropriate parameter.

Yoshikawa (1984, 1985) proposed a continuous measure that evaluates the kinematic quality of robotic mechanisms; he named it the *measure of manipulability* (MOM). MOM takes a continuous nonnegative scalar value and becomes equal to zero if and only if DOM becomes less than m. This measure can be considered as a distance from singularity. MOM has been used for kinematic analysis, design, and control of robotic mechanisms. In this section, we overview the definition and properties of MOM based on Yoshikawa's (1984) presentation.

The definition of MOM is given as follows.

Definition 3.3 (Measure of Manipulability)

A scalar value w given by the following equation is called the *measure of manipulability* at state $\boldsymbol{\theta}$ with respect to manipulation variable $\boldsymbol{r}$:

$$w \triangleq \sqrt{\det\{\boldsymbol{J}(\boldsymbol{\theta})\boldsymbol{J}^T(\boldsymbol{\theta})\}} \qquad 3.41$$

Note that this measure works for both redundant ($m < n$) and nonredundant ($m = n$) cases. It works even in the case of $m > n$, where MOM always becomes zero, since the robot obviously cannot make every motion that is described by the manipulation variable.

3.4.2 Properties

Relationship with Singular Values. In this section, we generally assume that $m \leq n$. The singular value decomposition of the Jacobian matrix is represented by (see Section 2.3.1):

$$\boldsymbol{J}(\boldsymbol{\theta}) = \boldsymbol{U\Sigma V}^T \qquad 3.42$$

where $U \in R^{m \times m}$ and $V \in R^{n \times n}$ are orthogonal matrices, and Σ is given by

$$\Sigma \triangleq \mathrm{diag}(\sigma_1, \cdots, \sigma_m) \in R^{m \times n} \qquad 3.43$$

$$\sigma_1 \geq \sigma_2 \geq \cdots \geq \sigma_m \geq 0$$

where the σ_is are called *singular values*. Considering that U and V are orthogonal matrices, substituting Eq. 3.42 into Eq. 3.41 results in

$$w = \sqrt{\det(\Sigma \Sigma^T)}$$
$$= \sigma_1 \sigma_2 \cdots \sigma_m \qquad 3.44$$

Equation 3.44 shows that MOM is nothing but the product of singular values. When the Jacobian matrix degenerates, one or more singular values become zero, and so does MOM.

Elliptic Geometry. The elliptic geometry of the singular value decomposition was discussed in Section 2.3.4. We can apply the same discussion to Eqs. 3.28 and 3.42, to derive the following results.

When $\delta\boldsymbol{\theta}$ takes all the values in a unit hypersphere of n-dimensional space, $\delta\boldsymbol{r}$ takes all the values in a rotated hyperellipsoid in m-dimensional space. The principal axes of the ellipsoid are $\sigma_i \, \boldsymbol{u}_i \, (i = 1, \cdots, m)$, where $\boldsymbol{u}_i \in R^m$ is the ith column vector of the orthogonal matrix U in Eq. 3.42. Yoshikawa (1984) named the ellipsoid the *manipulability ellipsoid* and suggested that we use it as a means for analysis, design, and control of robotic mechanisms.

From Eqs. 2.101 and 3.44, the volume of the manipulability ellipsoid is represented using MOM as follows:

$$\text{volume} = \frac{\pi^{m/2}}{\Gamma(1 + m/2)} \, \text{MOM} \qquad 3.45$$

Equation 3.45 means that the volume of the manipulability ellipsoid is proportional to MOM. The proportional coefficient is constant, and is determined only by m.

Nonredundant Case. When a robotic mechanism is nonredundant—namely, when $m = n$—, MOM is computed by an equation simpler than Eq. 3.41. From Eqs. 3.42 and 3.44 with $\det U = \pm 1$ and $\det V = \pm 1$, the following equation is easily derived:

$$w = |\det \boldsymbol{J}(\boldsymbol{\theta})| \qquad 3.46$$

Equation 3.46 was used to analyze the singularity of wrist mechanisms by Paul and Stevenson (1983).

Relationship with Condition Number. The condition number of a matrix, as discussed in Sections 2.3.2 and 2.3.3, has been used to evaluate the property of linear equations. Salisbury and Craig (1982) proposed that we use the condition number of the Jacobian matrix to evaluate workspace quality. The condition number of the Jacobian matrix becomes

$$\kappa = \frac{\sigma_1}{\sigma_m} \qquad 3.47$$

The inverse of the condition number has similar characteristics to MOM in that it is a nonnegative continuous scalar and becomes equal to zero if and only if the rank of the Jacobian matrix becomes less than m. The relationship between MOM and the condition number is clear when they are physically interpreted using the manipulability ellipsoid. MOM is proportional to the volume of the ellipsoid, whereas the condition number is the ratio of the lengths of the longest and shortest principal axes. In other words, MOM is related to the magnitude of the ellipsoid, whereas the condition number concerns the shape of the ellipsoid. Therefore, MOM prefers large ellipsoids and, on the other hand, the condition number prefers ellipsoids with spherelike shape.

Example 3.1

We shall discuss an example from Yoshikawa (1985) to illustrate the application of MOM.

A 2-DOF manipulator working in a plane is given in Fig. 3.8. The joint angles are defined in the figure. The manipulation variable is the

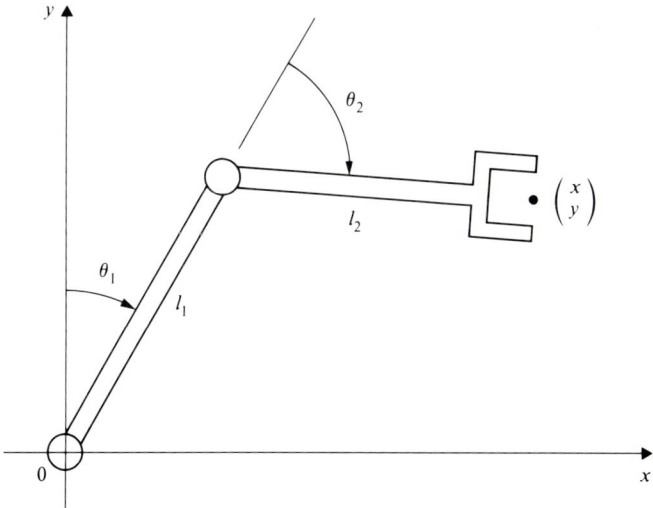

Figure 3.8 Planar 2-DOF manipulator.

position of the end-effector—namely, $\boldsymbol{r} = (x\ y)^T$. The Jacobian matrix is calculated by

$$\boldsymbol{J}(\boldsymbol{\theta}) = \begin{pmatrix} l_1 c_1 + l_2 c_{12} & l_2 c_{12} \\ -l_1 s_1 - l_2 s_{12} & -l_2 s_{12} \end{pmatrix} \qquad 3.48$$

where $c_1 = \cos\theta_1$, $s_1 = \sin\theta_1$, $c_{12} = \cos(\theta_1 + \theta_2)$, and $s_{12} = \sin(\theta_1 + \theta_2)$. MOM is computed using Eq. 3.46 as follows:

$$w = |\det \boldsymbol{J}(\boldsymbol{\theta})| = l_1 l_2 |\sin\theta_2| \qquad 3.49$$

Equation 3.49 indicates that the configuration of the manipulator is optimal in the sense of the measure of manipulability when $\theta_2 = \pm\pi/2$, regardless of θ_1 and the link lengths. When the constraint $l_1 + l_2 =$ constant is imposed on the link lengths, the optimal ratio becomes $l_1/l_2 = 1$, regardless of θ_1 and θ_2. Thus, we conclude that the optimal design of link lengths is $l_1 = l_2$ and the optimal configuration is $\theta_2 = \pm\pi/2$ from the perspective of MOM.

Summary

In this chapter, as a basis for the following chapters, we have discussed differential kinematics of robotic mechanisms.

Differential kinematics and statics have a tightly connected relationship, which is sometimes called *duality*. Using the mathematical results established in this chapter, we can discuss whether or not a robot manipulator can apply a given force to a given environment. This problem is called *force applicability*. A detailed discussion is given in Nakamura (1987).

The singularity of robotic mechanisms is defined. The measure of manipulability can be used as a mathematical tool to verify the distance from singularity. Singularity avoidance using kinematic redundancy will be discussed in Chapters 4 and 5.

Even for a redundant manipulator, singularity does exist, and the situation where the robot must pass singular points or their neighborhood occurs. Most control schemes have treated singularity as an exception, and have excluded it from their frameworks of discussions. It is important to establish a control scheme that works even at or near singular points. Chapter 9 provides a kinematic approach to this problem.

References

Craig, J. J. 1989. *Introduction to robotics: Mechanics and control*, second edition. Reading, MA: Addison-Wesley.

Denavit, J., and Hartenberg, R.S. 1955. A kinematic notation for lower-pair mechanisms based on matrices. *ASME Trans. J. Appl. Mech.* 77 (2): 215–221.

Hanafusa, H., Yoshikawa, T., and Nakamura, Y. 1981 (Kyoto). Analysis and control of articulated robot arms with redundancy. *Preprints 8th Triennial IFAC World Congress*, Vol. 14: pp. 78–83.

Hanafusa, H., Yoshikawa, T., and Nakamura, Y. 1983. Redundancy analysis of articulated robot arms and its utilization for tasks with priority. *Trans. Society of Instrument and Control Engineers* 19 (5): 421–426 *(in Japanese)*.

Khatib, O. 1987. A unified approach for motion and force control of manipulators: The operational space formulation. *IEEE J. Robotics Automat.* 3 (1): 43–53.

Luh, J. Y. S., Walker, M. W., and Paul, R. P. C. 1980. Resolved acceleration control of mechanical manipulators. *IEEE Trans. Automatic Contr.* 25 (3): 468–474.

Nakamura, Y. 1979. Control of articulated robot arms with redundancy. M.S. thesis, Kyoto University, Dept. of Precision Engineering *(in Japanese)*.

Nakamura, Y. 1987 (Los Angeles). Force applicability of robotic mechanisms. *Proc. 26th IEEE Conf. Decision and Control*, pp. 570–575.

Nakamura, Y., Yokokohji, Y., Hanafusa, H., and Yoshikawa, T. 1986 (Osaka). Unified recursive formulation of kinematics and dynamics of robot manipulators. *Proc. Japan–USA Symp. Flexible Automation*, pp. 53–60.

Nakamura, Y., Yokokohji, Y., Hanafusa, H., and Yoshikawa, T. 1987. Unified computation of kinematics and dynamics for robot manipulators. *Trans. Society of Instrument and Control Engineers* 23 (5): 491–498 *(in Japanese)*.

Orin, D. E., and Schrader, W. W. 1984. Efficient computation of the Jacobian for robot manipulators. *Int. J. Robotics Res.* 3 (4): 66–75.

Paul, R. P. 1981. *Robot manipulators: Mathematics, Programming, and Control.* Cambridge, MA: MIT Press.

Paul, R. P., and Stevenson, C. N. 1983. Kinematics of robot wrists. *Int. J. Robotics Res.* 1 (2): 31–38.

Pieper, D. L. 1968. The kinematics of manipulators under computer control. Ph.D. thesis, Stanford University, Dept. of Computer Science.

Salisbury, J. K., and Craig, J. T. 1982. Articulated hands: Force control and kinematic issues. *Int. J. Robotics Res.* 1 (1): 4–17.

Takase, K. 1976. Generalized decomposition and control of a motion of a manipulator. *Trans. Society of Instrument and Control Engineers* 12 (3): 62–68 *(in Japanese)*.

Uchiyama, M. 1979. Study on dynamic control of artificial arms—part 1. *Trans. Japanese Society of Mechanical Engineers.* C-45 (391): 314–322 *(in Japanese)*.

Whitney, D. E. 1969. Resolved motion rate control of manipulators and human prostheses. *IEEE Trans. Man-Machine Sys.* MMS-10 (2): 47–53.

Whitney, D. E. 1972. The mathematics of coordinated control of prostheses and manipulators. *J. Dyn. Sys., Meas., Contr.* 94 (4): 303–309.

Yoshikawa, T. 1984. Analysis and control of robot manipulators with redundancy. In *Robotics research*, eds. M. Brady and R. Paul, pp. 735–747. Cambridge, MA: MIT Press.

Yoshikawa, T. 1985. Manipulability of Robotic Mechanisms. *Int. J. Robotics Res.* 4 (2): 3–9.

CHAPTER 4

Local Optimization of Kinematic Redundancy

4.1 Introduction

Robot manipulators have usually been designed to have no more than 6 DOF, which is the minimum DOF needed to perform three-dimensional tasks. Although 6 DOF is sufficient from the conventional mechanical-design viewpoint, it has limited the potential application of robot manipulators. When a robot manipulator follows a given trajectory of the end-effector and needs to avoid hitting obstacles at the elbow, it requires more DOF than it does when working in a free workspace. Moreover, due to the nonlinearity of kinematics, there exist specific arm configurations where the DOF degenerate and effectively become equivalent to less than 6 DOF. These configurations are called *singular points*. Robot manipulators in nature are expected to be more flexible and adaptive than are the conventional mechanisms, like human arms.

Except for Sections 4.3.2. and 4.4.3, this chapter is adapted from, by permission, "Analysis and control of articulated robot arms with redundancy," by H. Hanafusa, T. Yoshikawa, and Y. Nakamura, In *Control Science and Technology for the Progress of Society (Proc. 8th Triennial World Congress of IFAC)*, ed. H. Akashi, Vol. 4: pp. 1927–1932, 1981; "Redundancy analysis of articulated robot arms and its utilization for tasks with priority," by H. Hanafusa, T. Yoshikawa and Y. Nakamura, *Trans. Society of Instrument and Control Engineers*, Vol. 19, No. 5, pp. 421–426, 1983 (*in Japanese*); "Task priority based redundancy

Whitney (1972) and Uchiyama (1979) pointed out that a kinematically redundant robot manipulator having more than 6 DOF can be a remedy to overcome singularity. The kinematic redundancy is also effective for enabling a robot manipulator to approach a workpiece from all directions or to trace a given spatial trajectory avoiding obstacles in a workspace (Freund 1977). Whitney (1969, 1972) discussed the redundancy of prosthetic arms. Although he suggested a criterion to minimize the integral of kinetic energy, he simplified the criterion due to the expected high computational cost, and instead proposed that we minimize the quadratic form of joint velocity instantaneously. This method implies that we perform motion resolution by means of the weighted pseudoinverse of the full-rank Jacobian matrix. Nakano and Ozaki (1974), on the other hand, proposed the *minimum potential energy criterion*, which imposes a constraint on the inverse kinematics of an anthropomorphic manipulator so that the elbow position stays as low as possible. The criteria by Whitney and by Nakano and Ozaki seem to have been intended to determine the solution uniquely, rather than to utilize the kinematic redundancy actively for a specific purpose.

Liegeois (1977) discussed the active utilization of redundancy, where the general solution of joint velocity was represented by means of the generalized inverse of the Jacobian matrix. He proposed that the gradient vector of a scalar function be used as the arbitrary vector. He also showed the numerical simulation of utilizing redundancy for keeping the joint angles within their mechanical limitations. Hanafusa, Yoshikawa, and Nakamura (1978) proposed a numerical algorithm to plan a joint trajectory of a redundant manipulator using the general solution expressed by the pseudoinverse of the Jacobian matrix, where the limitation of joint angles and torques and the geometric constraints imposed by obstacles were considered. Hanafusa, Yoshikawa, and Nakamura (1981) analyzed the kinematic redundancy of robot manipulators in light of the matrix theory, and discussed its utilization. Klein and Huang (1983) reviewed the pseudoinverse control of redundant manipulators, and suggested an algorithm to reduce the computational burden. Moreover, Klein (1984) showed by numerical simulation that a manipulator can avoid an obstacle by paying attention to the point that has the shortest distance from the obstacle. Konstantinov, Markov, and Nenchev (1981) also discussed the utilization of redundancy for avoiding mechanical limits of joint angles by means of the pseudoinverse of the Jacobian matrix. Benati, Marasso, and Tagliasco (1982), on the other hand, proposed a recursive algorithm for inverse kinematics of an anthropomorphic redundant manipulator.

---

control of robot manipulators," by Y. Nakamura and H. Hanafusa, In *Robotics research 2*, eds. H. Hanafusa and H. Inoue, pp. 155–162, Cambridge, MA: MIT Press, 1985; and "Task priority based redundancy control of robot manipulators," by Y. Nakamura and H. Hanafusa, *International Journal of Robotics Research*, Vol. 6, No. 2, pp. 3–15, 1987.

There are two possible approaches to the redundancy utilization: (1) locally optimal control, and (2) globally optimal control. Although the local approach that instantaneously determines the present utilization of redundancy based on only present information is computationally inexpensive, it lacks a guarantee of global optimality. On the other hand, the second approach guarantees global optimality, but it requires a high computation cost. Thus, the locally optimal control approach is suitable for real-time applications, such as sensor-based obstacle avoidance. The globally optimal control approach is better for off-line trajectory planning for tasks requiring the strict optimality, such as obstacle avoidance in a complicated workspace and energy minimization. Therefore, these two frameworks should be used properly, according to the situation.

In this chapter, we discuss the locally optimal control of redundancy. In Section 4.2, the concept of task priority is introduced into inverse kinematics of robot manipulators. A task is divided into subtasks according to the order of priority. Then, the joint motion is resolved such that the subtasks with lower priority are realized using redundancy or extra DOF not committed to satisfying the subtask requirements of higher priority. This procedure is mathematically formulated based on the Jacobian matrix and its pseudoinverse. In Section 4.3, we derive the inverse kinematics of redundant robots considering the order of priority. This formulation is based on the Jacobian matrix and is considered to be an extension of the resolved motion rate control (Whitney 1969). Numerical simulations are carried out in Section 4.4 to show the effectiveness of the formulation. Three examples are simulated. In the first example, the position of the end-effector has a priority higher than that of the end-effector's orientation. In the second example, we make an obstacle avoidance motion using the artificial potential function (Khatib and Le Maitre 1978) in order to describe the second priority task. In the third example, the artificial potential function is also used to avoid kinematic singularity. Experiments of obstacle avoidance are carried out in Section 4.5 using UJIBOT, a robot manipulator with 7 DOF developed to demonstrate the implementation and the actual effectiveness of the redundancy utilization. The robot is taught by an operator information about an obstacle as a desirable arm configuration. Operator intervention of this kind would be a simple but effective method for utilizing the global judgment of human operators.

4.2 Tasks with the Order of Priority

We sometimes find tasks where the position of the end-effector is more important than the orientation, or vice versa. For example, in arc welding, arc cutting, and shape measuring, although both position and orientation must be controlled, the position has more significance than the orientation. On the other hand, the position has less significance than the orientation in spray painting, in directing a camera-in-hand to objects, and so on. We consider

128 Chapter 4 Local Optimization of Kinematic Redundancy

these tasks to be composed of subtasks with different levels of significance, and call them *tasks with the order of priority*.

The problems of redundancy utilization can generally be formulated in the framework of tasks with the order of priority. When a redundant manipulator is required to follow a trajectory of the end-effector while avoiding obstacles or kinematic singular points, trajectory following is given first priority, and obstacle or singularity avoidance is given second priority.

For tasks with the order of priority, if it is impossible to perform all the subtasks completely because of the degeneracy or the shortage of DOF, it would be reasonable to perform the most significant subtasks preferentially and the less important subtasks as much as possible using the remaining DOF. Even for a 6-DOF manipulator, the subtask decomposition between position and orientation is advantageous, because it will enlarge the reachable workspace of the first-priority subtask by allowing incompleteness for the second-priority subtask. In the following section, redundancy utilization will be discussed based on the concept of tasks with the order of priority.

4.3 Inverse Kinematics Considering the Order of Priority

4.3.1 General Formulation

First, we assume that a task is composed of two subtasks with the order of priority. The first-priority subtask is specified using the first manipulation variable, $r_1 \in R^{m_1}$, and the second-priority subtask by the second manipulation variable, $r_2 \in R^{m_2}$. The kinematic relationships between the joint variable $\theta \in R^n$ and the manipulation variables are expressed by

$$r_i = f_i(\theta), \qquad (i = 1, 2) \qquad 4.1$$

Their differential relationships are given as follows:

$$\dot{r}_i = J_i(\theta)\dot{\theta}, \qquad (i = 1, 2) \qquad 4.2$$

where $J_i(\theta) \triangleq \partial f_i / \partial \theta \in R^{m_i \times n}$ is the Jacobian matrix of the ith manipulation variable.

With $m_1 < n$, Eq. 4.2 for $i = 1$ generally has an infinite variety of solutions of $\dot{\theta}$, whose general solution is obtained using the pseudoinverse of the Jacobian matrix as follows (see Theorem 2.5 in Section 2.4.3):

$$\dot{\theta} = J_1^\#(\theta)\dot{r}_1 + \{E_n - J_1^\#(\theta)J_1(\theta)\}y \qquad 4.3$$

where $J_1^\#(\theta) \in R^{n \times m_1}$ is the pseudoinverse of $J_1(\theta)$, $y \in R^n$ is an arbitrary vector, and $E_n \in R^{n \times n}$ indicates an identity matrix. If the exact solution does not exist, Eq. 4.3 covers all the least-squares solutions that minimize $\| \dot{r}_1 - J_1(\theta)\dot{\theta} \|$.

4.3 Inverse Kinematics Considering the Order of Priority

Now, substituting Eq. 4.3 into Eq. 4.2 for $i = 2$, we have the following equation:

$$J_2(E_n - J_1^\# J_1)y = \dot{r}_2 - J_2 J_1^\# \dot{r}_1 \qquad 4.4$$

If the exact solution of y exists for Eq. 4.4, it implies that the second manipulation variable can be realized. The exact solution, however, does not generally exist. We can obtain y that minimizes $\| \dot{r}_2 - J_2(\theta)\dot{\theta} \|$, in the same fashion as Eq. 4.3. That is,

$$y = \hat{J}_2^\#(\dot{r}_2 - J_2 J_1^\# \dot{r}_1) + (E_n - \hat{J}_2^\# \hat{J}_2)z \qquad 4.5$$
$$\hat{J}_2 \triangleq J_2(E_n - J_1^\# J_1)$$

where $z \in R^n$ is an arbitrary vector.

The solution $\dot{\theta}$ is obtained from Eqs. 4.3 and 4.5 as follows:

$$\begin{aligned}\dot{\theta} &= J_1^\# \dot{r}_1 + (E_n - J_1^\# J_1)\hat{J}_2^\#(\dot{r}_2 - J_2 J_1^\# \dot{r}_1) \\ &+ (E_n - J_1^\# J_1)(E_n - \hat{J}_2^\# \hat{J}_2)z\end{aligned} \qquad 4.6$$

Maciejewski and Klein (1985) proved that the second term of Eq. 4.6 is reduced to $\hat{J}_2^\#(\dot{r}_2 - J_2 J_1^\# \dot{r}_1)$. Their proof considered the symmetry and idempotency[4.1] of $E_n - J_1^\# J_1$; see the property 6 of pseudoinverses in Section 2.4.2. Hence, Eq. 4.6 becomes

$$\dot{\theta} = J_1^\# \dot{r}_1 + \hat{J}_2^\#(\dot{r}_2 - J_2 J_1^\# \dot{r}_1) + (E_n - J_1^\# J_1)(E_n - \hat{J}_2^\# \hat{J}_2)z \qquad 4.7$$

Equation 4.7 represents the inverse kinematic solution taking account of the priority of subtasks.

We defined the range space of the Jacobian matrix, $\mathcal{R}(J)$, as the manipulable space, and the null space of the Jacobian matrix, $\mathcal{N}(J)$, as the redundant space in Definitions 3.1 and 3.2, respectively. Figure 4.1 shows the general relationship between the manipulable and redundant spaces for the first and second manipulation variables. In Fig. 4.1, subspace A, subspace B, and subspace C are $S_{R1}^\perp$, $S_{R1} \cap (S_{R1} \cap S_{R2})^\perp$, and $S_{R1} \cap S_{R2}$, respectively, where $S_{Ri} = \mathcal{N}(J_i)$ and $S_*^\perp$ indicates the orthogonal complement of subspace S_*. Subspace A shows all the possible contributions of $\dot{\theta}$ to the first manipulation variable. Subspace B implies the contribution to the second manipulation variable without disturbing the first manipulation variable. Subspace C shows the remaining DOF that affect neither the first nor the second manipulation variables. Subspace C can be used for the third and higher manipulation variables if necessary. The first term in the right-hand side of Eq. 4.7 is the least-squares mapping of $\dot{r}_1$ onto subspace A. The second term implies the least-squares

---

[4.1] Matrix M is called *idempotent* if it satisfies $M^2 = M$.

Chapter 4 Local Optimization of Kinematic Redundancy

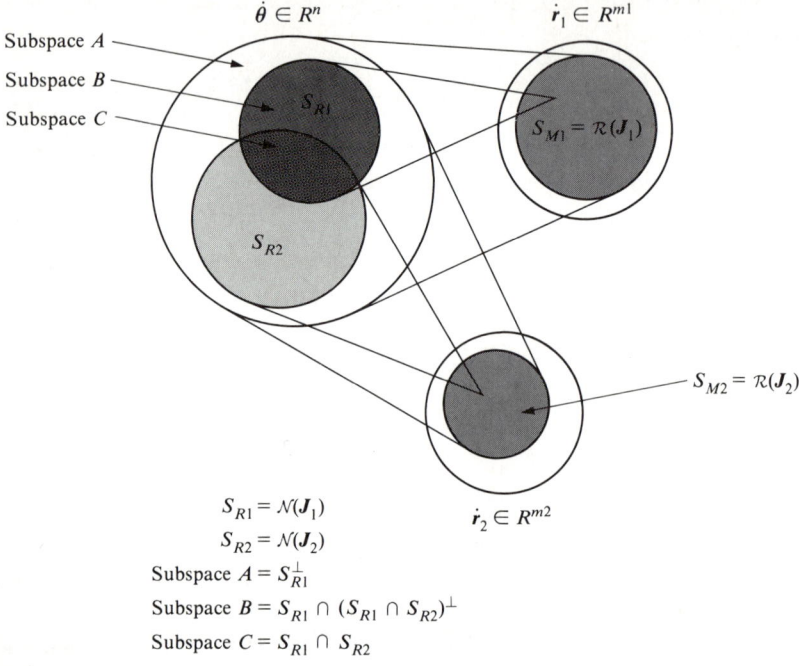

Figure 4.1 Manipulable and redundant spaces for the first and second manipulation variables.

mapping of $\dot{r}_2 - J_2 J_1^\# \dot{r}_1$ onto subspace B, where $\dot{r}_2 - J_2 J_1^\# \dot{r}_1$ is the modified desired value of the second manipulation variable due to the effect of the first term on the second manipulation variable. The third term is the orthogonal projection of the arbitrary vector z onto subspace C. If there is a third manipulation variable, the arbitrary vector z is determined in the same manner as y.

In the case of $r_2 = \theta$, Eq. 4.7 can be reduced to a simpler form using $J_2 = E_n$ as follows:

$$\dot{\theta} = J_1^\# \dot{r}_1 + (E_n - J_1^\# J_1)\dot{r}_2 \qquad 4.8$$

where $(E_n - J_1^\# J_1)^\# = E_n - J_1^\# J_1$ and the idempotency of $E_n - J_1^\# J_1$ are used. In Eq. 4.8, the term corresponding to the third term of Eq. 4.7 becomes intrinsically equal to zero, which means that zero DOF remains for the higher manipulation variables, because the second manipulation variable $r_2 = \theta$ requires all the DOF remaining after being used for r_1.

4.3.2 Using a Potential Function as the Second Manipulation Variable

Liegeois (1977) proposed that we utilize redundancy to avoid mechanical joint limits. He used Eq. 4.3 and determined the arbitrary vector $\boldsymbol{y}$ from the potential function that takes larger values as the manipulator approaches its mechanical joint limits. With the potential function $p(\boldsymbol{\theta})$ to be reduced, the potential approach is described by

$$\dot{\boldsymbol{\theta}} = \boldsymbol{J}_1^\# \dot{\boldsymbol{r}}_1 + (\boldsymbol{E}_n - \boldsymbol{J}_1^\# \boldsymbol{J}_1)(-k\frac{\partial p}{\partial \boldsymbol{\theta}})^T \qquad 4.9$$

where k is a positive constant. When $\dot{\boldsymbol{r}}_1 = \boldsymbol{O}$, Eq. 4.9 guarantees monotonous decrease of the potential function. Indeed,

$$\dot{p} = \frac{\partial p}{\partial \boldsymbol{\theta}} \dot{\boldsymbol{\theta}}$$
$$= -k\frac{\partial p}{\partial \boldsymbol{\theta}}(\boldsymbol{E}_n - \boldsymbol{J}_1^\# \boldsymbol{J}_1)(\frac{\partial p}{\partial \boldsymbol{\theta}})^T$$
$$= -k\frac{\partial p}{\partial \boldsymbol{\theta}}(\boldsymbol{E}_n - \boldsymbol{J}_1^\# \boldsymbol{J}_1)(\boldsymbol{E}_n - \boldsymbol{J}_1^\# \boldsymbol{J}_1)^T(\frac{\partial p}{\partial \boldsymbol{\theta}})^T$$
$$\leq 0 \qquad 4.10$$

where the idempotency and symmetry of $\boldsymbol{E}_n - \boldsymbol{J}_1^\# \boldsymbol{J}_1$ are used. Note that, for nonzero $\dot{\boldsymbol{r}}_1$, Eq. 4.9 does not necessarily mean the nonpositiveness of $\dot{p}(\boldsymbol{\theta})$ or the monotonous decrease of $p(\boldsymbol{\theta})$, although the second term of Eq. 4.9 still tends to reduce $p(\boldsymbol{\theta})$.

When the second manipulation variable is a scalar variable neglecting the unnecessary third term, Eq. 4.7 becomes

$$\dot{\boldsymbol{\theta}} = \boldsymbol{J}_1^\# \dot{\boldsymbol{r}}_1 + (\boldsymbol{E}_n - \boldsymbol{J}_1^\# \boldsymbol{J}_1)(-k_1 \boldsymbol{J}_2^T - k_2 \boldsymbol{J}_2^T \boldsymbol{J}_2 \boldsymbol{J}_1^\# \dot{\boldsymbol{r}}_1) \qquad 4.11$$

$$k_1 = -k_2 \dot{r}_2$$

$$k_2 = \| \boldsymbol{J}_2(\boldsymbol{E}_n - \boldsymbol{J}_1^\# \boldsymbol{J}_1) \|^{-2}$$

Choosing $\dot{r}_2 < 0$, both k_1 and k_2 become positive. Unless $\boldsymbol{J}_2(\boldsymbol{E}_n - \boldsymbol{J}_1^\# \boldsymbol{J}_1) = \boldsymbol{O}$, Eq. 4.11 guarantees the monotonous decrease of r_2 with the specified velocity $\dot{r}_2 < 0$, even if $\dot{\boldsymbol{r}}_1 \neq 0$.

If we choose the potential function $p(\boldsymbol{\theta})$ as the second manipulation variable, then $\boldsymbol{J}_2 = \partial p / \partial \boldsymbol{\theta}$ and, therefore, Eq. 4.9 can be considered to be an approximation of Eq. 4.11 obtained by assigning $k_1 = k$ and neglecting the second term in the right pair of parentheses.

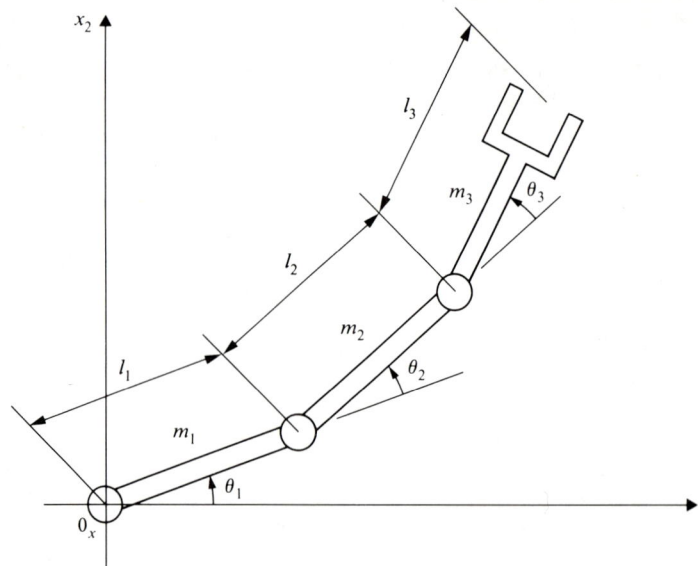

Figure 4.2 Planar 3-DOF manipulator.

4.4 Numerical Simulations

4.4.1 Position Prior to Orientation

The first numerical simulation is the case where the position of the end-effector is given the first priority, and the orientation has the second priority. Figure 4.2 and Table 4.1 show the numerical model of a planar 3-DOF manipulator. It is assumed that the mass of each link is uniformly distributed, and that the manipulator is constrained in a horizontal plane. Hence, the dynamics of the manipulator are represented by

$$T = A(\theta)\ddot{\theta} + B(\theta, \dot{\theta}) \qquad 4.12$$

Table 4.1. Parameters of a 3-DOF manipulator.

i	l_i (cm)	m_i (kg)
1	50.0	30.0
2	43.3	25.0
3	35.0	20.0

where $T, \theta, \dot{\theta}, \ddot{\theta}$ and $B(\theta, \dot{\theta}) \in R^3$, $A(\theta) \in R^{3\times 3}$. We call $A(\theta)$ an inertia matrix, and $B(\theta, \dot{\theta})$ indicates the effect of Coriolis and centrifugal forces.

We assign the first and second manipulation variables as follows:

$$r_1 = \begin{pmatrix} x_1 \\ x_2 \end{pmatrix} \qquad 4.13$$

$$r_2 = \cos(\theta_1 + \theta_2 + \theta_3) \qquad 4.14$$

where x_1 and x_2 represent the position of the end-effector in the Cartesian coordinates and r_2 is the orientation of the end-effector.

By differentiating Eq. 4.2 again, we obtain the following equation:

$$\ddot{r}_i = J_i(\theta)\ddot{\theta} + \dot{J}_i(\theta)\dot{\theta} \qquad 4.15$$

A feedback control scheme is designed so that the following equation represents the closed-loop characteristics:

$$\ddot{e}_i + G_{1i}\dot{e}_i + G_{2i}e_i = O \qquad 4.16$$

$$e_i \triangleq r_i^0(t) - r_i$$

where $r_i^0(t)$ is the desired trajectory of the ith manipulation variable, and e_i is the error vector. The positive scalar feedback coefficients, G_{1i} and G_{2i}, guarantee $e_i \to O$ as $t \to \infty$. From Eqs. 4.15 and 4.16, the joint acceleration should satisfy the following equation:

$$J_i(\theta)\ddot{\theta} = \ddot{r}_i^0(t) - \dot{J}_i(\theta)\dot{\theta} + G_{1i}\dot{e}_i + G_{2i}e_i$$

$$\triangleq h_i(\theta, \dot{\theta}, t) \qquad 4.17$$

Since Eq. 4.17 is similar to Eq. 4.2, the approach discussed in section 4.3 is straightforwardly applied. To simplify the computation, we apply Eq. 4.8 rather than Eq. 4.7 by solving Eq. 4.17 for $i = 2$ and regarding the solution $\ddot{\theta} = J_2^\# h_2$ as the desired acceleration of the second manipulation variable. That is,

$$\ddot{\theta} = J_1^\# h_1 + (E_3 - J_1^\# J_1)J_2^\# h_2 \qquad 4.18$$

We calculate the necessary joint torque to generate the acceleration by substituting Eq. 4.18 into Eq. 4.12.

Figure 4.3 shows the result of simulation. The desired trajectory $r_1^0(t)$ traces the lines connecting points A, B, and C. The motion is described by the third-order polynomials with respect to time. The desired trajectory $r_2^0(t)$ is constantly equal to zero. It is impossible to realize both $r_1^0(t)$ and $r_2^0(t)$ for $8 \leq t \leq 12$ (sec), because they mean outside of the workspace when both are required as the first-priority task. Since the solution by the inverse or pseudoinverse of the Jacobian matrix tends to cause unstable motion near

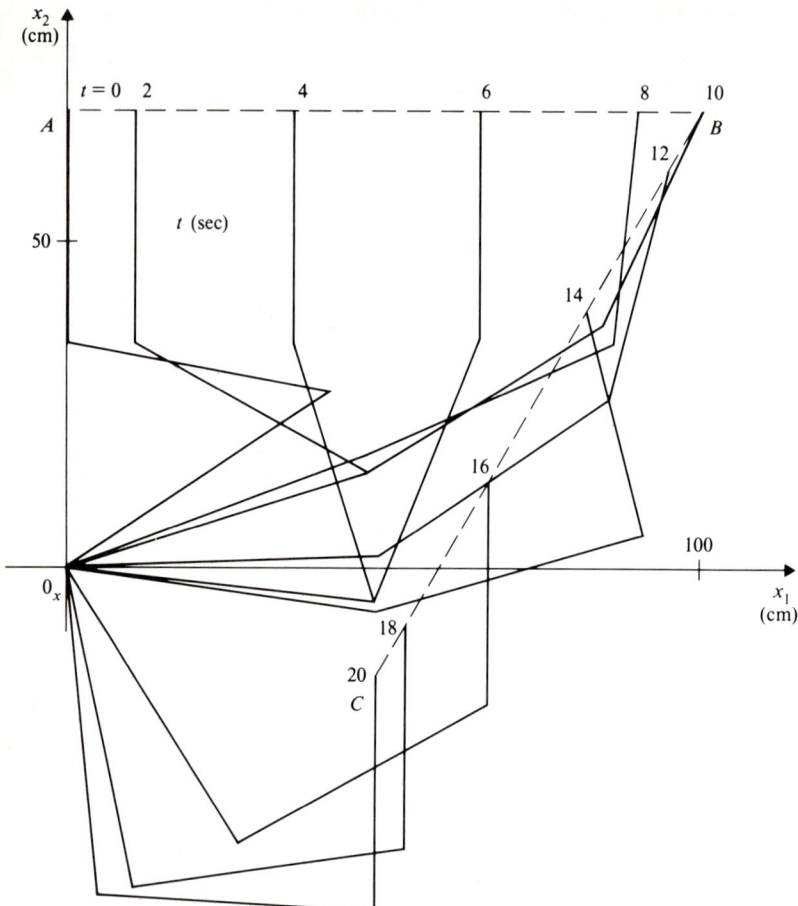

Figure 4.3 Simulation result: The position of the end-effector has a priority higher than that of the end-effector's orientation.

singularity (Nakamura and Hanafusa 1986), the robot manipulator would either stop or oscillate near $t = 8$ (sec) if the inverse or pseudoinverse solution of the both requirements were applied without consideration for the order of priority. Figure 4.3 shows that $\boldsymbol{r}_1^o(t)$ is always satisfied and only $r_2^o(t)$ is disturbed when both are not realizable. This example indicates that the inverse kinematic solution considering the order of priority is not only effective for redundant manipulators, but also useful for enlarging the workspace of nonredundant manipulators.

The feedback coefficients were chosen as $G_{11} = 0$ (1/sec), $G_{21} = 0$ (1/sec^2)[†4.2], $G_{12} = 20$ (1/sec), and $G_{22} = 100$ (1/sec^2). In the simulation

---

[†4.2] Since the first manipulation variable is subject to no disturbance, the open-loop control is sufficient for simulation purpose.

$r_2(t)$ is disturbed near $t = 14$ (sec). This disturbance occurs because we used Eq. 4.8 rather than Eq. 4.7 to reduce the computation cost. The disturbance would be much reduced if Eq. 4.7 were applied, and there would be no overshoot in the convergence process of $r_2(t)$ since the chosen G_{12} and G_{22} imply critical dumping.

4.4.2 Obstacle Avoidance Using the Potential Function

We apply local redundancy control to an obstacle-avoidance problem. Figure 4.4 shows the numerical model of a 4-DOF manipulator and an obstacle in the workspace. The parameters of the manipulator are given in Table 4.2, where $\theta_{i\max}$ is the mechanical limit of the ith joint—namely, $|\theta_i| \leq \theta_{i\max}$. The mass of each link is assumed to be uniformly distributed. The manipulator moves within a horizontal plane. Therefore, the dynamics equation takes the same form as Eq. 4.12, where $T, \boldsymbol{\theta}, \dot{\boldsymbol{\theta}}, \ddot{\boldsymbol{\theta}}$ and $\boldsymbol{B}(\boldsymbol{\theta}, \dot{\boldsymbol{\theta}}) \in R^4$, $\boldsymbol{A}(\boldsymbol{\theta}) \in R^{4\times 4}$.

The first manipulation variable $\boldsymbol{r}_1 \in R^3$ is represented by

$$\boldsymbol{r}_1 = \begin{pmatrix} x_1 \\ x_2 \\ \cos(\theta_1 + \theta_2 + \theta_3 + \theta_4) \end{pmatrix} \qquad 4.19$$

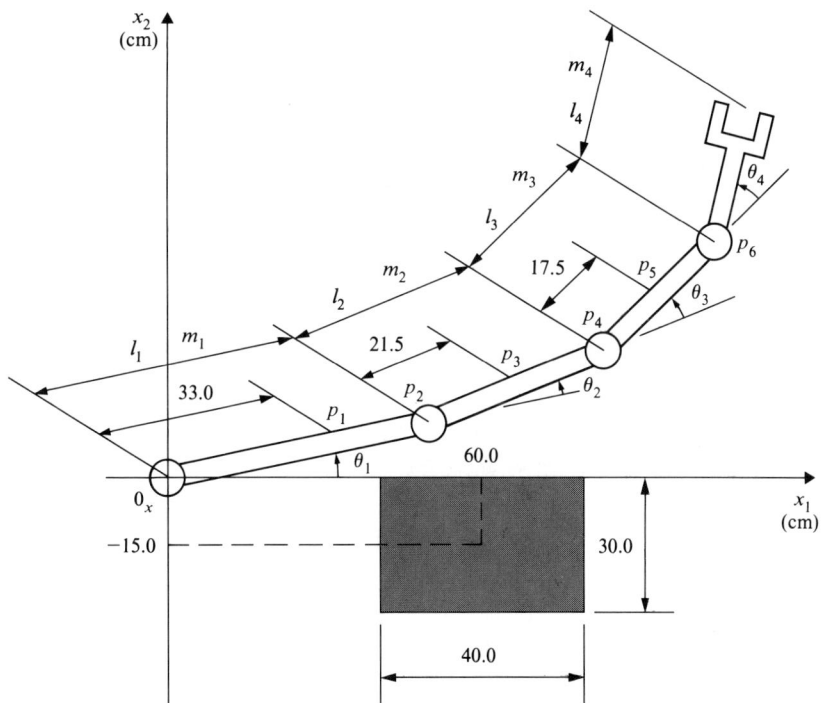

Figure 4.4 Planar 4-DOF manipulator and an obstacle.

Table 4.2. Parameters of 4-DOF manipulator.

i	I_i (cm)	m_i (kg)	$\theta_{i_{max}}$ (degree)
1	50.0	30.0	180
2	43.3	25.0	120
3	35.0	20.0	120
4	22.0	10.0	150

Khatib and Le Maitre (1978) proposed the use of artificial potential and dissipative functions in determining the joint torque for obstacle-avoidance problems when only the final goal of the end-effector is given and the trajectory is free. According to this method, the joint torque is computed as follows:

$$T = -(\frac{\partial P}{\partial \theta} + \frac{\partial D}{\partial \dot{\theta}})^T \qquad 4.20$$

where P and D are the artificial potential and dissipative functions, respectively. If the joint torque of Eq. 4.20 is applied, the following joint acceleration is generated from Eq. 4.12:

$$\ddot{\theta} = -A^{-1}(\theta)\{(\frac{\partial P}{\partial \theta} + \frac{\partial D}{\partial \dot{\theta}})^T + B(\theta, \dot{\theta})\} \qquad 4.21$$

We consider the acceleration in Eq. 4.21 as the acceleration of the second manipulation variable. Hence, from Eqs. 4.18 and 4.21, the joint acceleration considering both the first and second manipulation variables becomes

$$\ddot{\theta} = J_1^{\#} h_1 - (E_4 - J_1^{\#} J_1) A^{-1} \{(\frac{\partial P}{\partial \theta} + \frac{\partial D}{\partial \dot{\theta}})^T + B\} \qquad 4.22$$

We obtain the required joint torque by substituting $\theta, \dot{\theta}$ and Eq. 4.22 into Eq. 4.12.

The artificial potential and dissipative functions are defined by

$$P = P_O + P_J \qquad 4.23$$

$$P_O = k_O \sum_{i=1}^{6} (C_O(p_i) - 1)^{-1} \qquad 4.24$$

$$P_J = k_J \sum_{i=1}^{4} (\theta_{i_{max}}^2 - \theta_i^2)^{-1} \qquad 4.25$$

$$D = k_D \sum_{i=1}^{4} \frac{1}{2} \dot{\theta}_i^2 \qquad 4.26$$

where P_O and P_J are potential functions due to the obstacle and the joint limits, respectively. The $\boldsymbol{p}_i \triangleq (x_{1i} \; x_{2i})^T$ (cm) are the positions of points on the manipulator and are used to evaluate the distance between the manipulator and the obstacle. The locations of these points are indicated in Fig. 4.4. We use $C_O(\boldsymbol{p}_i) = 1$ to approximate the contour of the obstacle, and $C_O(\boldsymbol{p}_i)$ is given as follows (Khatib and Le Maitre 1978):

$$C_O(\boldsymbol{p}_i) = \left(\frac{x_{1i} - 60}{20}\right)^8 + \left(\frac{x_{2i} + 15}{15}\right)^8 \qquad 4.27$$

Figure 4.5 illustrates the results of simulation. Figure 4.5(a) was computed neglecting the second term of Eq. 4.22. In Fig. 4.5(b), we took the obstacle into account by including the second term of Eq. 4.22. The desired trajectory of the first manipulation variable $\boldsymbol{r}_1$ is the constant velocity motion along the line connecting points A and B with the fixed end-effector orientation. Figure 4.5(b) clearly shows that the inverse kinematics based on the order of priority is useful for obstacle avoidance. We chose k_O, k_J, and k_D in Eqs. 4.23 through 4.26 as $k_O = 5.0 \times 10^5$ (kgcm^2/sec^2), $k_J = 6.0 \times 10^4$ (kgcm^2/sec^2), and $k_D = 1.0 \times 10^4$ (kgcm^2/sec^2). The feedback coefficients for the first manipulation variable were zero—namely, $G_{11} = G_{21} = 0$.

4.4.3 Singularity Avoidance Using Potential Function[4.3]

Yoshikawa (1984) proposed using the measure of manipulability (see Section 3.4) as the potential function for avoiding singularity. The potential function is now given as follows:

$$p(\boldsymbol{\theta}) = -\sqrt{\det \boldsymbol{J}_1 \boldsymbol{J}_1^T} \qquad 4.28$$

The measure of manipulability, $\sqrt{\det \boldsymbol{J}_1 \boldsymbol{J}_1^T}$, is nonnegative and becomes zero only at singular points. Therefore, it can be considered as a kind of distance from singular points. Choosing Eq. 4.28 as a potential function is expected to keep a manipulator away from singularity. Note that minimizing the potential function implies not only avoiding the singularity, but also maintaining the kinematic ability of the manipulator as much as possible. We obtain the resultant motion of the manipulator by substituting Eq. 4.28 into Eq. 4.9.

A 3-DOF manipulator as shown in Fig. 4.3 is again used for numerical simulation. The link lengths are assumed to be $l_1 = 0.6$, $l_2 = 0.85$, and $l_3 = 0.2$ (m). The given task for the first manipulation variable is to move in the negative x_2 direction with constant velocity 0.01 (m/sec) for 10 seconds. The initial joint angles are $\boldsymbol{\theta}(t_0) = (180° \; -175° \; 0°)^T$.

Figure 4.6(a) shows the result of simulation when the second term of

---

[4.3]The content of Section 4.4.3 is from Yoshikawa, T. 1984."Analysis and control of robot manipulators with redundancy," in *Robotics research*, eds. M. Brady and R. Paul, pp. 735–747, Cambridge, MA: MIT Press. Reprinted with permission.

138 Chapter 4 Local Optimization of Kinematic Redundancy

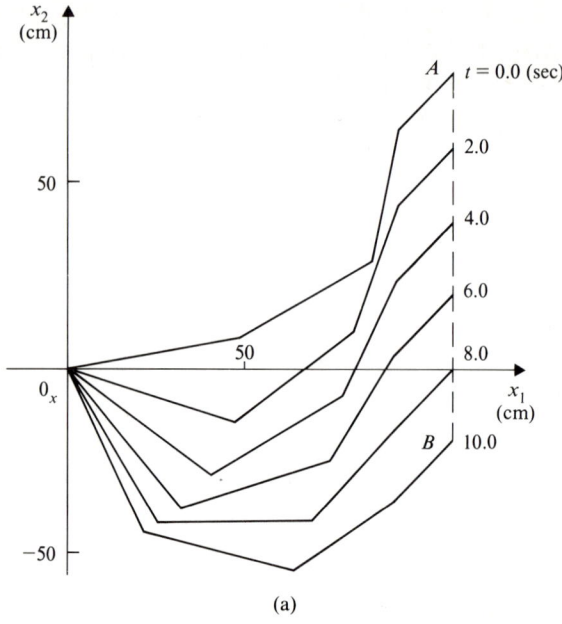

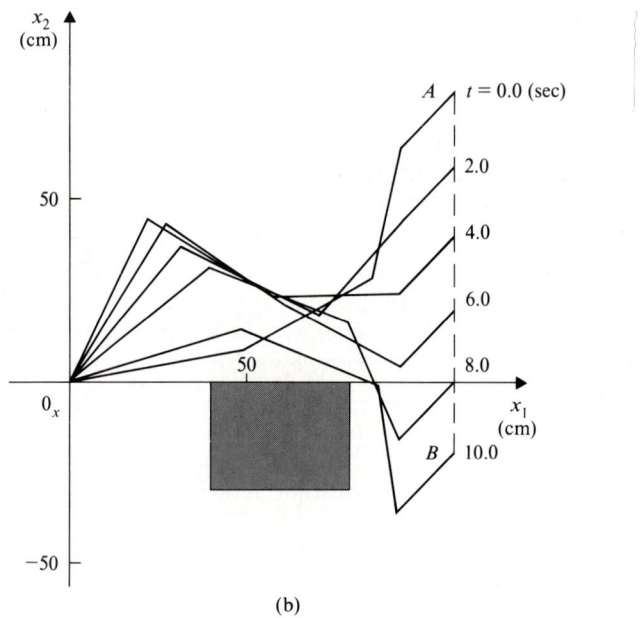

Figure 4.5 Simulation results of obstacle avoidance. (a) Without obstacle. (b) With obstacle.

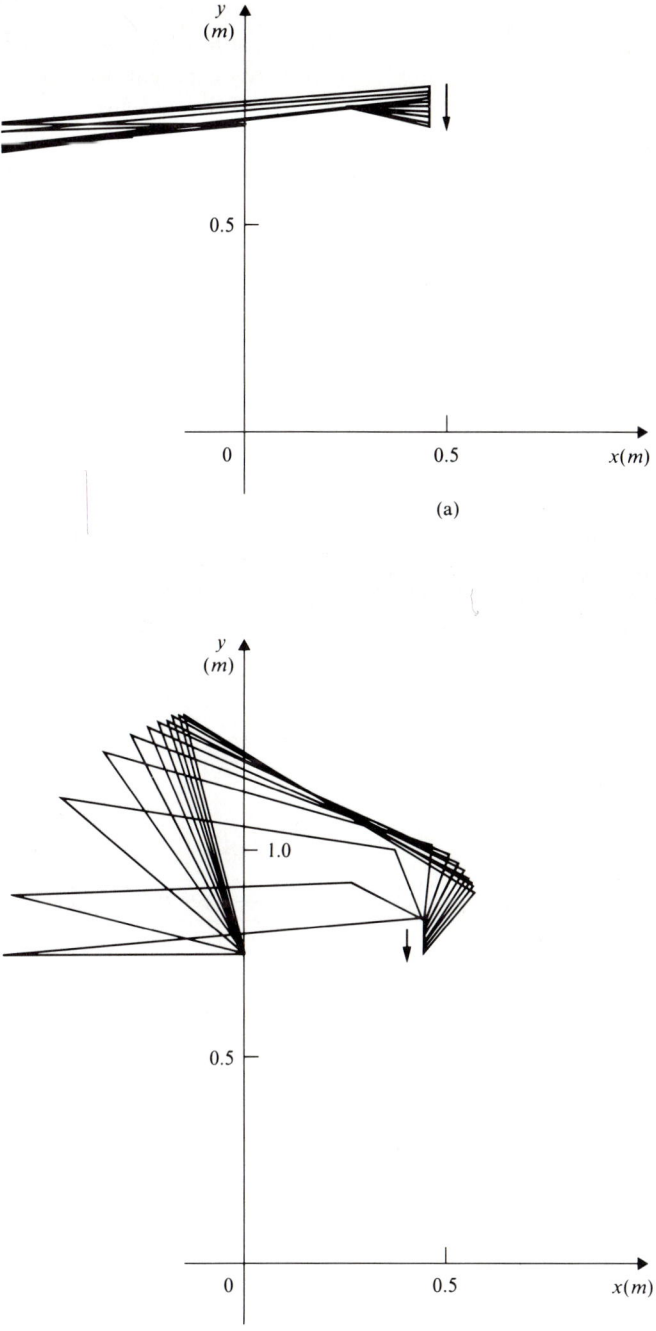

Figure 4.6 Simulation results of singularity avoidance. (a) Second term of Eq. 4.9 neglected. (b) Second term of Eq. 4.9 included. (From Yoshikawa (1984), reprinted with permission.)

Eq. 4.9 was neglected. This is nothing but the resolved motion rate control (Whitney 1969). Figure 4.6(b) shows the result when the second term was included in the same equation. The positive constant was chosen as $k = 5.0$ $(\text{m}^{-2}\text{sec}^{-1})$. Figure 4.7 shows the change of the measure of manipulability. Figure 4.7(a) and (b) correspond to the cases of Fig. 4.6(a) and (b).

From Figs. 4.6 and 4.7, we can see that the conventional resolved motion rate control has no ability for moving away from the singularity, whereas utilizing redundancy by the potential function quickly drove the manipulator away from the singular point and afterward maintained the maximum value of the measure of manipulability.

4.5 Experiments

4.5.1 UJIBOT, a Robot Manipulator with 7 DOF

Figure 4.8 shows the kinematic structure of UJIBOT, a robot manipulator with 7 DOF driven by DC servo motors. The specifications and dimensions are summarized in Table 4.3. The vector from the center of mass of the ith link to the ith joint, l_i, and the inertia tensor, I_i, are described in terms of the ith link coordinates fixed at the center of mass of the ith link. The $x, y,$

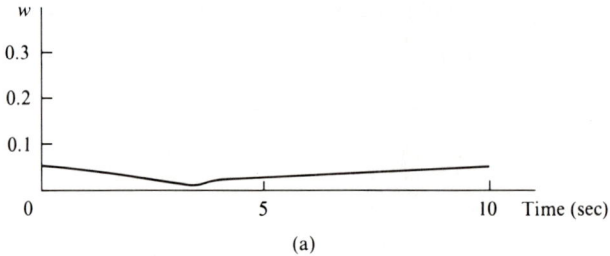

(a)

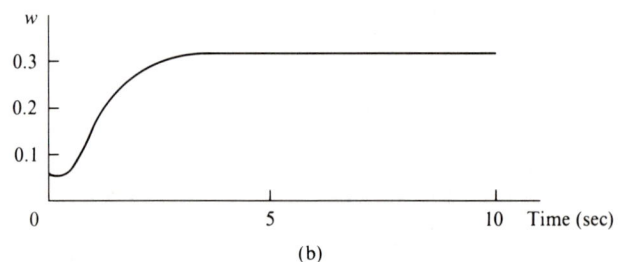

(b)

Figure 4.7 Change of the measure of manipulability. (a) Second term of Eq. 4.9 neglected. (b) Second term of Eq. 4.9 included. (From Yoshikawa (1984), reprinted with permission.)

4.5 Experiments 141

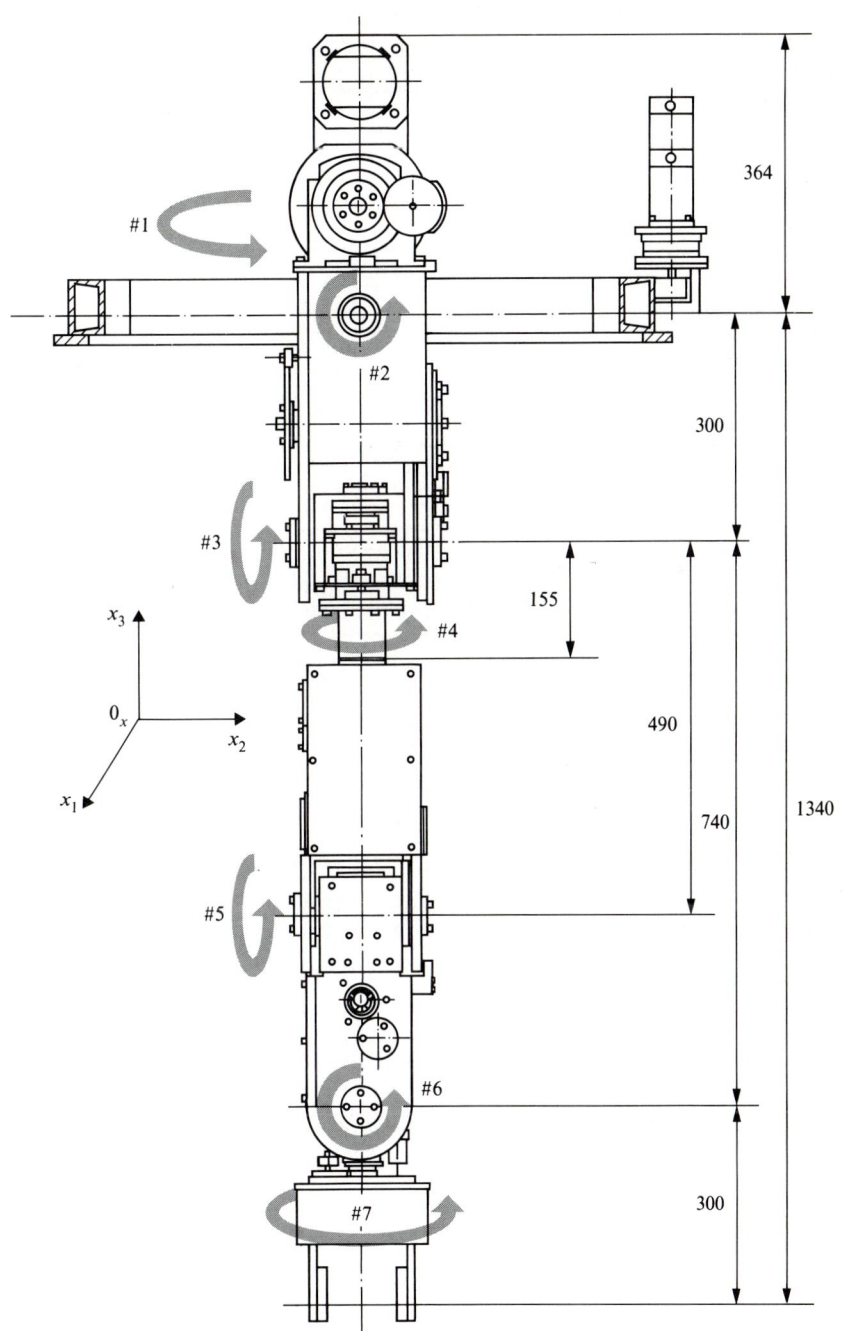

Figure 4.8 Distribution of DOF of UJIBOT, a robot manipulator with 7 DOF.

Table 4.3. Dimensions of UJIBOT.

	i		(Unit)	1	2
M o t o r	Rated	Output (W)		35	95
		Voltage (V)		24	55
		R.p.m.		3500	1100
		Torque (nm)		9.80×10^{-2}	9.02×10^{-1}
	Inductance		(vsec)	4.74×10^{-2}	4.54×10^{-1}
	Torque constant		(nm/A)	4.74×10^{-2}	4.51×10^{-1}
	Armature resistance		(Ω)	1.68	4.7
Reduction ratio				3160	2321.4
Inertia of motor and transmission			(kgm^2)	1.444×10^{-4}	1.800×10^{-1}
Mass of link			(kg)	49.75	43.16
Length of link			(m)	0.0	3.00×10^{-1}
Gravity center l_i		l_{i1}	(m)	-1.856×10^{-2}	3.33×10^{-3}
		l_{i2}	(m)	-3.39×10^{-2}	-2.62×10^{-3}
		l_{i3}	(m)	0.0	-1.427×10^{-2}
Inertia matrix I_i		l_{i11}	(kgm^2)	8.374	1.898×10^{-1}
		l_{i22}	(kgm^2)	7.873	1.877×10^{-1}
		l_{i33}	(kgm^2)	9.022×10^{-1}	1.6167
		l_{i12}	(kgm^2)	-1.036×10^{-1}	-3.17×10^{-4}
		l_{i23}	(kgm^2)	-9.79×10^{-2}	-5.284×10^{-3}
		l_{i31}	(kgm^2)	2.007×10^{-1}	-1.622×10^{-2}

and z axes of link coordinates are parallel with the $x, y,$ and z axes of the base Cartesian coordinates, respectively, when UJIBOT takes the configuration as shown in Fig. 4.8, which is defined as the home position—namely, $\boldsymbol{\theta} = \boldsymbol{O}$.

Since the mechanical reduction ratios are very large (as shown in Table 4.3), the inertia of links and the effects of Coriolis and centrifugal forces are negligible in the dynamics. Hence, the dynamics of UJIBOT are approximated by

$$\boldsymbol{V} = \boldsymbol{P}_U \ddot{\boldsymbol{\theta}} + \boldsymbol{Q}_U \dot{\boldsymbol{\theta}} + \boldsymbol{R}_U \sin \dot{\boldsymbol{\theta}} + \boldsymbol{S}_U \boldsymbol{C}(\boldsymbol{\theta}) \qquad 4.29$$

where $\boldsymbol{\theta} \in R^7$ and $\sin \dot{\boldsymbol{\theta}} \stackrel{\triangle}{=} (\sin \dot{\theta}_i \cdots \sin \dot{\theta}_i)^T$. Here, $\boldsymbol{V} \in R^7$ is the input voltage vector of DC motors, and $\boldsymbol{C}(\boldsymbol{\theta}) \in R^7$ is the gravity torque vector; $\boldsymbol{P}_U = \mathrm{diag}\,(P_{Ui}), \boldsymbol{Q}_U = \mathrm{diag}\,(Q_{Ui}), \boldsymbol{R}_U = \mathrm{diag}\,(R_{Ui})$ and $\boldsymbol{S}_U = \mathrm{diag}\,(S_{Ui}) \in R^{7 \times 7}$ are constant diagonal matrices given in Table 4.4. Figure 4.9 is the overall view of the experimental setup.

Table 4.3. (continued)

3	4	5	6	7
80	64	64	35	10
31.3	30.8	30.8	24	17
4000	4000	4000	3500	3200
1.96×10^{-1}	1.57×10^{-1}	1.57×10^{-1}	9.80×10^{-2}	2.94×10^{-2}
6.49×10^{-2}	6.44×10^{-2}	6.44×10^{-2}	4.74×10^{-2}	3.72×10^{-2}
6.47×10^{-2}	6.43×10^{-2}	6.43×10^{-2}	4.70×10^{-2}	3.72×10^{-2}
1.3	1.7	1.7	1.68	4.0
3703.7	2222.2	2133.3	853.3	110
1.627×10^{-4}	1.098×10^{-4}	1.018×10^{-4}	1.607×10^{-4}	2.040×10^{-5}
10.79	14.16	13.18	2.29	0.60
2.781×10^{-1}	2.119×10^{-1}	2.50×10^{-1}	6.853×10^{-2}	2.315×10^{-1}
7.75×10^{-3}	-1.80×10^{-3}	9.82×10^{-3}	0.0	0.0
-1.977×10^{-2}	1.432×10^{-1}	-1.84×10^{-3}	-1.189×10^{-2}	3.20×10^{-6}
3.90×10^{-3}	0.0	8.97×10^{-2}	1.580×10^{-2}	0.0
3.119×10^{-2}	2.01×10^{-2}	3.861×10^{-2}	1.03×10^{-3}	1.368×10^{-3}
2.456×10^{-2}	4.089×10^{-2}	1.465×10^{-2}	2.32×10^{-3}	1.377×10^{-3}
4.303×10^{-2}	2.466×10^{-1}	1.118×10^{-1}	4.40×10^{-3}	1.136×10^{-3}
2.952×10^{-3}	-5.23×10^{-4}	8.79×10^{-4}	0.0	0.0
5.540×10^{-3}	1.437×10^{-2}	5.476×10^{-3}	3.84×10^{-4}	2.15×10^{-5}
4.981×10^{-3}	-4.591×10^{-3}	-5.696×10^{-3}	0.0	0.0

4.5.2 Task Description and Control Scheme

The task is described as follows: reach for a tennis ball on a box, grasp it, return to the initial position, transport the ball along the x_2 axis at a constant rate of 0.1 (m/sec) for 6.58 (sec), stop, and release the ball into a tennis-ball can below, while avoiding another box that possibly could prevent the motion.

The first and second manipulation variables are chosen as follows:

$$r_1 = \begin{pmatrix} x_1 \\ x_2 \\ x_3 \end{pmatrix} \qquad 4.30$$

$$r_2 = \theta \qquad 4.31$$

where x_1, x_2, and x_3 are the positions of the end-effector in the base Cartesian coordinates. Since the acceleration term is much less than the velocity term

Table 4.4. Identified dynamic characteristics of UJIBOT.

i	$P_{Ui}(\text{Vsec}^2)$	$Q_{Ui}(\text{Vsec})$	$R_{Ui}(\text{V})$	$S_{Ui}(\text{V/Nm})$
1	3.095	7.034×10^1	6.010×10^{-1}	0
2	6.800	1.812×10^2	3.446×10^{-1}	7.306×10^{-4}
3	2.326	7.060×10^1	2.881×10^{-1}	3.386×10^{-3}
4	1.107	3.905×10^1	2.789×10^{-1}	6.264×10^{-3}
5	1.058	5.506×10^1	3.293×10^{-1}	1.007×10^{-2}
6	4.548	1.962×10^1	5.309×10^{-1}	1.188×10^{-1}
7	3.568×10^{-1}	2.843	5.065×10^{-1}	0

in Eq. 4.29 (see Table 4.4), velocity control is adopted as the control scheme. The velocity command for the manipulation variable is computed by

$$\dot{r}_i^* = \dot{r}_i^0(t) + G_i\{r_i^0(t) - r_i(t)\} \qquad 4.32$$

where $\dot{r}_i^*$ is the velocity command for the ith manipulation variable, $r_i^0(t)$ indicates the desired trajectory, and G_i is the scalar feedback coefficient. The values of G_i were determined experimentally. The joint velocity command $\dot{\theta}^*$ is calculated using Eq. 4.8 as follows:

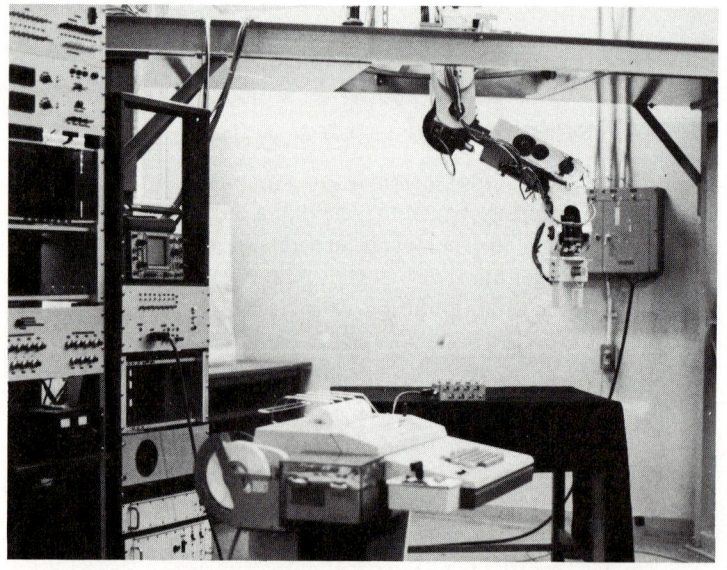

Figure 4.9 Overall view of the experimental setup.

$$\dot{\boldsymbol{\theta}}^* = \boldsymbol{J}_1^\# \dot{\boldsymbol{r}}_1^* + (\boldsymbol{E}_7 - \boldsymbol{J}_1^\# \boldsymbol{J}_1)\dot{\boldsymbol{r}}_2^* \qquad 4.33$$

The motor voltages are determined using Eq. 4.29 as follows:

$$\boldsymbol{V} = \boldsymbol{Q}_U \dot{\boldsymbol{\theta}}^* + \boldsymbol{R}_U \sin \dot{\boldsymbol{\theta}}^* + \boldsymbol{S}_U \boldsymbol{C}(\boldsymbol{\theta}) \qquad 4.34$$

where the acceleration term is neglected.

The desired trajectory is specified by the following equation:

$$\boldsymbol{r}_1^0(t) = \boldsymbol{r}_1(t_0) + (t - t_0) \begin{pmatrix} 0.0 \\ 0.1 \\ 0.0 \end{pmatrix} \quad (\text{m}) \qquad 4.35$$

$$\boldsymbol{r}_2^0(t) = \boldsymbol{\theta}_r \qquad 4.36$$

where $\boldsymbol{\theta}_r \in R^7$ is a constant joint variable vector, called the *reference configuration*, which was chosen intuitively by the operator such that the elbow of UJIBOT would stay away from the obstacle. Operator intervention of this kind would be a simple but effective way to utilize the global judgment of a human operator for obstacle avoidance. In the case of complicated obstacles, it would be necessary to represent the reference configuration as a function of time. Figure 4.10 shows the reference configuration $\boldsymbol{\theta}_r$. Note that inconsistency of the reference configuration with the desired trajectory of the end-effector is allowed. The inconsistency is automatically adjusted in the inverse kinematics, considering the order of priority.

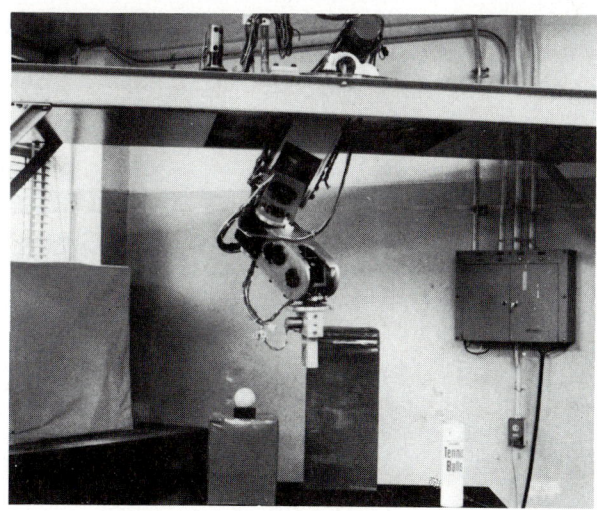

Figure 4.10 Reference configuration.

146   Chapter 4   Local Optimization of Kinematic Redundancy

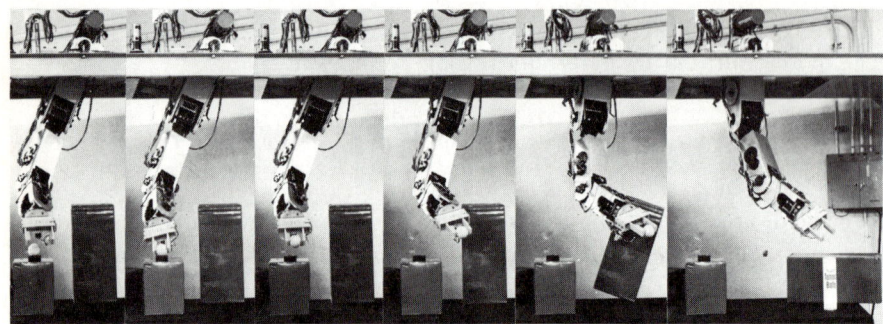

Figure 4.11 Experimental motion of UJIBOT without provisions for the obstacle.

4.5.3   Results and Discussion

Figure 4.11 shows the motion of UJIBOT when the second term of Eq. 4.33 was neglected ($G_1 = 3.0$ 1/sec, $G_2 = 0.0$ 1/sec); this is the resolved motion rate control by Whitney (1969). UJIBOT collided with the obstacle.

Figure 4.12 shows the motion of UJIBOT when the second term of Eq. 4.33 was included ($G_1 = 3.0$ 1/sec, $G_2 = 0.3$ 1/sec). UJIBOT successfully avoided colliding with the obstacle by utilizing redundancy. Figure 4.13 illustrates the motions projected onto the x_3-x_1 plane every 0.47 sec. The dashed lines represent the reference configuration. The broken lines indicate the obstacle. Observe that the elbow of UJIBOT avoided the obstacle as though it had been attracted by the reference configuration.

These experimental results proved that we can implement the inverse kinematics taking account of the order of priority in real control systems of redundant robots, and that the technique is effective in utilizing kinematic redundancy. The overall sampling time, including the real-time computation

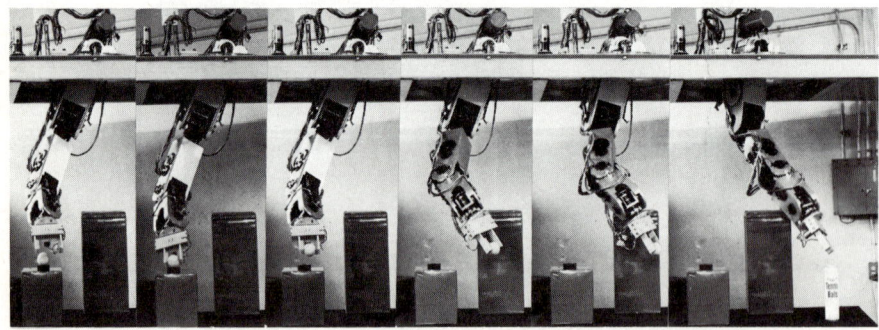

Figure 4.12 Experimental motion of UJIBOT with provisions for the obstacle.

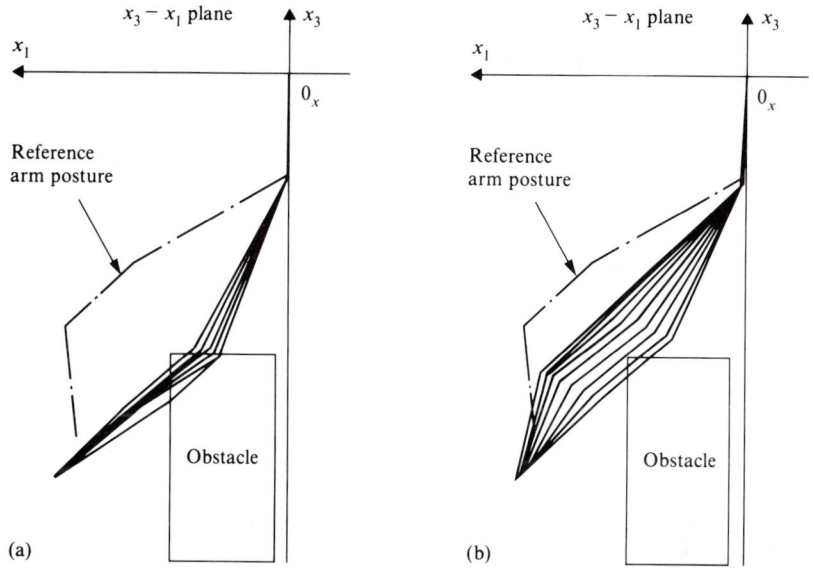

Figure 4.13 Motion of UJIBOT projected onto the x_3-x_1 plane. (a) Without provisions for the obstacle ($G_1 = 3.0, G_2 = 0.0$ 1/sec). (b) With provisions for the obstacle ($G_1 = 3.0, G_2 = 3.0$ 1/sec).

of the Jacobian matrix and its pseudoinverse and the evaluation of Eqs. 4.32 through 4.34, was 47 (msec) using a minicomputer NOVA 03 (Data General Corp.) with a floating-point processing unit.

Experiments were also performed to investigate the relationship between the feedback coefficients and the control performance. Figures 4.14 and 4.15 show the errors of the first and second manipulation variables: $e_1 \triangleq \| r_1^0(t) - r_1(t) \|$ and $e_2 \triangleq \| r_2^0 - r_2(t) \|$. From the figures, the following observations can be made:

1. Increasing G_1, as long as stability is maintained, reduces e_1, but hardly changes e_2.

2. Having G_2 greater than a certain value has no effect in improving e_2.

3. Increasing G_2 disturbs e_1. This effect is due to the coupling between the first and second manipulation variables through the dynamics. A more precise dynamic model and the control scheme based on it would be necessary to reduce such coupling problems.

148   Chapter 4   Local Optimization of Kinematic Redundancy

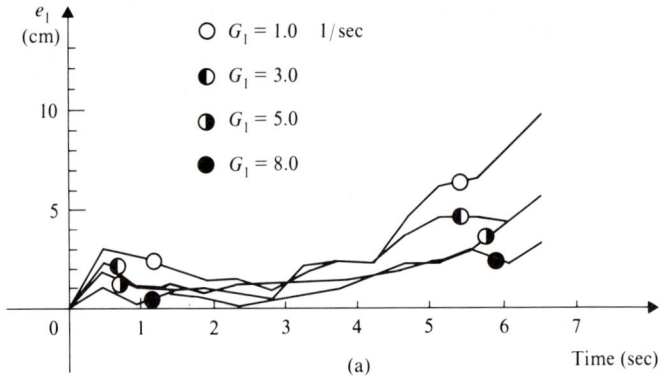

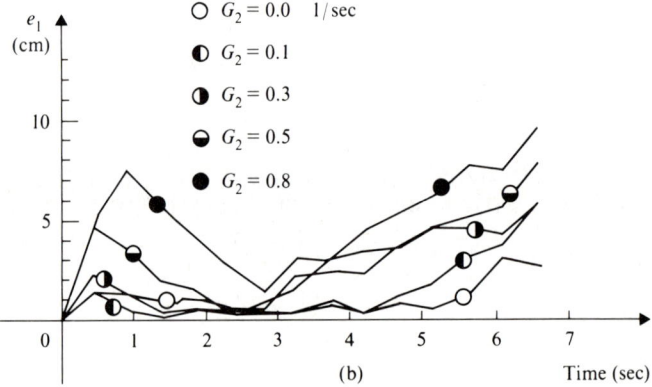

Figure 4.14 Errors of the first manipulation variable, $e_1 = \| \boldsymbol{r}_1^0(t) - \boldsymbol{r}_1(t) \|$. (a) G_1 changes, $G_2 = 0.3$ (1/sec). (b) $G_1 = 3.0$ (1/sec), G_2

Summary

The concept of task priority was introduced into the inverse kinematics of redundant manipulators. The inverse kinematic solution was derived taking account of the order of priority. This is a locally optimal solution that is suitable for real-time redundancy control.

Numerical simulations and experiments were performed to verify the effectiveness and the implementability of the proposed method for redundancy control.

We confirmed that dividing a task into subtasks with the order of priority is useful not only for actively utilizing the kinematic redundancy, but also for relaxing the shortage or the degeneracy of DOF even for nonredundant manipulators.

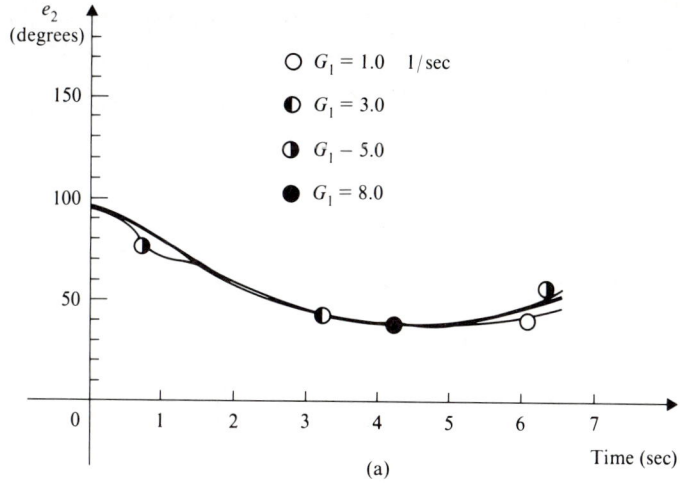

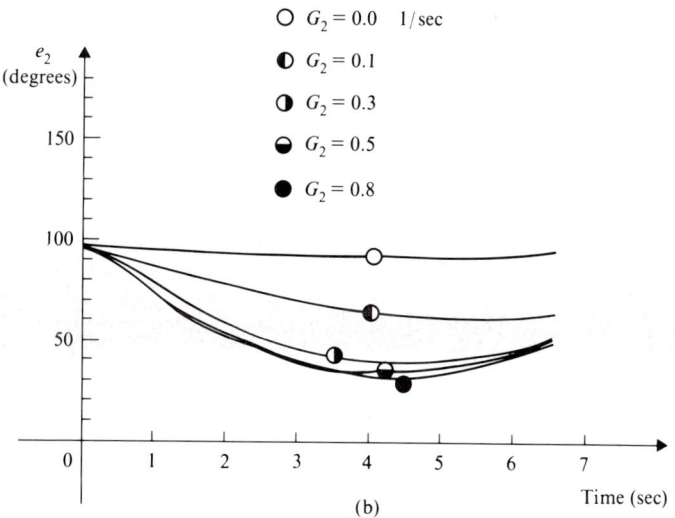

Figure 4.15 Errors of the second manipulation variable, $e_2 = \| \boldsymbol{r}_2^0 - \boldsymbol{r}_2(t) \|$. (a) G_1 changes, $G_2 = 0.3$ (1/sec). (b) $G_1 = 3.0$ (1/sec), G_2 changes.

The obstacle-avoidance problem was solved in two different ways: (1) the potential function of the obstacle was used to determine the motion of the second manipulation variable, and (2) the reference configuration given by an operator based on global judgment were used to specify the motion of the second manipulation variable. Although the first method may work even for fairly complex obstacles, it requires an environmental model and needs additional computation. Since the second method provides a robot manipulator with direct information about how to avoid obstacles, it is computationally less expensive, if intervention by a human operator is acceptable.

References

Benati, M., Morasso, P., and Tagliasco, V. 1982. The inverse kinematic problem for anthromorphic manipulator arms. *J. Dyn. Sys., Meas., Contr.* 104 (1):110–113.

Boullion, T. L., and Odell, P. L. 1971. *Generalized inverse matrices.* New York: Wiley-Interscience.

Freund, E. 1977 (Tokyo). Path control for a redundant type of industrial robot. *Proc. 7th Int. Symp. Industr. Robots*, pp. 107–114.

Hanafusa, H., Yoshikawa, T., and Nakamura, Y. 1978. Control of articulated robot arms with redundancy. *Proc. 21st Joint Automatic Contr. Conf. in Japan*, pp. 237–238 *(in Japanese)*.

Hanafusa, H., Yoshikawa, T., and Nakamura, Y. 1981 (Kyoto). Analysis and control of articulated robot arms with redundancy. In *Control Science and Technology for the Progress of Society (Proc. 8th Triennial World Congress of IFAC)*, ed. H. Akashi, vol. 4: pp. 1927–1932.

Hanafusa, H., Yoshikawa, T., and Nakamura, Y. 1983. Redundancy analysis of articulated robot arms and its utilization for tasks with priority. *Trans. Society of Instrument and Control Engineers* 19 (5): 421–426. *(in Japanese)*.

Khatib, O. 1985 (St. Louis). Real-time obstacle avoidance for manipulators and mobile robots. *Proc. 1985 Int. Conf. Robotics and Automation*, pp. 500–505.

Khatib, O., and Le Maitre, J.-F. 1978 (Udine, Italy). Dynamic control of manipulators operating in a complex environment. *Proc. 3rd Int. CISM-IFToMM Symp.*, pp. 267–282.

Klein, C. A. 1985. Use of redundancy in the design of robotic systems. In *Robotics research 2*, eds. H. Hanafusa and H. Inoue, pp. 207–214. Cambridge, MA: MIT Press.

Klein, C. A., and Huang, C. H. 1983. Review of pseudoinverse control for use with kinematically redundant manipulators. *IEEE Trans. Sys., Man, Cyber.* SMC-13 (3): 245–250.

Konstantinov, M. S., Markov, M. D., and Nenchev, D. N. 1981 (Tokyo). Kinematic control of redundant manipulators. *Proc. 11th Int. Symp. Industr. Robots*, pp. 561–568.

Ligeois, A. 1977. Automatic supervisory control of the configuration and behavior of multibody mechanisms. *IEEE Trans. Sys., Man, Cyber.* SMC-7 (12): 868–871.

Luh, J. Y. S., Walker, M. W., and Paul, R. P. C. 1980. Resolved acceleration control of mechanical manipulators. *IEEE Trans. Automatic Contr.* 25 (3): 468–474.

Maciejewski, A. A., and Klein, C. A. 1985. Obstacle avoidance for kinematically redundant manipulators in dynamically varying environments. *Int. J. Robotics Res.* 4 (3): 109–117.

Nakamura, Y., and Hanafusa, H. 1985. Task priority based redundancy control of robot manipulators. In *Robotics research 2*, eds. H. Hanafusa and H. Inoue, pp. 155–162. Cambridge, MA: MIT Press.

Nakamura, Y., and Hanafusa, H. 1986. Inverse kinematic solutions with singularity robustness for robot manipulator control. *J. Dyn. Sys., Meas., Contr.* 108 (3): 163–171.

Nakamura, Y., Hanafusa, H., and Yoshikawa, T. 1987. Task-priority based redundancy control of robot manipulators. *Int. J. Robotics Res.* 6 (2): 3–15.

Nakano, E., and Ozaki, S. 1974 (Tokyo). Coorperative control of a pair of anthropomorphous manipulators—MELARM. *Proc. 4th Int. Symp. Industr. Robots*, pp. 250–260.

Uchiyama, M. 1979. Study on dynamic control of artificial arms—part 1. *Trans. Japanese Society of Mechanical Engineers.* C-45: 314–322 *(in Japanese)*.

Whitney, D. E. 1969. Resolved motion rate control of manipulators and human prostheses. *IEEE Trans. Man–Machine Sys.* MMS-10 (2): 47–53.

Whitney, D. E. 1972. The mathematics of coordinated control of prostheses and manipulators. *J. Dyn. Sys., Meas., Contr.* 94 (4): 303–309.

Yoshikawa, T. 1984. Analysis and control of robot manipulators with redundancy. In *Robotics research*, eds. M. Brady and R. Paul, pp. 735–747. Cambridge, MA: MIT Press.

CHAPTER 5

Global Optimization of Kinematic Redundancy

5.1 Introduction

The redundancy control problem of a robot manipulator has been discussed in the framework of how to utilize the null space of the Jacobian matrix instantaneously (Ligeois 1977; Hanafusa, Yoshikawa and Nakamura 1981, 1983; Nakamura, Hanafusa, and Yoshikawa, 1987; Klein and Huang 1983; Yoshikawa 1984). Although the gradient vector of a scalar function that evaluates global performance was used in these formulations, the essence of the formulation is that we utilize redundancy as well as possible without paying attention to its global affect. Therefore, the method should be called *locally optimal control of redundancy*. The formulation has the advantage of real-time computation, but unfortunately it lacks the guarantee of global optimality.

This chapter is adapted from, by permission, Nakamura, Y. and Hanafusa, H. 1985. "Task priority based redundancy control of robot manipulators." In *Robotics research 2*, eds. H. Hanafusa and H. Inoue, pp. 155–162, Cambridge, MA: MIT Press; Nakamura, Y. and Hanafusa, H. 1985."Optimal Redundancy Control of Articulated Robot Arms," *Trans. Society of Instrument and Control Engineers*, Vol. 21, No. 5, pp. 501–507 *(in Japanese)*; and Nakamura, Y. and Hanafusa, H. 1987. "Optimal Redundancy Control of Robot Manipulators," *International Journal of Robotics Research*, Vol. 6, No. 1, pp. 32–42.

Practically, we sometimes find tasks that require global optimality. It would be difficult to find a feasible solution of joint trajectory for obstacle-avoidance problems if local optimal control were applied to a robot manipulator in a workspace with complicated obstacles. For space applications where energy is limited and expensive, it would be necessary to plan the minimum-energy motion for performing a given task by utilizing redundancy. Singularity avoidance using kinematical redundancy can best be solved by planning the joint trajectories farthest from singular points. These problems should be discussed in the framework of globally optimal control.

A few papers have been published on globally optimal control of redundancy. Whitney (1969, 1972) suggested that the integrated value of kinetic energy be used as a global criterion for redundancy utilization, but only the computational difficulty was pointed out. Uchiyama, Shimizu and Hakomori (1985) proposed that a performance index of the integral type be used to determine the joint trajectory when the specified constraint on the end-effector is not sufficient for unique determination of the joint trajectory. They also proposed a computational scheme to find the solution, where the trajectory was approximated by a polynomial of finite dimension and its coefficients were found by a minimum-value search using random numbers. Strictly speaking, the scheme cannot guarantee global optimality in the sense of the performance index because of the approximation of trajectory. Furthermore, if the specified trajectory of the end-effector is of α DOF and the order of polynomial is β, we have to search the minimum value in $\alpha(\beta+1)$-dimensional space, which requires a huge amount of computation.

In this chapter, we discuss globally optimal control of redundancy. The problem is formulated based on Pontryagin's maximum principle. We propose a computational algorithm that reduces the computation to a minimum-value search problem in a space of as many dimensions as the degrees of redundancy, if only kinematics is considered. If dynamics is taken into account, the search space becomes twice as many dimensions as the degrees of redundancy. Several numerical examples will be given to show the effectiveness of the proposed globally optimal control scheme.

5.2 Optimal Control Problem of Redundancy

The optimal control problem of redundancy is formulated in the framework of tasks with the order of priority. Suppose that the first manipulation variable $r_1 \in R^{m_1}$ represents the constrained variable such as the position and the orientation of an end-effector, whose desired trajectory is described as a function of time. Also suppose that the second manipulation variable, $r_2 \in R^{m_2}$, $m_2 = 1$, represents a performance index of the integral type, which evaluates the performance of redundancy utilization. That is,

$$r_1(t) = f_1(\theta) \qquad 5.1$$

$$r_2 = \int_{t_0}^{t_1} p(\boldsymbol{\theta}, t) dt \qquad 5.2$$

where $\boldsymbol{\theta} \in R^n$, $n > m_1$, is the joint variable vector of a redundant robot manipulator. The initial time and the final time are represented by t_0 and t_1, respectively.

The problem is to find the joint trajectory $\boldsymbol{\theta}(t)$, $t_0 \leq t \leq t_1$, that minimizes r_2 among joint trajectories realizing the desired trajectory of the first manipulation variable $\boldsymbol{r}_1(t)$. Note that the determination of the initial joint variable $\boldsymbol{\theta}(t_0)$ is also involved in the problem, which has not been discussed in locally optimal control of redundancy, although it is fundamental and important in trajectory-planning problems.

5.3 Application of Pontryagin's Maximum Principle

In this section, the optimal control problem of redundancy is formulated based on Pontryagin's maximum principle (see Section 2.6). By differentiating Eq. 5.1 with respect to time, we derive the following equation:

$$\dot{\boldsymbol{r}}_1(t) = \boldsymbol{J}_1(\boldsymbol{\theta})\dot{\boldsymbol{\theta}} \qquad 5.3$$

where $\boldsymbol{J}_1(\boldsymbol{\theta}) \in R^{m_1 \times n}$ is the Jacobian matrix. Now, suppose that $\dot{\boldsymbol{\theta}}$ satisfying Eq. 5.3 exists. Then, Eq. 5.3 is equivalently transformed to

$$\dot{\boldsymbol{\theta}} = \boldsymbol{J}_1^{\#} \dot{\boldsymbol{r}}_1(t) + (\boldsymbol{E}_n - \boldsymbol{J}_1^{\#} \boldsymbol{J}_1) \boldsymbol{y} \qquad 5.4$$

where $\boldsymbol{J}_1^{\#}(\boldsymbol{\theta}) \in R^{n \times m_1}$ is the pseudoinverse of the Jacobian matrix and $\boldsymbol{y} \in R^n$ is an arbitrary vector (Theorem 2.5 in Section 2.4.3). $\boldsymbol{E}_n \in R^{n \times n}$ is an identity matrix. Equation 5.4 explicitly represents redundancy by the second term of the right-hand side, compared with Eqs. 5.1 and 5.3.

By replacing Eq. 5.1 with Eq. 5.4, we represent the optimal control problem of redundancy as follows:

$$\dot{\boldsymbol{\theta}} = \boldsymbol{J}_1^{\#} \dot{\boldsymbol{r}}_1(t) + (\boldsymbol{E}_n - \boldsymbol{J}_1^{\#} \boldsymbol{J}_1) \boldsymbol{y}$$
$$\triangleq g(\boldsymbol{\theta}, t, \boldsymbol{y}) \qquad 5.5$$

$$r_2 = \int_{t_0}^{t_1} p(\boldsymbol{\theta}, t) dt \qquad 5.6$$

where $\boldsymbol{g} \in R^n$. Although Eq. 5.5 represents the kinematic relationship between $\boldsymbol{\theta}(t)$ and $\boldsymbol{r}_1(t)$, it can be regarded as the system equation of a nonlinear time-varying dynamical system with $\boldsymbol{\theta}$ as the state vector and $\boldsymbol{y}$ as the input vector. Now, considering Eqs. 5.5 and 5.6 as an ordinary optimal control problem of a dynamical system, we apply Pontryagin's maximum principle to the problem.

Concerning the boundary conditions, at the left-hand endpoint ($t = t_0$), $\boldsymbol{\theta}(t_0)$ should satisfy $\boldsymbol{r}_1(t_0) = \boldsymbol{f}_1(\boldsymbol{\theta}(t_0))$. On the other hand, $\boldsymbol{\theta}(t_1)$ is completely free at the right-hand endpoint ($t = t_1$) because $\boldsymbol{\theta}(t_1)$ necessarily satisfies $\boldsymbol{r}_1(t_1) = \boldsymbol{f}_1(\boldsymbol{\theta}(t_1))$, if $\boldsymbol{\theta}(t)$ is governed by Eq. 5.5 with the initial condition of $\boldsymbol{r}_1(t_0) = \boldsymbol{f}_1(\boldsymbol{\theta}(t_0))$. Hence, this is a fixed-time optimal control problem with a variable left-hand endpoint and a free right-hand endpoint.

We can use Theorem 2.17 in Section 2.6.6 to solve this problem. From Eqs. 2.277, 2.278, 2.283, 2.286, 2.287, and 2.356, the Hamiltonian for our fixed-time and free end-state problem becomes as follows:

$$H(\boldsymbol{\psi}, \boldsymbol{\theta}, t, \boldsymbol{y}) = -p + \boldsymbol{\psi}^T \boldsymbol{g} \qquad 5.7$$

where $\boldsymbol{\psi} \in R^n$ is an adjoint vector. Note that we used $\psi_0(t) = -1$ in Eq. 5.7. Also note that we use H to represent the same Hamiltonian represented in Section 2.6 by H_0. From Eq. 2.355 in Theorem 2.17, we choose $\boldsymbol{y}(t)$ so as to maximize the Hamiltonian Eq. 5.7 at every moment t. The optimal joint trajectory $\boldsymbol{\theta}(t)$ is then yielded by the solution of the following differential equations:

$$\dot{\boldsymbol{\theta}} = (\frac{\partial H}{\partial \boldsymbol{\psi}})^T \qquad 5.8$$

$$\dot{\boldsymbol{\psi}} = -(\frac{\partial H}{\partial \boldsymbol{\theta}})^T \qquad 5.9$$

where Eq. 5.8 is equivalent to Eq. 5.5. Note that Eqs. 5.8 and 5.9 were obtained from Eqs. 2.288 and 2.289 by neglecting the trivial first component of each equation.[†5.1]

5.4 Boundary Conditions

Since Eqs. 5.8 and 5.9 have $2n$ entries, the boundary conditions of the same number are necessary. The self-evident boundary condition is $\boldsymbol{r}_1(t_0) = \boldsymbol{f}_1(\boldsymbol{\theta}(t_0))$; $\boldsymbol{\psi}(t_1) = \boldsymbol{O}$ is the boundary condition derived from the condition of a free right-hand endpoint (see Eq. 2.354 in Section 2.6.6). The former has m_1 boundary conditions and the latter has n boundary conditions. The remaining $n - m_1$ boundary conditions are to be obtained from the transversality condition at the left-hand endpoint.

From Theorem 2.14 and Definition 2.8 in Section 2.6.4, the transversality condition is that $\boldsymbol{\psi}(t_0)$ should orthogonally intersect with the manifold $\boldsymbol{r}_1(t_0) = \boldsymbol{f}_1(\boldsymbol{\theta}(t_0))$, which means a set of points where $\boldsymbol{\theta}$ must be located at

---

[†5.1] The first component of Eq. 2.288 is equivalent to Eq. 2.284 and that of Eq. 2.289 implies $\dot{\psi}_0 = 0$.

$t = t_0$. The normal vectors of the manifold consist of the column vectors of $(\partial \boldsymbol{f}_1/\partial \boldsymbol{\theta})^T$. Therefore, with the transversality condition there must exist a vector $\boldsymbol{q} \in R^{m_1}$ that satisfies

$$\boldsymbol{J}_1(\boldsymbol{\theta}(t_0))^T \boldsymbol{q} = \boldsymbol{\psi}(t_0) \qquad 5.10$$

Generally speaking, a necessary and sufficient condition for the existence of the solution of a linear equation $\boldsymbol{Ma} = \boldsymbol{b}$, $\boldsymbol{M} \in R^{n \times m}$, is $(\boldsymbol{E}_n - \boldsymbol{MM}^\#)\boldsymbol{b} = \boldsymbol{O}$ (Rao and Mitra 1971). Considering $(\boldsymbol{MM}^\#)^T = \boldsymbol{MM}^\#$ and $(\boldsymbol{M}^T)^\# = (\boldsymbol{M}^\#)^T$, Eq. 5.10 is transformed into

$$\{\boldsymbol{E}_n - \boldsymbol{J}_1^\#(\boldsymbol{\theta}(t_0))\boldsymbol{J}_1(\boldsymbol{\theta}(t_0))\}\boldsymbol{\psi}(t_0) = \boldsymbol{O} \qquad 5.11$$

Equation 5.11 can also be directly obtained by replacing $\boldsymbol{g}_1$ and $\boldsymbol{x}$ in Eq. 2.323 with $\boldsymbol{f}_1$ and $\boldsymbol{\theta}$. When $\boldsymbol{J}_1(\boldsymbol{\theta}(t_0))$ is full rank—namely, rank $\boldsymbol{J}_1(\boldsymbol{\theta}(t_0)) = m_1$— the rank of the coefficient matrix of Eq. 5.11 is $n - m_1$. Hence, Eq. 5.11 has $n - m_1$ independent boundary conditions. Equation 5.11 implies that the initial value of the adjoint vector $\boldsymbol{\psi}(t_0)$ must be an element of the orthogonal complement of the redundant space of the first manipulation variable.

Finally, the boundary conditions are summarized as follows:

L.E. : $\quad \boldsymbol{r}_1(t_0) = \boldsymbol{f}_1(\boldsymbol{\theta}(t_0))$,

$\qquad \{\boldsymbol{E}_n - \boldsymbol{J}_1^\#(\boldsymbol{\theta}(t_0))\boldsymbol{J}_1(\boldsymbol{\theta}(t_0))\}\boldsymbol{\psi}(t_0) = \boldsymbol{O} \qquad 5.12$

R.E. : $\quad \boldsymbol{\psi}(t_1) = \boldsymbol{O}$

where L.E. and R.E. represent the left-hand ($t = t_0$) and right-hand ($t = t_1$) endpoints, respectively. Equation 5.12 shows that the problem has been reduced to a two-point boundary value problem (see Section 2.6.7) with n conditions at each endpoint.

To solve multipoint boundary value problems, researchers have proposed the *initial value adjusting method* (see, for example, Ojika and Kasue 1979), which we shall explain briefly for the case of two-point boundary value problems, as follows. First, we estimate the unknown left-hand endpoint boundary values. Second, we solve the differential equations as an ordinary initial value problem by combining estimates of unknown left-hand endpoint boundary values with the known left-hand endpoint boundary values. Next, depending on the difference between the known right-hand endpoint boundary values and their computed values, we modify the estimated left-hand endpoint boundary values. Then, we solve the differential equations as the initial value problem again. These processes are repeated until the boundary values converge to the given values at both endpoints. Figure 5.1 illustrates the procedure conceptually.

If we solve the two-point boundary value problem of Eq. 5.12 based on the initial value adjusting method, we have to repeat the estimation and modification of n conditions, which requires a lot of computation.

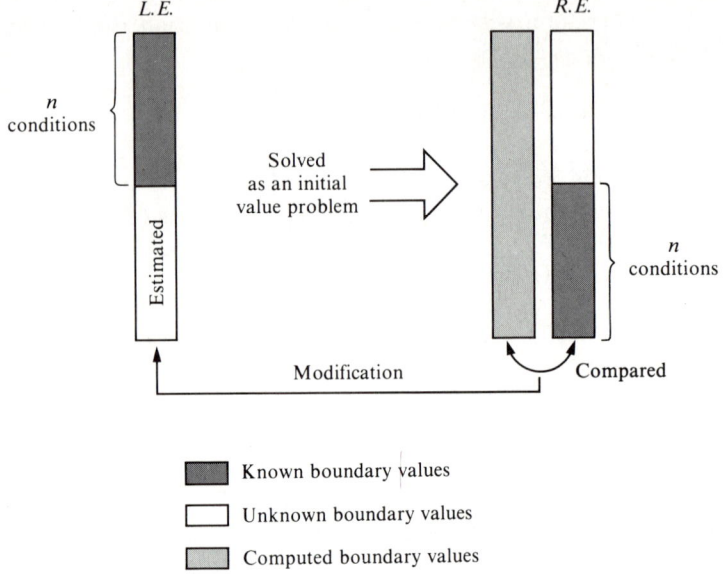

Figure 5.1 Initial value adjusting method for two-point boundary value problems with n conditions at each of the endpoints.

Now, we reduce the computational complexity by considering the particularity of our optimal control problem. As we stated in Section 5.3, if $\boldsymbol{\theta}(t_0)$ satisfies $\boldsymbol{r}_1(t_0) = \boldsymbol{f}_1(\boldsymbol{\theta}(t_0))$, then $\boldsymbol{\theta}(t_1)$ necessarily satisfies $\boldsymbol{r}_1(t_1) = \boldsymbol{f}_1(\boldsymbol{\theta}(t_1))$, as long as $\boldsymbol{\theta}(t)$ is governed by Eq. 5.5. The reverse is also true. Therefore, the boundary conditions of Eq. 5.12 are replaced by

$$\text{L.E.}: \quad \{\boldsymbol{E}_n - \boldsymbol{J}_1^{\#}(\boldsymbol{\theta}(t_0))\boldsymbol{J}_1(\boldsymbol{\theta}(t_0))\}\boldsymbol{\psi}(t_0) = \boldsymbol{O}$$
$$\text{R.E.}: \quad \boldsymbol{r}_1(t_1) = \boldsymbol{f}_1(\boldsymbol{\theta}(t_1)) \qquad\qquad 5.13$$
$$\boldsymbol{\psi}(t_1) = \boldsymbol{O}$$

Equation 5.13 means that the problem can be regarded as a two-point boundary value problem with $n - m_1$ conditions at the left-hand endpoint and $n + m_1$ conditions at the right-hand endpoint. Therefore, if we solve Eqs. 5.8 and 5.9 conversely from t_1 to t_0, the number of variables whose boundary values should be estimated and modified in the initial value adjusting method is reduced to $n - m_1$. That is, the optimal redundancy control problem of Eqs. 5.1 and 5.2 has been reduced to a minimum-value search problem of as many dimensions as the degrees of redundancy. Especially in the case of one degree of redundancy, $n - m_1$ is equal to one, and then our optimal problem results in a

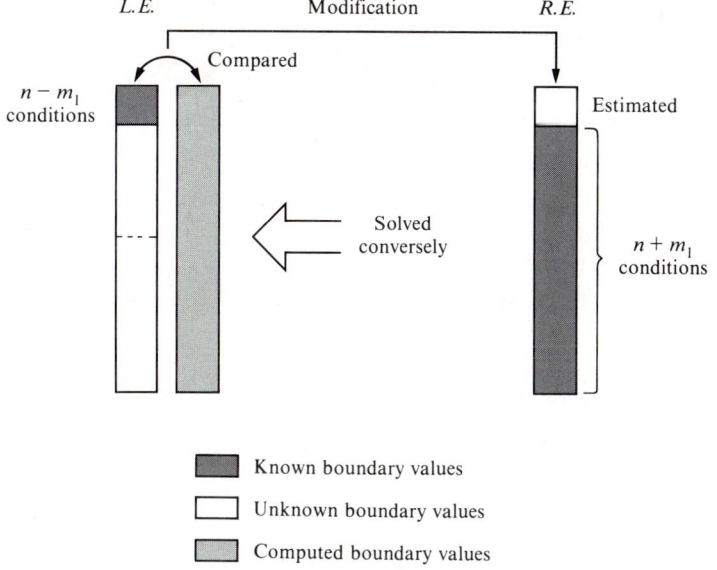

Figure 5.2 Modified initial value adjusting method, where the differential equations are solved conversely and the estimation and modification are carried out for $n - m_1$ conditions.

one-dimensional minimum-value search problem. Figure 5.2 illustrates the computational procedure conceptually based on Eq. 5.13.

5.5 An Example of Performance Index

In this section, a representative example of performance indices will be given and the concrete forms of Eqs. 5.8 and 5.9 will be derived. The representative example follows:

$$r_2 = \int_{t_0}^{t_1} \{kp_0(\boldsymbol{\theta}) + \dot{\boldsymbol{\theta}}^T \dot{\boldsymbol{\theta}}\} dt \qquad 5.14$$

where k is a nonnegative scalar. The second term of the integrand of Eq. 5.14 guarantees the existence of the solution. In addition, the second term makes it easy to determine the $\boldsymbol{y}(t)$ that maximizes the Hamiltonian. For example, by choosing $p_0(\boldsymbol{\theta})$ as a potential function that gives large values in the neighborhood of obstacles, we can deal with the obstacle avoidance problem. If we choose a function of the measure of manipulability (Yoshikawa 1984) as $p_0(\boldsymbol{\theta})$, we can plan a singularity-free trajectory. This technique will be discussed in Section 5.7.1.

The Hamiltonian Eq. 5.7 for the performance index Eq. 5.14 becomes

$$H = -kp_0 - g^T g + \psi^T g$$
$$= -(g - \frac{1}{2}\psi)^T(g - \frac{1}{2}\psi) + \frac{1}{4}\psi^T \psi - kp_0 \qquad 5.15$$

In Eq. 5.15, the second and third terms of the right-hand side have nothing to do with $y(t)$ explicitly, because $y(t)$ is included only in $g(\theta, t, y)$. Therefore, the $y(t)$ that minimizes $\| g - \frac{1}{2}\psi \|$ will maximize the Hamiltonian of Eq. 5.15. The g is given by Eq. 5.5, and such a $y(t)$ is obtained from Theorem 2.5 in Section 2.4.3 as follows:

$$y = (E_n - J_1^\# J_1)^\# \{-J_1^\# \dot{r}_1(t) + \frac{1}{2}\psi\} \qquad 5.16$$

Considering the property of pseudoinverses $(E_n - J_1^\# J_1)^\# = E_n - J_1^\# J_1$ and $(E_n - J_1^\# J_1)J_1^\# = O$, Eq. 5.16 results in the following simpler equation:

$$y = \frac{1}{2}(E_n - J_1^\# J_1)\psi \qquad 5.17$$

Note that the integrand of Eq. 5.14 makes the Hamiltonian quadratic with respect to g and simplifies the computation of optimal value of y. Equation 5.17 means that the half of the orthogonal projection of the adjoint vector ψ onto the redundant space (Definition 3.2 in Section 3.3.2) of the first manipulation variable is the optimal input for Eqs. 5.5 and 5.14. From Eqs. 5.5, 5.9, 5.15, and 5.17, we find that the optimal trajectory $\theta(t)$ for the performance index Eq. 5.14 is governed by the following differential equations:

$$\dot{\theta} = g \qquad 5.18$$

$$\dot{\psi} = (\frac{\partial g}{\partial \theta})^T (2g - \psi) + k(\frac{\partial p_0}{\partial \theta})^T \qquad 5.19$$

$$g = J_1^\# \dot{r}_1(t) + \frac{1}{2}(E_n - J_1^\# J_1)\psi \qquad 5.20$$

In Section 5.7, numerical examples of the performance index of Eq. 5.14 will be shown and solved using Eqs. 5.18, 5.19, and 5.20.

5.6 Consideration of Dynamics

We have discussed optimal redundancy control considering only kinematics. However, when a trajectory is required to minimize the energy used in performing tasks, the dynamics of robot manipulators must be taken into consideration. In this section, we shall extend the results obtained so far to the optimal redundancy control problem considering dynamics.

5.6 Consideration of Dynamics

The dynamics of robot manipulators is generally represented by the following equation:

$$T = A(\theta)\ddot{\theta} + B(\theta,\dot{\theta}) + C(\theta) \qquad 5.21$$

where $A(\theta) \in R^{n \times n}$ is an inertia matrix, $B(\theta,\dot{\theta}) \in R^n$ is a torque vector caused by centrifugal and Coriolis forces, $C(\theta) \in R^n$ is a gravity torque vector, and $T \in R^n$ is a joint torque vector.

When the dynamics is described by a second-order differential equation, such as Eq. 5.21, dynamical redundancy is explicitly represented as follows by differentiation of Eq. 5.3 once more with respect to time:

$$\ddot{\theta} = J_1^\#(\ddot{r}_1(t) - \dot{J}_1\dot{\theta}) + (E_n - J_1^\# J_1)y \qquad 5.22$$

Now, let us introduce a vector $x \triangleq (\theta^T \dot{\theta}^T)^T, \in R^{2n}$, by which Eqs. 5.22 and 5.21 are represented as follows:

$$\dot{x} = Q(x,t) + R(x)y \qquad 5.23$$

$$Q(x,t) \triangleq \begin{pmatrix} \dot{\theta} \\ J_1^\#(\ddot{r}_1(t) - \dot{J}_1\dot{\theta}) \end{pmatrix} \in R^{2n}$$

$$R(x) \triangleq \begin{pmatrix} O \\ E_n - J_1^\# J_1 \end{pmatrix} \in R^{2n \times n}$$

$$T(x,t,y) = U(x,t) + V(x)y \qquad 5.24$$

$$U(x,t) \triangleq A J_1^\#(\ddot{r}_1(t) - \dot{J}_1\dot{\theta}) + B + C \in R^n$$

$$V(x) \triangleq A(E_n - J_1^\# J_1) \in R^{n \times n}$$

Let the following performance index be a representative one:

$$r_2 = \int_{t_0}^{t_1} \{kp_0(x) + T^T T\} dt \qquad 5.25$$

where k is a nonnegative scalar. We can obtain the optimal trajectory that minimizes the energy cost by setting $k = 0$.

Considering that x is a state vector and y is an input vector in Eqs. 5.23, 5.24, and 5.25, we can apply Pontryagin's maximum principle in the same way as in Section 5.3. The Hamiltonian for the performance index Eq. 5.25 then becomes

$$H = -kp_0 - T^T T + \psi^T(Q + Ry)$$
$$= -kp_0 - (U + Vy)^T(U + Vy) + \psi^T(Q + Ry) \qquad 5.26$$

where the adjoint vector is extended to $\boldsymbol{\psi} \in R^{2n}$. Since Eq. 5.26 is quadratic with respect to $\boldsymbol{y}$, the $\boldsymbol{y}$ that maximizes the Hamiltonian is readily computed as follows:

$$\boldsymbol{y} = \boldsymbol{V}^{\#}(-\boldsymbol{U} + \frac{1}{2}\boldsymbol{S}\boldsymbol{\psi}) \qquad 5.27$$

$$\boldsymbol{S} \triangleq (\boldsymbol{O} \quad \boldsymbol{A}^{-1}) \in R^{n \times 2n}$$

Equation 5.27 is derived, considering that the inertia matrix $\boldsymbol{A}(\boldsymbol{\theta})$ is symmetric and positive definite. Again, note that choosing the performance index as Eq. 5.25 simplifies the computation of the optimal value of $\boldsymbol{y}$.

Consequently, the optimal trajectory for the performance index Eq. 5.25 is governed by the following differential equations:

$$\dot{\boldsymbol{x}} = \boldsymbol{Q}(\boldsymbol{x},t) + \boldsymbol{R}(\boldsymbol{x})\boldsymbol{y} \qquad 5.28$$

$$\dot{\boldsymbol{\psi}} = -(\frac{\partial H}{\partial \boldsymbol{x}})^{T} \qquad 5.29$$

where $\boldsymbol{y}$ is the optimal input vector given by Eq. 5.27.

In addition, based on ideas similar to the discussion in Section 5.4, we can readily show that the optimal redundancy control problem taking account of dynamics is reduced to a two-point boundary value problem with $2(n - m_1)$ conditions at the left-hand endpoint and $2(n + m_1)$ conditions at the right-hand endpoint. Accordingly, if the degree of redundancy $n - m_1$ is equal to 1, the optimal problem results in a two-dimensional minimum-value search problem.

5.7 Numerical Examples

Numerical examples of the optimal redundancy control problem are examined in this section. The performance index of Eq. 5.14 is assumed for an anthropomorphic robot manipulator with 7 DOF and a total length of 1.5 m, shown in Fig. 5.3.

As the first manipulation variable, we choose the position of the end-effector described in three-dimensional Cartesian coordinates, and the orientation described by Euler angles. Accordingly, $\boldsymbol{r}_1 \in R^6$, and then the degree of redundancy at nonsingular points is equal to one. The defined trajectory of $\boldsymbol{r}_1(t)$ is a straight-line trajectory 0.5 m in length, to be traced at a constant speed of 0.5 m/sec with the orientation fixed. Therefore, $t_0 = 0$ sec and $t_1 = 1$ sec. Figure 5.4 shows the motion $\boldsymbol{\theta}_0(t)$ of the robot manipulator computed by resolved motion rate control (Whitney 1969) from an initial angle $\boldsymbol{\theta}_0(t_0)$, satisfying $\boldsymbol{r}_1(t_0) = \boldsymbol{f}_1(\boldsymbol{\theta}_0(t_0))$. In Fig. 5.4, $\boldsymbol{r}_1(t)$ is displayed by the motion of the end-effector.

5.7 Numerical Examples   163

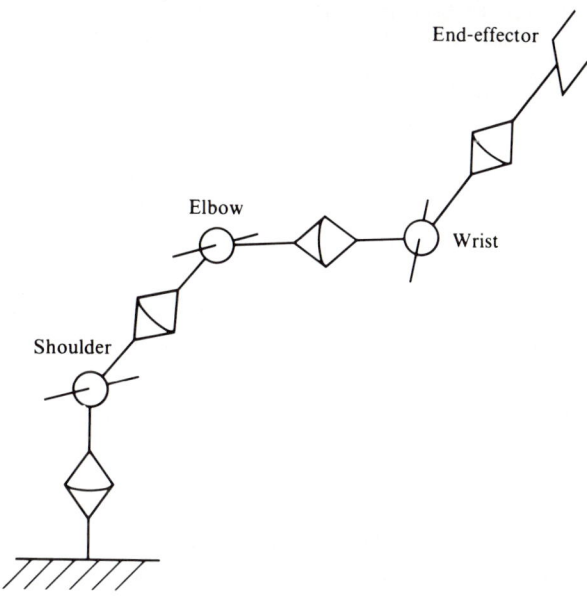

Figure 5.3 An anthropomorphic robot manipulator with 7 DOF.

Our optimal problem is to find the final angle $\boldsymbol{\theta}(t_1)$ that satisfies $\boldsymbol{r}_1(t_1) = \boldsymbol{f}_1(\boldsymbol{\theta}(t_1))$ and fulfills the transversality condition at $t = t_0$ after the backward integration from $t = t_1$ to $t = t_0$ of Eqs. 5.18, 5.19, and 5.20. When the position and the orientation of the end-effector are chosen as the first manipulation variable, the redundancy of the anthropomorphic manipulator as shown in Fig. 5.3 can be parametrized by a rotational angle of the elbow around the axis connecting the shoulder and the wrist. In other words, we can uniquely distinguish $\boldsymbol{\theta}(t_1)$ satisfying $\boldsymbol{r}_1(t_1) = \boldsymbol{f}_1(\boldsymbol{\theta}(t_1))$ by means of the rotational angle of the elbow.

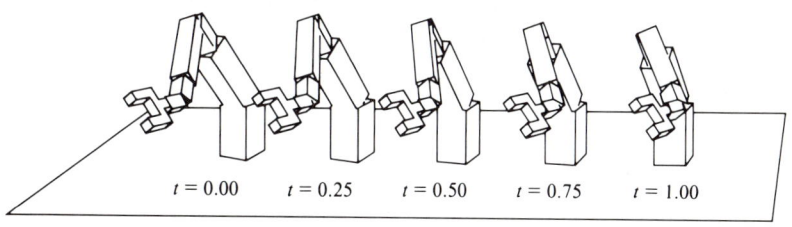

Figure 5.4 Desired motion of the end-effector (computed by resolved motion rate control).

Now, we define the following angle:

$$\alpha \triangleq \int_0^T \|\dot{\boldsymbol{\theta}}_e\| \, d\tau \qquad 5.30$$

$$\dot{\boldsymbol{\theta}}_e \triangleq (\boldsymbol{E}_7 - \boldsymbol{J}_1^\#(\boldsymbol{\theta}_e)\boldsymbol{J}_1(\boldsymbol{\theta}_e))\boldsymbol{e} \qquad 5.31$$

$$\boldsymbol{e} \triangleq (0\ 0\ 1\ 0\ 0\ 0\ 0)^T \qquad 5.32$$

where τ is imaginary time, $\boldsymbol{\theta}_e(0)$ is equal to the final joint angle $\boldsymbol{\theta}_0(t_1)$ of Fig. 5.4, and $\boldsymbol{e}$ is a joint velocity vector. Equation 5.32 means that the third joint rotates at a unity velocity with the other joints fixed.

Since $\dot{\boldsymbol{\theta}}_e$ is an orthogonal projection of $\boldsymbol{e}$ onto the redundant space of the first manipulation variable, $\dot{\boldsymbol{\theta}}_e$ can be regarded as an approximate solution when the robot manipulator is required to move at a velocity of $\boldsymbol{e}$ under the constraint that the position and the orientation of the end-effector are fixed at $\boldsymbol{r}_1(t_1)$. It can be readily understood from Fig. 5.3 that, when the rotational angle of the elbow monotonously increases or decreases with the end-effector fixed, so does that of the third joint. Our next claim is that the third element of $\dot{\boldsymbol{\theta}}_e$ is nonnegative. Indeed,

$$\begin{aligned} \boldsymbol{e}^T \dot{\boldsymbol{\theta}}_e &= \boldsymbol{e}^T (\boldsymbol{E}_7 - \boldsymbol{J}_1^\# \boldsymbol{J}_1) \boldsymbol{e} \\ &= \boldsymbol{e}^T (\boldsymbol{E}_7 - \boldsymbol{J}_1^\# \boldsymbol{J}_1)^T (\boldsymbol{E}_7 - \boldsymbol{J}_1^\# \boldsymbol{J}_1) \boldsymbol{e} \\ &\geq 0 \end{aligned} \qquad 5.33$$

where the properties of pseudoinverses, $\boldsymbol{J}_1^\# \boldsymbol{J}_1 \boldsymbol{J}_1^\# = \boldsymbol{J}_1^\#$ and $(\boldsymbol{J}_1^\# \boldsymbol{J}_1)^T = \boldsymbol{J}_1^\# \boldsymbol{J}_1$, are used. Therefore, $\dot{\boldsymbol{\theta}}_e$ always drives the elbow in one direction with the end-effector fixed. In the other words, α has a one-to-one correspondence with the rotational angle of the elbow. Accordingly, we can use α in place of the actual rotational angle of the elbow to specify $\boldsymbol{\theta}(t_1)$, satisfying $\boldsymbol{r}_1(t_1) = \boldsymbol{f}_1(\boldsymbol{\theta}(t_1))$. It was verified by the computation that the elbow makes a round from $\alpha = 0°$ to $\alpha = 633°$.

5.7.1 Minimizing the Integral of Pseudokinetic Energy

First, we consider the following performance index, which is the simplest case ($k = 0$) of Eq. 5.14:

$$r_2 = \int_{t_0}^{t_1} \dot{\boldsymbol{\theta}}^T \dot{\boldsymbol{\theta}} \, dt \qquad 5.34$$

This criterion was suggested by Whitney (1969), but not solved. The integrand of the criterion is considered as a pseudokinetic-energy.

Figure 5.5(a) shows the relationship between α of every 30° from 0° to 660° and $\|\{E_7 - J_1^\#(\theta(t_0))J_1(\theta(t_0))\}\psi(t_0)\|$, the Euclidean norm of the transversality condition of Eq. 5.11 to be computed after the backward integration from $t = t_1$ to $t = t_0$ of Eqs. 5.18, 5.19, and 5.20. Since it is a necessary condition for the optimal trajectory that the Euclidean norm is equal to zero, we find from the figure that the optimal trajectory will be given from $\alpha = 120$, 300, 390 or 510°.

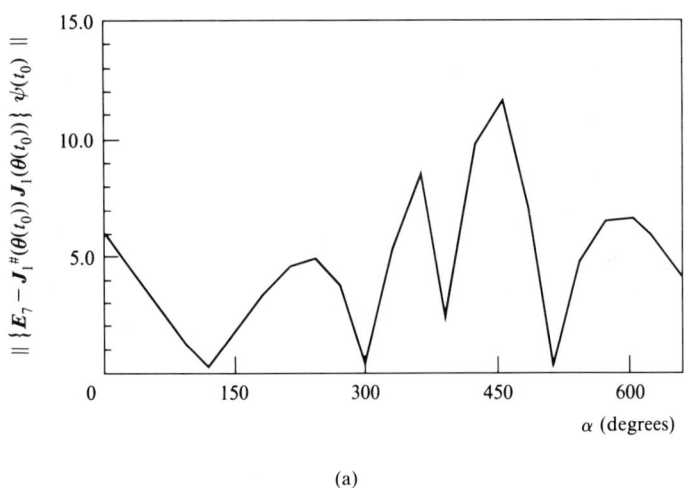

(a)

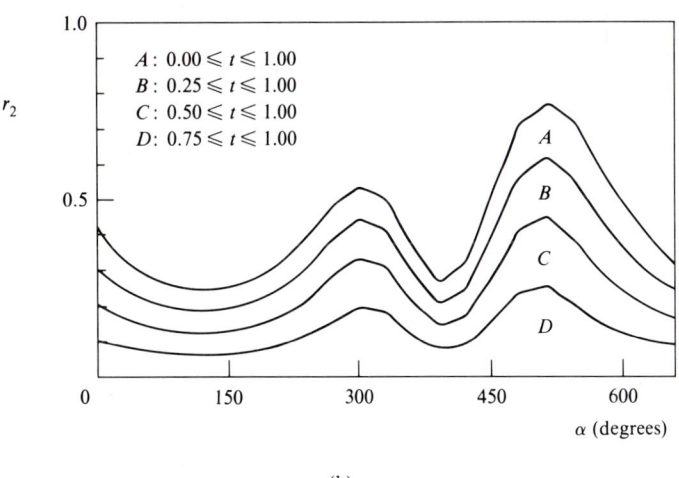

(b)

Figure 5.5 Computation of the optimal solution for performance index Eq. 5.34. (a) $\|\{E_7 - J_1^\#(\theta(t_0))J_1(\theta(t_0))\}\psi(t_0)\|$ versus α. (b) r_2 versus α.

166   Chapter 5   Global Optimization of Kinematic Redundancy

Figure 5.5(b) shows the relationship between α and the performance index r_2 of Eq. 5.34 computed after the backward integration from $t = t_1 = 1.0$ sec to $t = 0.75, 0.50, 0.25$, and 0.0 sec. We find from the figure that $\alpha = 300°$ and $\alpha = 510°$ are in the neighborhood of the local maxima, and that $\alpha = 120°$ and $\alpha = 390°$ are in the neighborhood of the local minima of the performance index. Based on the results, local minimum search was performed using the Newton's method around $\alpha = 120°$ and $\alpha = 390°$, which found that $\alpha = 115°$ yields the optimal trajectory.

Based on this experience, we can make the following remarks concerning the practical computation of the optimal trajectory:

1. Since the conditions given by Pontryagin's maximum principle are necessary conditions and, therefore, the transversality condition is satisfied at the local maxima and minima of the performance index, we have to compute the performance index directly to find the optimal solution. In addition, although the curve of the Euclidean norm of the transversality condition is so steep (as seen in Fig. 5.5a) that it is not easy to search local minima, the curve of the performance index is relatively gentle (as seen in Fig. 5.5b). Therefore, from the viewpoint of practical computation, the transversality condition is not very useful, and it is appropriate to compute the performance index to search for local minima.

2. From Fig. 5.5(b), the local minima of the performance index seem to change continuously, as t_0 changes from $t_0 = 0.75$ sec to $t_0 = 0.0$ sec. Therefore, the following recursive procedure would generally be sufficient

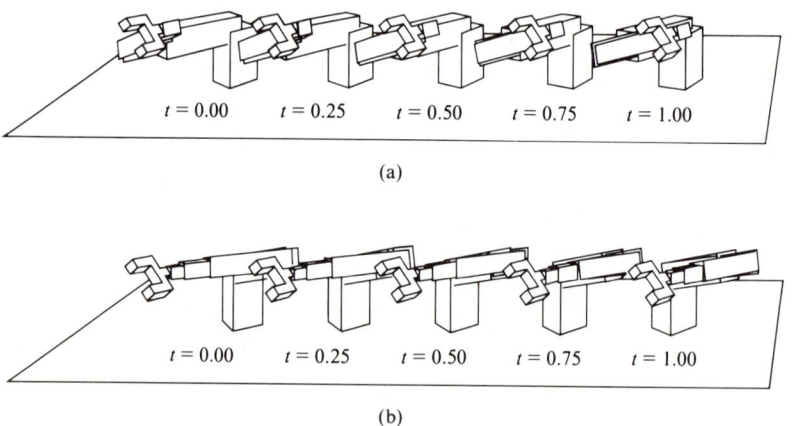

Figure 5.6 Motions for performance index Eq. 5.34. (a) Optimal (global minimum) trajectory ($\alpha = 115°$). (b) Local minimum trajectory ($\alpha = 394°$).

for reducing computation. First, the interval from t_1 to t_0 is divided into small intervals: $t^1 (= t_1)$, t^2, t^3, $\cdots$, $t^l (= t_0)$. Second, the local minima for the integration interval from t^1 to t^{k+1} are searched, starting at the local minima for the integration interval from t^1 to t^k. This procedure is repeated until $k = l - 1$.

Figure 5.6 shows the motions of the robot manipulator corresponding to the optimal trajectory at $\alpha = 115°$ and to another local minimum at $\alpha = 394°$. Figure 5.7 shows the motions corresponding to the global maximum at $\alpha = 570°$ and another local maximum at $\alpha = 300°$. The performance index values of $\alpha = 115°$ and $\alpha = 394°$ were similar. This might be because the two motions are nearly symmetric with respect to the line connecting the shoulder and the wrist. In addition, note that the nonoptimal trajectories in Figs. 5.6 and 5.7 are also optimal if their initial joint variables $\boldsymbol{\theta}(t_0)$ are given as the boundary condition in the place of $\boldsymbol{r}_1(t_0) = \boldsymbol{f}_1(\boldsymbol{\theta}(t_0))$.

5.7.2 Planning the Joint Trajectory Farthest from Singular Points

Next, we discuss the trajectory-planning problem taking manipulability into account. As the measure to evaluate manipulability, Yoshikawa (1984) defined the measure of manipulability w (Definition 3.3 in Section 3.4.1), which is calculated for the first manipulation variable as follows:

$$w = \sqrt{\det(\boldsymbol{J}_1 \boldsymbol{J}_1^T)} \qquad 5.35$$

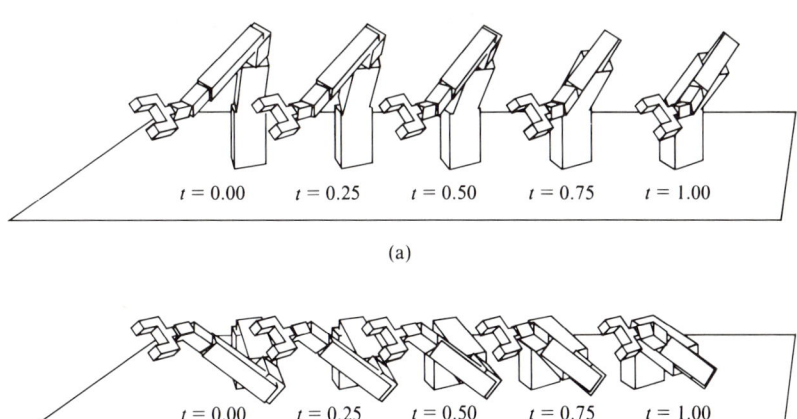

Figure 5.7 Motions for performance index Eq. 5.34. (a) Global maximum trajectory ($\alpha = 570°$). (b) Local maximum trajectory ($\alpha = 300°$).

168   Chapter 5 Global Optimization of Kinematic Redundancy

The measure of manipulability becomes zero only at singular points. In the neighborhood of singular points, the measure of manipulability increases as a robot manipulator moves away from singular points. Yoshikawa (1984) used the measure in the local redundancy control for avoiding the singularity (see Section 4.4.3). The local control, however, does not always guarantee

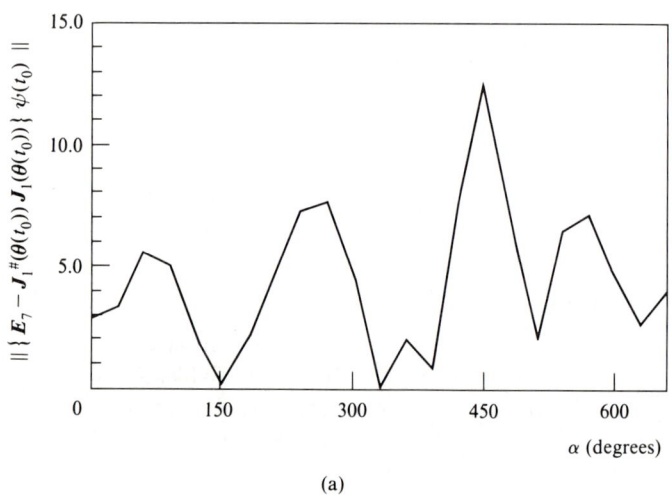

(a)

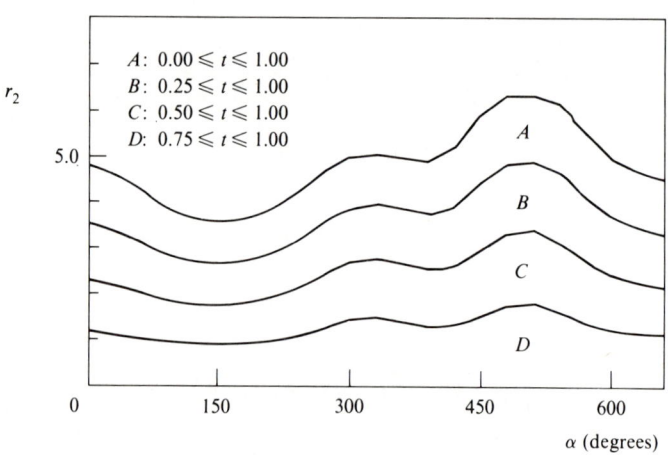

(b)

Figure 5.8 Computation of the optimal solution for performance index Eq. 5.36. (a) $\| \{ \boldsymbol{E}_7 - \boldsymbol{J}_1^{\#}(\boldsymbol{\theta}(t_0)) \boldsymbol{J}_1(\boldsymbol{\theta}(t_0)) \} \boldsymbol{\psi}(t_0) \|$ versus α. (b) r_2 versus α.

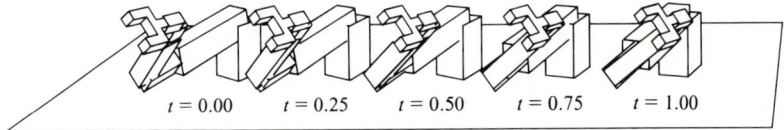

Figure 5.9 Optimum (global minimum) trajectory for performance index Eq. 5.36.

singularity-free motion. We propose formulating the problem in the framework of the globally optimal redundancy control. The performance index is given as follows:

$$r_2 = \int_{t_0}^{t_1} \{\frac{1}{\sqrt{\det(\boldsymbol{J}_1 \boldsymbol{J}_1^T)}} + \dot{\boldsymbol{\theta}}^T \dot{\boldsymbol{\theta}}\} dt \qquad 5.36$$

If the trajectory approaches the singularity, the first term of the integrand of Eq. 5.36 increases infinitely. Therefore, Eq. 5.36 will naturally offer a singularity-free motion by finding the joint trajectory farthest from singular points in the sense of Eq. 5.36

Figure 5.8 shows the relationships between α and the transversality condition, and between α and the performance index. Based on the figure, we performed local minimum search around $\alpha = 150°$ by the Newton's method, which proved that $\alpha = 148°$ gives us the optimal trajectory.

Figure 5.9 shows the optimum motion of the robot manipulator. It is noteworthy that the motion resembles one we often choose for our arm in doing similar tasks.

In computing the preceding examples, the *Versatile Simulation Library for Articulated Robot Manipulators* (VSLAM) written in FORTRAN, was developed and used at Kyoto University Data Processing Center.

Summary

A method for solving the globally optimal redundancy control problem of robot manipulators was proposed in a general form based on Pontryagin's maximum principle. Using this method, we can obtain the optimal trajectory if that trajectory exists.

Concerning boundary conditions, we found that the problem can be reduced to a two-point boundary value problem, with $n - m_1$ conditions at the left-hand endpoint and $n + m_1$ conditions at the right-hand endpoint, if we consider only kinematics. This means that the optimal trajectory can be computed by a minimum-value search in a space of as many dimensions as the degrees of redundancy. It was also proved that, if dynamics is also taken into consideration, the globally optimal redundancy control problem results in a minimum-value search in a space twice as many dimensions as the degrees of redundancy.

To verify the effectiveness of the proposed method, we used two numerical examples: (1) an optimal problem to minimize the integral of joint rate, and (2) an optimal problem to find a singularity-free motion using the measure of manipulability. Some remarks were also made regarding the practical computation of optimal solution.

Recent Related Research. After this research, several different approaches to global optimization of kinematic redundancy have been made.

Suh and Hollerbach (1987) compared local and global optimization of joint torque. The global problem is the case where $k = 0$ in Eq. 5.25. By using the calculus of variations, the authors obtained the formulae equivalent to Eqs. 5.28 and 5.29 without using the adjoint variable ψ.

Kazerounian and Wang (1988) also applied the calculus of variations to the global optimization of joint rates, which is the same problem as that in Section 5.7, and analytically derived that the solution must be governed by differential equation Eq. 5.22 with $y = O$. This result implies that the global joint-rate optimization produces the local least squares solutions of the joint accelerations with appropriate boundary conditions.

Pontryagin's maximum principle was derived from the variational method. Therefore, the technique used in these papers is mathematically very similar to one we discussed in this chapter. The main difference lies in the fact that Pontryagin's maximum principle is applicable even when the input variable must be an entry of a closed set, whereas the classical variational method works only for open-set input variables. Therefore, the discussions in this chapter can be extended readily to the global optimization problems for which we have to take account of the limits of joint torques, although these problems cannot be solved using the calculus of variations.

The other difference is that these methods result in the second-order differential equation of the state variable (joint angles and velocities for joint-torque optimization, joint angles for joint-rate optimization), whereas the method in this chapter results in the first-order differential equation of both the state and adjoint variables. Since the dimension of the adjoint variable is equal to that of the state variable, the computational complexity of numerical integration is exactly the same. The methods that use the calculus of variations may enable more analytical intuition because they use fewer variables. Another difference lies in the treatment of boundary conditions. The transversality condition of Pontryagin's maximum principle has the simpler geometry that the adjoint variable should be orthogonal to the manifold of the state variable allowed at the boundary (see Sections 2.5.2 and 2.6.4).

References

Hanafusa, H., Yoshikawa, T., and Nakamura, Y. 1981 (Kyoto). Analysis and control of articulated robot arms with redundancy. In *Control Science and*

Technology for the Progress of Society, (Proc. 8th Triennial IFAC World Congress), ed. H. Akashi, Vol. 4: pp. 1927–1932.

Hanafusa, H., Yoshikawa, T., and Nakamura, Y. 1983. Redundancy analysis of articulated robot arms and its utilization for tasks with priority. *Trans. Society of Instrument and Control Engineers* 19 (5): 421–426 *(in Japanese)*.

Kazerounian, K., and Wang, Z. 1988. Global versus local optimization in redundancy resolution of robotic manipulators. *Int. J. Robotics Res.* 7 (5): 3–12.

Klein, C. A., and Huang, C. H. 1983. Review of pseudoinverse control for use with kinematically redundant manipulators. *IEEE Trans. Sys., Man, Cyber.* SMC-13 (3): 245–250.

Ligeois, A. 1977. Automatic supervisory control of the configuration and behavior of multibody mechanisms. *IEEE Trans. Sys., Man, Cyber.* SMC-7 (12): 868–871.

Nakamura, Y., and Hanafusa, H. 1985a. Task priority based redundancy control of robot manipulators. In *Robotics research 2*, eds. H. Hanafusa and H. Inoue, pp. 155-162. Cambridge, MA: MIT Press.

Nakamura, Y., and Hanafusa, H. 1985b. Optimal redundancy control of articulated robot arms. *Trans. Society of Instrument and Control Engineers* 21 (5): 501–507 *(in Japanese)*.

Nakamura, Y., and Hanafusa, H. 1987. Optimal redundancy control of robot manipulators. *Int. J. Robotics Res.* 6 (1): 32–42.

Nakamura, Y., Hanafusa, H., and Yoshikawa, T. 1987. Task-priority based redundancy control of robot manipulators. *Int. J. Robotics Res.* 6 (2): 3–15.

Ojika, T., and Kasue, Y. 1979. Initial-value adjusting method for the solution of nonlinear multipoint boundary-value problems. *J. Math. Anal. Appl.* 69: 359–371.

Pontryagin, L. S., Boltyanskii, V. G., Gamkrelidze, R. V., and Mishchenko, E. F. 1962. *The mathematical theory of optimal processes.* (trans. from Russian by K. N. Trirogoff; ed. L. W. Neustadt). New York: Wiley.

Rao, C. R., and Mitra, S. K. 1971. *Generalized inverse of matrices and its applications.* New York: Wiley.

Suh, K. C., and Hollerbach, J. M. 1987 (Raleigh). Local versus global torque optimization of redundant manipulators. *Proc. 1987 IEEE Int. Conf. Robotics and Automation*, pp. 619–624.

Uchiyama, M., Shimizu, K., and Hakomori, K. 1985. Performance evaluation of manipulators using the Jacobian and its application to trajectory planning. In *Robotics research 2*, eds. H. Hanafusa and H. Inoue, pp. 447–454. Cambridge, MA: MIT Press.

Whitney, D. E. 1969. Resolved motion rate control of manipulators and human prostheses. *IEEE Trans. Man-Machine Sys.* MMS-10 (2): 47–53.

Whitney, D. E. 1972. The mathematics of coordinated control of prostheses and manipulators. *J. Dyn. Sys., Meas., Contr.* 94 (4): 303–309.

Yoshikawa, T. 1984. Analysis and control of robot manipulators with redundancy. In *Robotics research*, eds. M. Brady and R. Paul, pp. 735–747. Cambridge, MA: MIT Press.

CHAPTER 6

Redundancy in Multirobot Coordination

6.1 Introduction

Advanced applications sometimes require multiple robotic manipulators to perform a single task in a coordinated manner. Grasping and manipulation by a multifingered robot hand are important technologies for using robots for fine applications. These problems have substantially the same physical characteristics and should be discussed in a unified way as the problem of coordination of multiple robotic mechanisms.

This chapter is adapted from, by permission, Nakamura, Y., Nagai, K., and Yoshikawa, T. 1986. "Mechanics of coordinative manipulation by multiple robotic mechanisms," *Journal of Robotic Society of Japan*, Vol. 4, No. 5, pp. 489–498 *(in Japanese)*; Nakamura, Y., Nagai, K., and Yoshikawa, T. 1987 (Raleigh). "Mechanics of coordinative manipulation by multiple robotic mechanisms," *Proc. 1987 IEEE International Conference on Robotics and Automation*, pp. 991–998; Nakamura, Y. 1988 (Minneapolis). "Contact stability measure and optimal finger force control of multi-fingered robot hands," *Proc. 1988 USA–Japan Symposium on Flexible Automation*, pp. 523–528; and Nakamura, Y., Nagai, K., and Yoshikawa, T. 1989. "Dynamics and stability in coordination of multiple robotic mechanisms," *International Journal of Robotics Research*, Vol. 8, No. 2, pp. 44–61.

For two manipulators, Nakano et al. (1974) adopted a scheme to control one as a master (by position control) and the other as a follower (by force control). Kurono (1975) proposed changing compliance depending on the directions. Mason (1981) sought to determine the forces of two manipulators without specifying either one as a master or a follower. Uchiyama, Hakomori, and Shimizu (1983) proposed a method to obtain the forces of manipulators based on the maintaining force and the restoring force. For three manipulators Takase (1985) suggested controlling two positional variables of each arm by position control, and one positional variable and three orientational variables by force control.

Regarding the grasping and manipulation by a robot hand, many papers have been published. Hanafusa and Asada (1977) proposed a method to grasp an object such that the potential energy in the elastic fingers is minimized. Salisbury (1982) and Salisbury and Craig (1982) discussed the contact condition between the fingers and an object, and suggested that the internal forces should be determined so that the net forces have positive magnitude in the inner normal direction at the contact points on the object surface. Hanafusa, Kobayashi, and Terasaki (1983) and Kobayashi (1985) discussed a necessary and sufficient condition to manipulate an object kinematically by considering the DOF of both fingers and contact points. Nguyen (1986a, 1986b, 1986c) discussed the force-closure grasp of polyhedra and the stability when each finger's stiffness is controlled. Fearing (1986) proposed a method for stable grasping of two-dimensional polygons. The problems of friction and stability of a soft-finger contact model were investigated by Cutkosky (1985) and by Cutkosky and Wright (1986). Jameson and Leifer (1986) also studied the stability of a frictional point contact model and of the frictional soft-finger contact model. Frictionless enveloping grasping was studied by Trinkle, Abel, and Paul (1985). Hanafusa, Yoshikawa, Nakamura, and Nagai (1985) defined the magnitude of the grasping force of three-fingered robot hands as the area of a triangle made by internal force vectors and proposed a performance criterion to determine the internal forces based on it. Kerr and Roth (1986) approximated the frictional constraints by linear constraints, and proposed that we compute the optimal internal forces for the constraints and the joint torque constraints by a linear programming method. Kumar and Waldron (1987) discussed suboptimal algorithms for distributing finger forces.

Although stability is one of the main problems discussed in most papers, Jameson and Leifer (1986) have mentioned that "grasping stability" has at least two meanings. One is the ability to return to the static equilibrium position when the object position is perturbed. This ability should be termed "object stability." Object stability has been discussed as the problem of restoring forces (Hanafusa and Asada 1977; Nguyen 1986c; Jameson and Leifer 1986; Fearing 1986). The other meaning is the ability to maintain contact when the object is subjected to disturbing forces. This ability should be termed "contact stability." Hanafusa, Yoshikawa, Nakamura, and Nagai (1985) and Kerr and Roth (1986) believe that hard squeezing of an object with fingers will make the contacts more resistant to slipping due to disturbing forces.

The propriety of the asymmetric control schemes where one manipulator is controlled as a master and the other as a follower seems not to be very clear. Concerning object stability of a robot hand, the static restoring force has often been discussed, but no clear mathematical approach has been made in a dynamic sense, apparently because the analysis and synthesis of grasping and manipulation by robot fingers has been done statically. Moreover, the relationship between contact stability and large grasping forces has been discussed only intuitively. These facts indicate the need for studies on dynamic coordination of multiple robotic mechanisms.

In this chapter, we discuss the mechanics involved in coordinating multiple robotic mechanisms by considering the dynamics of an object and explicitly describing the resultant force and the internal forces applied to it. The resultant force is the vector sum of the forces applied to an object, and directly contributes to motion control of the object. The internal forces represent the elements of force vectors that are canceled within the object and do not contribute to motion control.

First, we discuss the coordinative manipulation by the resultant force, taking into consideration the dynamics of the object. This allows us to attain object stability in the dynamic sense. Second, we discuss the resolution of the resultant force into the individual net forces of robotic mechanisms. The main problem here is to determine internal forces. Hanafusa, Yoshikawa, Nakamura, and Nagai (1985) and Kerr and Roth (1986) proposed keeping contact stability by assigning large internal forces. However, such forces imply low-stability grasping, since even small position errors or disturbances may cause a large disturbing force and moment (Cutkosky and Wright 1986).

It is obvious that large grasping forces are not appropriate for grasping breakable objects. We define the minimizing internal forces as the internal forces that give the minimum net forces under static frictional constraints, and we develop their computational algorithm. We obtain the minimizing internal forces by solving, at most, $\Sigma_{j=1}^{m} \binom{m}{j} \Sigma_{i=0}^{j} \binom{j}{i}$ sets of algebraic equations, if they exist (m is the number of robotic mechanisms). To answer the fundamental question of whether a robot hand can grasp the object with the given finger placement, we have to solve all of the above algebraic equations in the worst case.

Third, contact stability against unexpected disturbing forces is discussed, and the measure of contact stability is defined as the mathematical tool for evaluating contact stability. The optimal forces are defined as the end-effector forces that generate the given resultant force and have the specified measure of contact stability. It is also shown that the optimal forces and the optimal internal forces are obtained using algorithms similar to those for the minimum forces and the minimizing internal forces. Finally, numerical examples of the computation of the minimum forces and of the minimizing internal forces are given.

6.2 Coordinative Manipulation by Multiple Robotic Mechanisms

6.2.1 Nomenclature

$O_o - x_o y_o z_o$	Absolute coordinates
$O - xyz$	Object coordinates fixed at the mass center of the object
$r \in R^3$	Vector from the origin of the absolute coordinates to the mass center of the object (m)
$\omega \in R^3$	Rotational velocity vector of the object (rad/sec)
m_o	Mass of the object (kg)
$I \in R^{3 \times 3}$	Inertia tensor of the object represented by the object coordinates (kgm^2)
m	Number of the robotic mechanisms
$p_i \in R^3$	Position vector of the ith contact point represented by the object coordinates (m)
$f_i \in R^3$	Force applied to the object at the ith contact point (N)
$Q \in R^6$	Resultant force and moment at the center of mass of the object (N, N/m)
$\phi \in R^6$	Position and orientation of the object with respect to the absolute coordinates (m, rad)
μ_i	Maximum static frictional coefficient between the object and the ith robotic mechanism
$r_i \in R^3$	Position vector of the ith contact point represented by the absolute coordinates (m)
$\tau_i \in R^{k_i}$	Generalized driving force of the ith robotic mechanism
k_i	DOF of the ith robotic mechanism
$E_i \in R^{i \times i}$	Unit matrix
$g \in R^3$	Gravity acceleration (m/sec^2)

Vectors are assumed to be represented by the object coordinates unless otherwise specified.

6.2.2 Basic Equations

We derive the basic equations for coordinative manipulation of a rigid object by m robotic mechanisms as shown in Fig. 6.1. To simplify the problem, we assume that

1. A robotic mechanism makes a frictional point contact with the object.

2. A contact point does not move on the object by the change in the relative orientation between the object and the end-effector.

The first assumption allows us to consider only forces at the contact points.

6.2 Coordinative Manipulation by Multiple Robotic Mechanisms

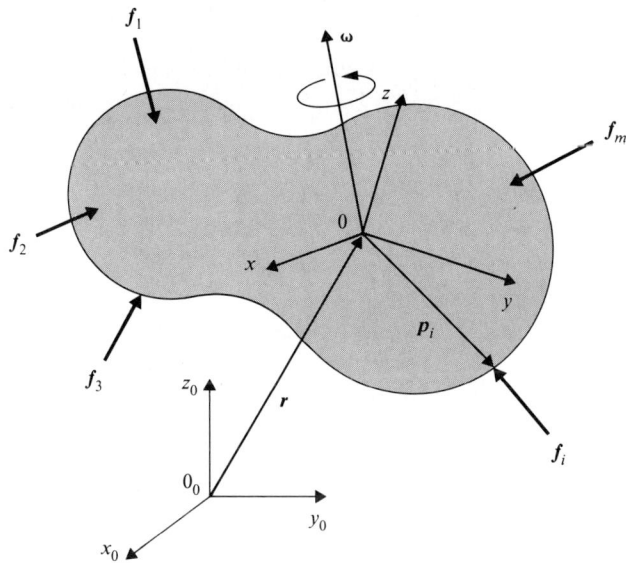

Figure 6.1 Mechanical model of coordinative manipulation of a rigid object by m robotic mechanisms.

This assumption is common and reasonable, and enables us to simplify the discussion. The second assumption means that the contact point of the end-effector is not a point on a smooth curved surface like spherical contact, but rather a vertex of a polyhedron or a cone. Although this assumption is not necessary for our dynamic problem, without it we would have to consider an additional kinematic problem, since the contact points would change in accordance with the motion of the object. This kinematic problem forms a nonholonomic constraint problem between the object and the end-effector, studied in detail by Kerr and Roth (1986).

The resultant force $\boldsymbol{f}_o$ and moment $\boldsymbol{n}_o$ of the external forces applied to the object are represented by

$$\boldsymbol{f}_o = \sum_{i=1}^{m} \boldsymbol{f}_i + m_o \boldsymbol{g} \qquad 6.1$$

$$\boldsymbol{n}_o = \sum_{i=1}^{m} \boldsymbol{p}_i \times \boldsymbol{f}_i \qquad 6.2$$

The motion equations of the object are represented by the following Newton and Euler equations:

$$m_o \ddot{\boldsymbol{r}} = \boldsymbol{f}_o \qquad 6.3$$

$$I\dot{\omega} + \omega \times (I\omega) = n_o \qquad 6.4$$

Equations 6.1 through 6.4 are summarized by the following equations:

$$I_o\ddot{\phi} + Q_o = Q \qquad 6.5$$

$$\ddot{\phi} = (\ddot{r}^T \quad \dot{\omega}^T)^T \in R^6$$

$$I_o = \begin{pmatrix} m_o E_3 & O \\ O & I \end{pmatrix} \in R^{6 \times 6}$$

$$Q_o = (-m_o g^T \quad \{\omega \times (I\omega)\}^T)^T$$

where

$$Q = WF \in R^6 \qquad 6.6$$

$$F = (f_1^T \quad f_2^T \quad \cdots \quad f_m^T)^T \in R^{3m}$$

$$W = \begin{pmatrix} E_3 & E_3 & \cdots & E_3 \\ P_1 & P_2 & \cdots & P_m \end{pmatrix} \in R^{6 \times 3m}$$

$$P_i = \begin{pmatrix} 0 & -p_{i3} & p_{i2} \\ p_{i3} & 0 & -p_{i1} \\ -p_{i2} & p_{i1} & 0 \end{pmatrix} \in R^{3 \times 3}$$

$$p_i = (p_{i1} \quad p_{i2} \quad p_{i3})^T \in R^3$$

6.2.3 Static Frictional Constraints

Suppose that the shape of the object is represented by

$$S(x, y, z) = 0 \qquad 6.7$$

where $S(x, y, z) > 0$ denotes the inside of the object and $S(x, y, z)$ is assumed to be at least once differentiable at the contact points. Then, the inner unit normal vector at contact point p_i is given by

$$e_{Ni} = \frac{\text{grad } S(p_i)}{\|\text{grad } S(p_i)\|} \in R^3 \qquad 6.8$$

where $\|*\|$ denotes the Euclidean norm of vector $*$. The force applied at the ith contact point is constrained as shown in Fig. 6.2 by the maximum static frictional condition

$$e_{Ni}^T f_i \geq \eta_i \|f_i\| \qquad 6.9$$

$$\eta_i = \frac{1}{\sqrt{1 + \mu_i^2}}$$

6.2 Coordinative Manipulation by Multiple Robotic Mechanisms   179

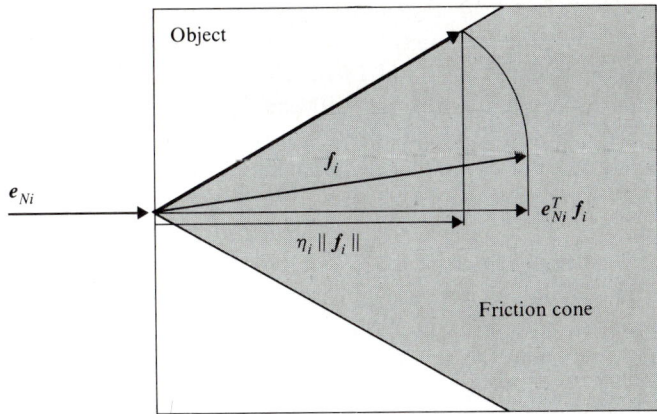

Figure 6.2 Static frictional constraints.

6.2.4 Dynamic Coordination and Object Stability

The main problem of grasping by robot fingers is to stabilize an object within the fingers. We call this stability *object stability*. Researchers attained object stability by controlling each finger's stiffness (Hanafusa and Asada 1987; Nguyen 1986a, 1986c). The main concern in this approach was to design the stiffness of the fingers so that the total stiffness would generate the restoring forces for any positional perturbation of the object. However, dynamic behavior, including dynamic stability, has not been considered so far; instead, grasping and manipulation have been discussed statically. It is also not self-evident whether stiffness control of each finger is the unique solution to object stability. We address these problems here by discussing the synthesis of the resultant force, considering the object dynamics.

Let the desired trajectory of the object be given by $\phi_d(t) \in R^6$. If the resultant force

$$Q = Q_o + I_o\{\ddot{\phi}_d + K_1(\dot{\phi}_d - \dot{\phi}) + K_2(\phi_d - \phi)\} \qquad 6.10$$

is applied to the object, then the object's motion is governed by the equation

$$(\ddot{\phi}_d - \ddot{\phi}) + K_1(\dot{\phi}_d - \dot{\phi}) + K_2(\phi_d - \phi) = 0 \qquad 6.11$$

where K_1 and $K_2 \in R^{6 \times 6}$ are constant matrices that guarantee asymptotic stability. Equation 6.11 means that $\phi(t)$ converges to $\phi_d(t)$, which means that the object is dynamically stable. From Eq. 6.5, the orientational element of ϕ becomes $\int_{t_o}^{t} \omega \, dt$, whose physical meaning is not clear. If we adopt the orientational representation by a 3×3 orthogonal matrix, we can obtain ω and $\dot{\omega}$ by differentiating it, and can compute $\phi_d - \phi$ from the difference between the desired orientational matrix and the actual one (see Section 3.2.4).

Researchers often have designed Trajectory control systems of single robot manipulators by using an equation similar to Eqs. 6.10 and 6.11 (Takase 1976 and 1985; Luh, Walker, and Paul 1980; Khatib 1985; Yoshikawa 1986). Note that the control variables in Eqs. 6.10 and 6.11 are the resultant force and moment, whereas they are joint torques in trajectory control of single manipulators. In addition to dynamic manipulation of the object, Eqs. 6.10 and 6.11 have the following physical meaning.

In Eq. 6.10 $I_o K_2$ is the stiffness matrix, and its inverse implies the compliance matrix. The stiffness of an object is not obtained by stiffness control of individual robotic mechanisms (Hanafusa and Asada 1977; Hanafusa, Kobayashi, and Terasaki 1983; Kobayashi 1985; Nguyen 1986a, 1986c). Rather, the object stiffness is specified explicitly by the resultant forces. It is noteworthy that there exists redundancy in determining the individual forces from a specified resultant force and moment Q, since the individual forces have nine elements even for three robotic mechanisms, whereas Q has only six. This redundancy is reserved as the DOF in determining the internal forces, which will be discussed in detail in Sections 6.4 and 6.5. Note that, when each finger is stiffness controlled to obtain an object stiffness, the finger forces are determined uniquely only by the object's displacement, which means that zero DOF is left for adjustment of internal forces. On the other hand, when Eq. 6.10 is used to determine finger forces, there is the possibility in this control scheme to make the net finger forces smaller by choosing appropriate internal forces for the same object stiffness.

6.2.5 *Requirement for Force Controller*

To manipulate an object dynamically, we must generate the resultant force as shown in Eq. 6.10. The distribution of the resultant force into individual end-effector forces will be discussed in the following sections. In this subsection, the basic requirement for the force controller is clarified.

So that contact with constraint surfaces will be maintained, force control should be done in the constrained directions (namely, the normal directions of the constraint surfaces), whereas position control to move the end-effector should be done in the unconstrained directions (namely, the tangential directions of constraint surfaces) (Mason 1981). The force control based on this formulation is usually termed a *hybrid position/force control* (Raibert and Craig 1981; Yoshikawa 1986). To make sure that the constrained directions and the unconstrained directions span the orthogonal complement spaces to each other in six-dimensional linear space (Mason 1981), it is an important assumption of hybrid position/force control that the end-effector does not move in the force-controlled directions. Therefore, the force controller in this formulation requires only to keep static equilibrium in the constrained directions.

6.2 Coordinative Manipulation by Multiple Robotic Mechanisms

In dynamic coordinative manipulation the force control is required in all directions of a six-dimensional space. An object is manipulated by the work done by the end-effector forces.[6.1] In other words, the force control should be done in the direction in which the end-effector moves. Therefore, the force control is different from hybrid position/force control. The force controller must consider both static and dynamic equilibrium. Static equilibrium is needed to generate the desired end-effector force at the contact point. On the other hand, dynamic equilibrium serves to maintain contact with the object by accelerating the end-effector to follow the motion of the contact point. The force control is mathematically formulated as follows.

From Fig. 6.1, the acceleration of the contact point is

$$\ddot{\boldsymbol{r}}_i = \ddot{\boldsymbol{r}} + \dot{\boldsymbol{\omega}} \times \boldsymbol{p}_i + \boldsymbol{\omega} \times (\boldsymbol{\omega} \times \boldsymbol{p}_i) \qquad 6.12$$

where $\ddot{\boldsymbol{r}}$ and $\dot{\boldsymbol{\omega}}$ are the accelerations of the object caused by the resultant force of Eq. 6.10, and are represented from Eqs. 6.10 and 6.11 by

$$\begin{aligned} (\ddot{\boldsymbol{r}}^T \quad \dot{\boldsymbol{\omega}}^T) &= \ddot{\boldsymbol{\phi}} \\ &= \ddot{\boldsymbol{\phi}}_d + \boldsymbol{K}_1(\dot{\boldsymbol{\phi}}_d - \dot{\boldsymbol{\phi}}) + \boldsymbol{K}_2(\boldsymbol{\phi}_d - \boldsymbol{\phi}) \qquad 6.13 \\ &= \boldsymbol{I}_o^{-1}(\boldsymbol{Q} - \boldsymbol{Q}_o) \end{aligned}$$

Let τ_{ip} be the generalized driving force of the ith robotic mechanism required to generate acceleration $\ddot{\boldsymbol{r}}_i$ when the mechanism is not making contact with the object. We compute τ_{ip} considering the dynamic characteristics of the robotic mechanism, such as the inertia tensor and the Colioris, centrifugal, and gravitational forces. Also let τ_{if} be the generalized driving force of the ith robotic mechanism needed to apply force $\boldsymbol{f}_i$ statically to the object. We compute τ_{if} from the static equilibrium condition by multiplying the transposed Jacobian matrix of the ith robotic mechanism to $\boldsymbol{f}_i$. According to d'Alembert's principle, the generalized driving force τ_i to manipulate the object is the sum of τ_{ip} and τ_{if}. The fundamental structure of the control system for coordinative manipulation is synthesized as shown in Fig. 6.3.

Note that this discussion does not mean that an arbitrary pair of $\boldsymbol{f}_i$ and $\ddot{\boldsymbol{r}}_i$ can be realized. As shown in Eqs. 6.12 and 6.13, $\ddot{\boldsymbol{r}}_i$ is constrained by the fact that $\ddot{\boldsymbol{r}}_i$ is the result of $\boldsymbol{f}_i$s. We could say that τ_{if} is the generalized driving force for force control, and that τ_{ip} is the generalized driving force for accompanying position control. These considerations are summarized by the following theorem.

[6.1] The total sum of the internal force elements of the end-effector forces does not do work, which will be discussed in Section 6.4.1. However, the internal force element, as well as the noninternal force element, of each end-effector force does work at each contact point.

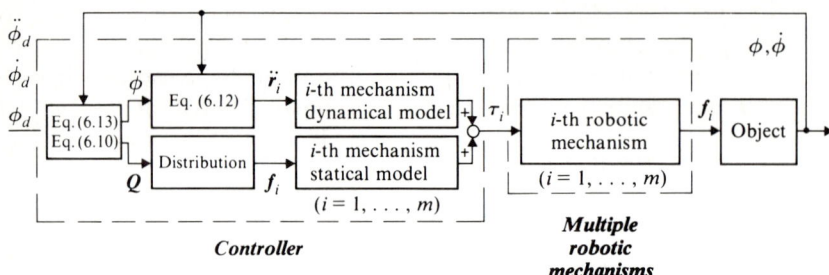

Figure 6.3 Requirement for force controller: Feedforward paths for position and force control.

Theorem 6.1 (Force Controller)

If a robotic mechanism is force controlled in the directions in which the end-effector works, it must accompany position-control so that the trajectory will be the result of the force control.

Figure 6.3 shows only the feedforward paths of position control and force control. Generally speaking, the control system needs a feedback loop for each of position control and force control (Raibert and Craig 1981; Yoshikawa 1986). In hybrid position/force control, the position control and the force control are decoupled in nature, and, therefore, their closed loops are designed independently. In our current problem, however, they are coupled and need to be designed considering the dependency. Further study of the synthesis of feedback control systems must be done.

Note that the discussion above does not mean that an object cannot be manipulated without consideration of the dynamic equilibrium. When robotic mechanisms are controlled only by the static equilibrium, a part of driving force computed from the static equilibrium is used to accelerate the end-effector of the robotic mechanism, and the object motion will be disturbed.

6.3 Coordinative Manipulability

Can the robotic mechanisms manipulate the object arbitrarily? This is a basic question. To answer it, we have to consider the static frictional constraints and the fact that a robotic mechanism can push but not pull the object. In this section, we derive a method to verify whether or not the given robotic mechanisms can generate an arbitrary acceleration $\ddot{\phi}$ under the static frictional constraints of Eq. 6.9.

Since the inertia tensor is positive definite in Eq. 6.5, generating an arbitrary $\ddot{\phi}$ is equivalent to generating an arbitrary Q. The grasp that enables exertion of an arbitrary force was named *force-closure grasp* by Ohwovoriole (1980), and has been discussed in detail for the two-dimensional case (Nguyen

1986c). Our result provides a method to investigate for any number of fingers in the three-dimensional case, whether or not a three-dimensional grasp is force-closure.

Suppose that two linearly independent resultant forces Q^1 and Q^2 can be generated under the constraints of Eq. 6.9; that is,

$$Q^j = W F^j \qquad \text{for } j = 1, 2 \qquad 6.14$$

$$F^j = (f_1^{jT} \quad f_2^{jT} \quad \cdots \quad f_m^{jT})^T \in R^{3m}$$

$$e_{Ni}^T f_i^j \geq \eta_i \|f_i^j\|$$
$$\text{for } i = 1, \ldots, m \quad j = 1, 2 \qquad 6.15$$

Then, for arbitrary nonnegative scalars k_1 and k_2,

$$Q^0 = k_1 Q^1 + k_2 Q^2 \qquad 6.16$$

can also be generated as follows. Let

$$F^0 = k_1 F^1 + k_2 F^2$$
$$= (f_1^{0T} \quad f_2^{0T} \quad \cdots \quad f_m^{0T})^T \qquad 6.17$$

The following inequality, derived from Eq. 6.15, implies that F^0 can be generated under static frictional constraints:

$$\begin{aligned} e_{Ni}^T f_i^0 &= e_{Ni}^T (k_1 f_i^1 + k_2 f_i^2) \\ &\geq \eta_i (k_1 \|f_i^1\| + k_2 \|f_i^2\|) \\ &\geq \eta_i (\|k_1 f_i^1 + k_2 f_i^2\|) \\ &= \eta_i \|f_i^0\| \end{aligned} \qquad 6.18$$

From Eqs. 6.14, 6.16, and 6.17, F^0 obviously yields the resultant force Q^0. This fact shows that, if $\pm Q^1$ and $\pm Q^2$ can be generated, any linear combination of Q^1 and Q^2 can be generated. Accordingly, the following theorem is given.

Theorem 6.2 (Coordinative Manipulability)

Let Q^j ($j = 1, \ldots, 6$) be linearly independent resultant forces. Then the robotic mechanisms can generate an arbitrary acceleration of the object if and only if Q^j and $-Q^j$ ($j = 1, \ldots, 6$) can be generated under the static frictional constraints.

Theorem 6.2 shows that we can conclude the coordinative manipulability by checking whether all 12 resultant forces can be generated. Since an object should preferably be grasped by force-closure, the theorem is used to tell whether given contact points are appropriate. To examine the coordinative manipulability by using the theorem, we must check whether each resultant force can be generated. The computational algorithm to do this is given in the next section. In addition, a necessary condition for generating six linearly independent resultant forces is that W be full rank in Eq. 6.14.

6.4 Computation of Minimum Forces

6.4.1 Internal Force in Coordinative Manipulation

The internal force in grasping and manipulation by robot hands was discussed based on Eq. 6.6. The internal force can be specified by elements in the null space of W. In this subsection, we give a physical meaning of the internal force, as well as its alternative definition.

Let the robotic mechanisms apply f_i ($i = 1, \ldots, m$) to the object, and let the virtual displacement of the object be represented by $\delta\phi = (\delta r^T \quad \delta\Omega^T)^T$, where $\delta\Omega \in R^3$ denotes the orientational virtual displacement, and $\dot{\Omega} = \omega$. The total sum of the virtual works made by all robotic mechanisms becomes

$$\delta w = \sum_{i=1}^{m}(\delta\Omega \times p_i + \delta r)^T f_i \qquad 6.19$$

Equation 6.19 is simplified to

$$\delta w = \delta\phi^T W F \qquad 6.20$$

The following equation is a sufficient and necessary condition to make δw equal to zero for an arbitrary $\delta\phi$:

$$W F = 0 \qquad 6.21$$

Accordingly, the internal forces are a set of f_i such that the total sum of the virtual works done by the fingers is zero for any and all virtual displacements of the object.

Definition 6.1 (Internal Forces)

The *internal forces* are defined as a set of f_i ($i = 1, \ldots, m$) such that the total sum of the virtual works done by the fingers results in zero for any and all virtual displacements of the object.

6.4.2 Minimizing Internal Forces

Hanafusa, Yoshikawa, Nakamura, and Nagai (1985) and Kerr and Roth (1986) proposed a technique to determine the optimal internal forces by considering the maximum value of f_i and the maximum torques of the finger joints, respectively. As mentioned in the latter paper, their approaches were aimed at determining "how hard to squeeze an object with the fingers in order to ensure that the object is grasped stably." Accordingly, the larger the available forces or torques are for the robotic mechanisms, the larger the grasping forces these approaches choose. However, with the large grasping forces, even a small error of the direction of force that is introduced by the measurement error or the geometrical uncertainty may cause a large disturbing force and moment at the mass-center of the object. We often experience this situation when we grasp a wet bar of soap: Hard squeezing results in slipping and misgrasping. Cutkosky and Wright (1986) pointed out that "increasing the gripping force reduced the chance of slipping but also made the grip less stable with respect to disturbances." These facts suggest that it is an appropriate principle for determining internal forces, from the viewpoint of stable grasping, to reduce the grasping forces as long as they have enough contact stability.

In this section, we discuss the extreme problem of obtaining the minimum forces under the static frictional constraints. The solution of this problem does not guarantee contact stability. This can be shown by the following example: When a rigid body is stationally floating in gravitation-free space, the minimum finger forces needed to maintain the body are obviously zero. With this zero-force grasp, we will lose contact if any disturbing forces are exerted. Although the solution of this extreme problem seems impractical the minimum forces have a significant physical meaning by themselves. An object can be grasped or manipulated with the specified resultant force only when the minimum forces exist. The algorithms to be developed in this section are extended in Section 6.5 to obtain the minimum forces, taking the contact stability into consideration.

Definition 6.2 (Minimum Force, Minimizing Internal Forces)

The minimum norm force F among all the forces that generate the specified resultant force Q under the static frictional constraints of Eq. 6.9 is called the *minimum force*, and the internal forces that yield the minimum force are called the *minimizing internal forces*.

To cope with the uncertainty of the coefficient of maximum static friction, we should regard

$$\mu_{ai} = \frac{1}{C}\mu_{ei} \qquad\qquad 6.22$$

as the effective coefficient of maximum static friction, where μ_{ei} is the critically estimated coefficient of maximum static friction and $C \geq 1$ is the safety factor. Since Q can be generated if and only if the minimizing internal forces exist, we can investigate the coordinative manipulability of Theorem 6.2 by verifying the existence of the minimizing internal forces for $\pm Q^j$ ($j = 1,\ldots,6$).

6.4.3 Application of Kuhn–Tucker Theorem

The problem of computing the minimum force and the minimizing internal forces is formulated as follows: For the specified resultant force Q, obtain F that satisfies

$$\min \ \|F\| \qquad 6.23$$

under the constraints

$$Q = WF \qquad 6.24$$

$$e_{Ni}^T f_i \geq \eta_i \|f_i\| \qquad 6.25$$

We assume that each finger has enough joints and can always apply an arbitrary force at each fingertip. If an F exists that satisfies Eq. 6.24,[6.2] we can write it as

$$F = F_o + Ay \qquad y \in R^b \qquad 6.26$$

$$F_o = W^\# Q$$
$$= (f_{o1}^T \ f_{o2}^T \ \cdots \ f_{om}^T)^T$$
$$f_{oi} \in R^3$$

$$A = (A_1^T \ A_2^T \ \cdots \ A_m^T)^T$$
$$A_i \in R^{3 \times b}$$

where $W^\# \in R^{3m \times 6}$ is the pseudoinverse of W, A is the matrix of the orthonormals of the null space of W,[6.3] and $b = 3m - \text{rank } W$ is the dimension of the null space of W. The first term of Eq. 6.26 is an element of the orthogonal complement of the null space of W, and the second term is an element of the null space of W. The internal force corresponds to the second term of Eq. 6.26. The constraints of Eq. 6.25 can be represented by the following $2m$ inequalities:

$$g_i(y) \leq 0 \qquad \text{for } i = 1, \ldots, 2m \qquad 6.27$$

$$g_i(y) = \begin{cases} (f_{oi} + A_i y)^T B_i (f_{oi} + A_i y) & \text{for } i = 1, \ldots, m \\ -e_{N\,i-m}^T (f_{o\,i-m} + A_{i-m} y) & \text{for } i = m+1, \ldots, 2m \end{cases}$$

$$B_i = \eta_i^2 \ E_3 - e_{Ni} e_{Ni}^T$$

[6.2] We verify the existence of F by checking that $(E_6 - WW^\#)Q = 0$.

[6.3] We can produce matrix A by collecting and orthonormalizing the linearly independent column vectors of $(E_{3m} - W^\# W)$.

Equation 6.23 can also be represented by

$$\min_{y} h(y) \qquad 6.28$$

$$h(y) = (F_o + Ay)^T (F_o + Ay)$$

From Eqs. 6.27 and 6.28, the problem of finding the minimizing internal force has been reduced to minimizing a quadratic function with linear and quadratic constraints.

To obtain the y that yields the minimizing internal forces, we apply Kuhn-Tucker Theorem of Section 2.5.4. We prepare a function $\varepsilon(y, \lambda)$ by using Lagrange multipliers $\lambda = (\lambda_1 \; \lambda_2 \; \cdots \; \lambda_{2m})^T \in R^{2m}$ as follows:

$$\varepsilon(y, \lambda) = h(y) + \lambda^T g(y) \qquad 6.29$$

$$g(y) \triangleq (g_1(y) \; g_2(y) \cdots g_{2m}(y))^T \qquad 6.30$$

From Theorem 2.10 in Section 2.5.4., a necessary condition for y^o to yield the local minimum of $h(y)$ under the condition $g_i(y) \leq 0$ is that $\lambda^o \geq 0$ ($\lambda^o = (\lambda_1^o \; \lambda_2^o \; \cdots \; \lambda_{2m}^o)^T$, $\lambda_i^o \geq 0$, $i = 1, \ldots, 2m$) satisfies the following equations:[†6.4]

$$\left. \frac{\partial \varepsilon}{\partial y} \right|_{y=y^o, \lambda=\lambda^o} = 0 \qquad 6.31$$

$$\left. \lambda^T g \right|_{y=y^o, \lambda=\lambda_o} = 0 \qquad 6.32$$

It is well known that Eqs. 6.31 and 6.32 become the necessary and sufficient conditions for the global minimum if $\varepsilon(y, \lambda)$ is a convex function with respect to y (Mine 1966).

6.4.4 Convexity of $\varepsilon(y, \lambda)$

Substituting $g_i(y)$ and $h(y)$ into Eq. 6.29 and using the fact that F_o and Ay are orthogonal to each other, we obtain

$$\varepsilon(y, \lambda) = y^T A^T (B_o + E_{3m}) Ay + (2 F_o^T B_o - e_{N_o}^T) Ay \\ + F_o^T (B_o + E_{3m}) F_o - e_{N_o}^T F_o \qquad 6.33$$

[†6.4] In our original problem of Eqs. 6.23, 6.24, and 6.25, we had both equality and inequality constraints. In the representation by Eqs. 6.27 and 6.28, we have explicitly eliminated the equality constraints. Therefore, the equality constraints are not included in Eqs. 6.29. The λ and g in Eqs. 6.29, 6.30, 6.31, and 6.32 correspond to μ and r in Theorem 2.10.

$$B_o = \begin{pmatrix} \lambda_1 B_1 & \cdots & O \\ \vdots & \ddots & \vdots \\ O & \cdots & \lambda_m B_m \end{pmatrix} \in R^{3m \times 3m}$$

$$e_{No} = (\lambda_{m+1} e_{N1}{}^T \quad \cdots \quad \lambda_{2m} e_{Nm}{}^T)^T \in R^{3m}$$

Equation 6.33 shows that $\varepsilon(y, \lambda)$ is a quadratic function of y. Convexity of the quadratic function is equivalent to the positive definiteness of the second-order term. If the second-order term is positive definite, the quadratic function has a unique local minimum and it is the global minimum. If it is positive semi-definite, the quadratic function has an infinite number of local minima. If it is neither positive definite nor positive semi-definite, then the quadratic function does not have the global minimum. The second-order term of Eq. 6.33 is expanded as follows:

$$y^T A^T (B_o + E_{3m}) A y$$
$$= y^T [\sum_{i=1}^{m} A_i{}^T \{(1 + \lambda_i \eta_i{}^2) E_3 - \lambda_i e_{Ni} e_{Ni}{}^T\} A_i] y \qquad 6.34$$

In Eq. 6.34, $A^T (B_o + E_{3m}) A$ involves λ_i, and cannot be positive definite for arbitrary $\lambda_i \ (\geq 0)$ and arbitrary geometry of grasping. Therefore, we have to verify directly the positive definiteness of $A^T (B_o + E_{3m}) A$ for obtained λ_i. That is, if obtained $\lambda_i \ (\geq 0)$ makes $A^T (B_o + E_{3m}) A$ positive definite, it guarantees the concavity of $\varepsilon(y, \lambda)$ with respect to y and provides the globally minimum solution.

6.4.5 *Computation of the Minimizing Internal Forces*

We give a procedure for computing the minimizing internal forces according to Eqs. 6.31 and 6.32. Substituting Eq. 6.33 into Eq. 6.31 yields

$$y^o = -\{A^T (B_o + E_{3m}) A\}^{-1} A^T (B_o F_o - \frac{1}{2} e_{No}) \qquad 6.35$$

Note that, if $\varepsilon(y, \lambda)$ is convex, $A^T (B_o + E_{3m}) A$ becomes positive definite and, therefore, nonsingular. Equation 6.32 is calculated as follows:

$$\sum_{i=1}^{2m} \lambda_i^o g_i(y^o) = 0 \qquad 6.36$$

Since $\lambda_i^o \geq 0$ and $g_i(y^o) \leq 0$, Eq. 6.36 is equivalent to

$$\lambda_i^o \begin{cases} = 0 & \text{for } g_i(y^o) < 0 \\ \geq 0 & \text{for } g_i(y^o) = 0 \end{cases} \quad (i = 1, \ldots, 2m) \qquad 6.37$$

Equation 6.37 implies that $\lambda_i^o = 0$ when the inequality for i is satisfied inside and that $\lambda_i^o > 0$ only when the inequality for i is satisfied on the boundary of the friction cone.

Note that $g_i(\boldsymbol{y}^o) \leq 0$ ($i = m+1, \ldots, 2m$) is satisfied on the boundary only when $\boldsymbol{f}_{i-m} = 0$. Therefore, when $\boldsymbol{f}_{i-m} \neq 0$ ($i = m+1, \ldots, 2m$), $g_i(\boldsymbol{y}^o) \leq 0$ must be satisfied inside the inequality constraint. In this case, we simplify Eqs. 6.35 and 6.37 by using $\lambda_i^o = 0$ ($i = m+1, \ldots, 2m$) as follows:

$$\boldsymbol{y}^o = -\{\boldsymbol{A}^T(\boldsymbol{B}_o + \boldsymbol{E}_{3m})\boldsymbol{A}\}^{-1}\boldsymbol{A}^T\boldsymbol{B}_o\boldsymbol{F}_o \qquad 6.38$$

$$\lambda_i^o \begin{cases} = 0 & \text{for } g_i(\boldsymbol{y}^o) < 0 \\ \geq 0 & \text{for } g_i(\boldsymbol{y}^o) = 0 \end{cases} \quad (i = 1, \ldots, m)$$

Assuming $\boldsymbol{f}_i \neq 0$ ($i = 1, \ldots, m$), the minimizing internal forces and the minimum forces are obtained by the following procedure.[6.5]

Algorithm 6.1

1. $k = 0$.

2. $j = 1$.

3. Assume k robotic mechanisms are on the boundaries of the friction cones and the others are inside them. Take a new guess from among the $\binom{m}{k}$ combinations, about which mechanisms are on the boundaries. Compute $\boldsymbol{y}^o$ according to Eq. 6.38 by setting $\lambda_i^o = 0$ for $m - k$ robotic mechanisms assumed to be inside the constraints. Obtain the rest of λ_i^o by solving k equations $g_i(\boldsymbol{y}^o) = 0$ ($1 \leq i \leq m$) for the robotic mechanisms assumed to be on the boundary of the constraints.

4. If the obtained λ_i^o satisfies $\lambda_i^o \geq 0$ and $2m$ inequalities of Eq. 6.27, and makes $\boldsymbol{A}^T(\boldsymbol{B}_o + \boldsymbol{E}_{3m})\boldsymbol{A}$ positive definite, then go to step 7.

5. If $j = \binom{m}{k}$ and $k < m$, then $k+1 \to k$ and go to step 2. If $j = \binom{m}{k}$ and $k = m$, then go to step 8.

6. $j + 1 \to j$ and go to step 3.

7. Compute $\boldsymbol{y}^o$ by substituting λ_i^o ($i = 1, \ldots, m$) into Eq. 6.38. $\boldsymbol{A}\boldsymbol{y}^o$ is the minimizing internal forces. The minimum forces are obtained by substituting $\boldsymbol{y}^o$ into Eq. 6.26. End.

8. The resultant force $\boldsymbol{Q}$ cannot be generated under the constraints. End.

[6.5] We may have $\boldsymbol{f}_i = 0$ as the minimum force although $\boldsymbol{f}_i \neq 0$ is assumed. This is because $\lambda_i^o = 0$ can be satisfied even when $g_i(\boldsymbol{y}^o) = 0$. There is no problem in treating $\boldsymbol{f}_i = 0$ as the solution.

The global solution for the general case where $f_i = 0$ is allowed can be computed according to the following procedure.

Algorithm 6.2

1. $m' = m$.

2. $j = 1$.

3. Assume $f_i \neq 0$ for m' robotic mechanisms, and $f_i = 0$ for the rest. Take a new guess from among the $\binom{m}{m'}$ combinations. Investigate whether the robotic mechanisms for which $f_i = 0$ is assumed satisfy $f_{oi} \in \mathcal{R}(A_i)$ (where $\mathcal{R}(A_i)$ is the range space of A_i). If they do not, then go to step 5.[†6.6]

4. Apply Algorithm 6.1 for m' robotic mechanisms for which $f_i \neq 0$ is assumed. If the global solutions are obtained, then go to step 7.

5. If $j = \binom{m}{m'}$ and $m' > 1$, then $m' - 1 \to m'$ and go to step 2. If $j = \binom{m}{m'}$ and $m' = 1$, then go to step 8.

6. $j + 1 \to j$ and go to step 3.

7. The minimizing internal forces and the minimum forces have been obtained for m' robotic mechanisms for which $f_i \neq 0$ is assumed. For $m - m'$ robotic mechanisms for which $f_i = 0$ is assumed, $-f_{oi}$ becomes the minimizing internal force and $f_i = 0$ becomes the minimum force. End.

8. The resultant force Q cannot be generated under the constraints. End.

When resultant force Q cannot be generated, step 4 must be repeated $\sum_{m'=1}^{m} \binom{m}{m'}$ times before we arrive at step 8. The loop of Algorithm 6.1 needs to be repeated at most $\sum_{i=0}^{m'} \binom{m'}{i}$ times for each computation of step 4 in Algorithm 6.2. Therefore, in the worst cases, we have to solve a set of algebraic equations $\sum_{j=1}^{m} \binom{m}{j} \sum_{i=0}^{j} \binom{j}{i}$ (26 for $m = 3$) times. However, the case where some robotic mechanisms satisfy $f_i = 0$ is very rare, because this implies that the mechanisms are not needed and $f_i \neq 0$ of the mechanisms results in the larger norm of F. Also, the case where some robotic mechanisms are simultaneously on the boundary of the constraints is considered to be rare in practice. Therefore, in most cases, the global solution would be found among the cases of $j = m$ and $i = 0, 1$; that is, $\binom{m}{0} + \binom{m}{1} = m + 1$ cases.

[†6.6] Since $f_i = 0$ holds from Eq. 6.26 only when $f_{oi} \in \mathcal{R}(A_i)$, step 4 can be skipped when $f_{oi} \in \mathcal{R}(A_i)$ is not satisfied.

6.5 Contact Stability and Optimal Forces

6.5.1 Measure of Contact Stability

We obtained Object stability by generating the resultant force specified by Eq. 6.10, where it was assumed that the end-effectors maintain contact with an object. Contact stability is the ability of the end-effectors to maintain contact with an object without slipping. The dynamic-equilibrium condition considered in Section 6.2.5 will serve for contact stability by accelerating the end-effectors to track the contact points, and it will be enough in the ideal situation where neither parameter ambiguity nor any unexpected disturbing force exists. In this section, we investigate contact stability against unexpected disturbing forces.

Previous works (Hanafusa, Yoshikawa, Nakamura, and Nagai 1985; Kerr and Roth 1986) were motivated by the idea that strong squeezing will ensure contact stability. We have already mentioned that hard squeezing may make gripping less stable because of measurement error or geometrical disturbances. Another problem is that hard squeezing may cause large slipping acceleration, since once slipping happens, the friction cone suddenly shrinks by the change of friction coefficient from static to kinetic. Therefore, all the force elements inside of the static friction cones but outside of the kinetic friction cones will become the accelerating forces, which would be larger in the case of hard squeezing. The motivation of our approach is that, for the mechanism to be more resistant to disturbing forces caused by measurement error or positional uncertainty, and to avoid generating large slipping accelerations even if slipping occurs, the end-effectors should apply as weak forces as possible so long as contact stability is maintained for a class of disturbing forces. As preparation, we develop an evaluation method of contact stability.

Our problem is schematically represented by Fig. 6.4. To generate the resultant force specified by Eq. 6.10, each end-effector applies some force inside the friction cone. With the force controller described in Section 6.2.5, the end-effectors can follow the motion of the contact points expected as a result of the resultant force. However, if unexpected external disturbing forces are exerted, the contacts of the end-effectors may not be maintained. We have to clarify the extent of disturbing forces that the contacts can be maintained without slipping.

Since dynamic equilibrium is maintained by applying τ_{ip} (Section 6.2.5), this problem is reduced to the static problem related to τ_{if}, so we can assume without loss of generality that the end-effectors are stationary, and are not subject to the gravitational force. Since $\ddot{\theta}_i = 0$ is satisfied and no gravity exists, from this assumption, the dynamics of the ith robotic mechanism are described by

$$\tau_{if} = I_i \ddot{\theta}_i + J_i^T f_i \qquad 6.39$$

192   Chapter 6   Redundancy in Multirobot Coordination

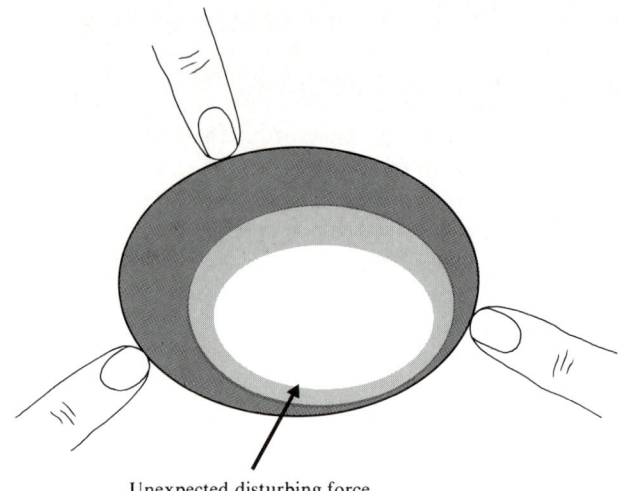

Figure 6.4 Can the contacts be maintained if unexpected disturbing forces are applied?

where $I_i \in R^{k_i \times k_i}$ is the inertia matrix, $\theta_i \in R^{k_i}$ is the generalized coordinate, and $J_i = \partial r_i / \partial \theta_i \in R^{3 \times k_i}$ is the Jacobian matrix of the ith robotic mechanism. We also assume that

$$\tau_{if} = J_i^T f_{id} \qquad 6.40$$

is generated by the joints and, at the initial time, the end-effector exerts $f_i = f_{id}$, which is included in the friction cone.[†6.7]

First, we discuss what will happen to $\ddot{\theta}_i$ and f_i if the object is suddenly accelerated by a disturbing force. Assume that the acceleration of the object at the contact point caused by the disturbing force is represented by $\ddot{r}_i$. Now, we suppose that the end-effector can follow the motion of the contact point. Then, from $\dot{\theta}_i = 0$, the following equation is satisfied:

$$\ddot{r}_i = J_i \ddot{\theta}_i + \dot{J}_i \dot{\theta}_i = J_i \ddot{\theta}_i \qquad 6.41$$

Considering the positive definiteness of the inertia matrix I_i and using Eqs. 6.40 and 6.41, the following equations are derived from Eq. 6.39:

[†6.7]See Nakamura (1987) for a necessary and sufficient condition for a robot mechanism to be able to exert f_{id} by generating τ_{if} of Eq. 6.40.

6.5 Contact Stability and Optimal Forces

$$f_i = f_{id} + f_{ia} \qquad 6.42$$

$$f_{ia} = -(J_i I_i^{-1} J_i^T)^{-1} \ddot{r}_i \qquad 6.43$$

where rank$(J_i) = 3$ is assumed, and $J_i I_i^{-1} J_i^T \in R^{3 \times 3}$ becomes positive definite. Equation 6.42 shows the force exerted on the object when the contact point is accelerated by $\ddot{r}_i$. Since the end-effector starts to slip when f_i is on the boundary of the friction cone, it can follow the motion of the contact point so long as f_i is in the friction cone. The relationship between contact stability and $\ddot{r}_i$ is shown schematically in Fig. 6.5.

Now, we consider the following spherical set of contact-point accelerations:

$$\{\ddot{r}_i \; ; \; \|\ddot{r}_i\| \leq a\} \qquad 6.44$$

where a is a positive scalar constant. Then, the set of f_{ia} made by all the possible $\ddot{r}_i$ is

$$\{f_{ia} \; ; \; f_{ia}^T (J_i I_i^{-1} J_i^T)^T (J_i I_i^{-1} J_i^T) f_{ia} \leq a^2\} \qquad 6.45$$

Since $J_i I_i^{-1} J_i^T$ is positive definite, from Eqs. 6.42 and 6.45 the set of endpoints of f_i becomes an ellipsoid with its center at the tip of f_{id}. The lengths of the principal axes are $2a/\sigma_{ij}$ $(j = 1, 2, 3)$, where $\sigma_{ij} > 0$ is the singular value of $J_i I_i^{-1} J_i^T$ (see Section 2.3.1). Let a_c be the minimum value of a that makes the ellipsoid contact with the friction cone. Then a_c implies the maximum radius of the sphere set of $\ddot{r}_i$ whose element never causes slipping. This relationship is shown in Fig. 6.6. Therefore, the following definition is given.

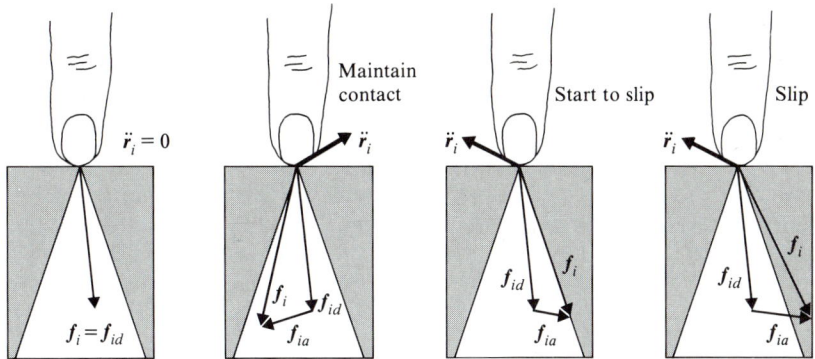

Figure 6.5 Contact stability and contact-point acceleration.

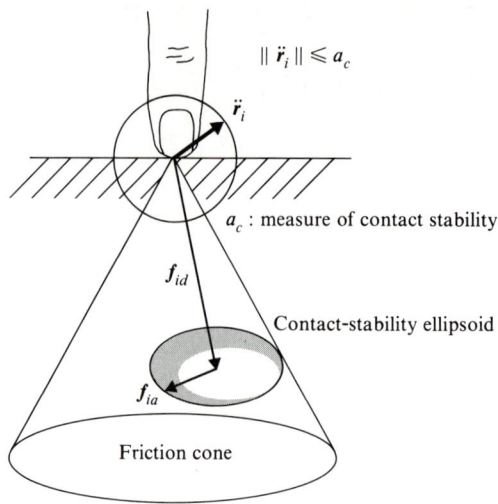

Figure 6.6 Measure of contact stability and contact-stability ellipsoid.

Definition 6.3 (Contact Stability Measure)

The minimum value of the radius of the sphere set of contact-point acceleration that makes the ellipsoid set of f_i defined by Eqs. 6.42 and 6.45 contact the friction cone is called the *measure of contact stability*, and the corresponding ellipsoid set of f_i is called the *contact-stability ellipsoid*.

The contact-stability ellipsoid is equivalent to the generalized inertia ellipsoid (Asada 1983) when $f_{id} = 0$ and $a_c = 1$.

Next, we discuss the relationship between squeezing and the object acceleration by disturbing forces. Suppose $\omega = 0$, there is no gravity, and the resultant force of f_{id} is zero. Then, from Eqs. 6.5, 6.6, and 6.42 we obtain

$$I_o \ddot{\phi} = W F_a + Q_{\text{dis}} \qquad 6.46$$

where $F_a = (f_{1a}{}^T \ f_{2a}{}^T \ \cdots \ f_{ma}{}^T)^T$, and Q_{dis} is the resultant force of disturbing forces. Using $\omega = 0$ and the definition of W in Eq. 6.6, we derive, from Eq. 6.12,

$$\ddot{r}_c = W^T \ddot{\phi} \qquad 6.47$$

where $\ddot{r}_c = (\ddot{r}_1^T \ \ddot{r}_2^T \ \cdots \ \ddot{r}_m^T)^T$. By using Eqs. 6.42, 6.43, 6.46, and 6.47, we get

$$\ddot{\phi} = (I_o + W I_r W^T)^{-1} Q_{\text{dis}} \qquad 6.48$$

$$I_r = \begin{pmatrix} (J_1 I_1^{-1} J_1^T)^{-1} & \cdots & O \\ \vdots & \ddots & \vdots \\ O & \cdots & (J_m I_m^{-1} J_m^T)^{-1} \end{pmatrix} \in R^{3m \times 3m}$$

In Eq. 6.48, $I_o + W I_r W^T$ is the combined inertia matrix of the object and the robotic mechanisms. From Eq. 6.48, we see that the object acceleration does not depend on how hard the robotic mechanisms squeeze the object. In other words, hard squeezing is effective for contact stability, but has no advantage for reducing the object acceleration caused by the disturbing forces. Once the acceleration causes the change of the object velocity and position, and if disturbing forces are not exerted thereafter, the object behavior is totally determined by the resultant force control strategy of Eq. 6.10.

6.5.2 Optimal Forces

In Section 6.4, the internal forces were determined so as to minimize the end-effector forces. We mentioned that the minimum force is worth computing to investigate the fundamental characteristics of grasping, but it is not practical for control because it does not have enough contact stability. In this subsection, the optimal forces are defined as the end-effector forces that generate the specified resultant force and have the specified measure of contact stability.

When the measure of contact stability is specified, the set of endpoints of all the possible force vectors that have at least the specified measure of contact stability make the elliptic cone as shown in Fig. 6.7. Since this elliptic cone seems to be too complicated to consider in the computation of force optimization, we propose that it be approximated by the following circular cone:

$$e_{Ni}^T \overline{f}_i \geq \eta_i \|\overline{f}_i\| \qquad 6.49$$

$$\overline{f}_i = f_i - \overline{f}_{oi} \qquad 6.50$$

$$\overline{f}_{oi} = \frac{\sqrt{\mu_i^2 + 1}}{\mu_i} \frac{a}{\sigma_{i3}} e_{Ni} \qquad 6.51$$

The relationship between this circular cone and the friction cone is shown in Fig. 6.8. The surface of this cone is parallel to the friction cone. The direction of the constant vector $\overline{f}_{oi}$ is the same as that of the internal unit normal vector at the contact point, and its length is determined so that the distance between the surfaces of two cones is equal to one half of the longest principal axis of the contact-stability ellipsoid. In Eq. 6.51, σ_{i3} is the smallest singular value of $J_i I_i^{-1} J_i^T \in R^{3 \times 3}$.

We call the circular cone defined by Eqs. 6.49 through 6.51 the *contact-stability cone*, based on which the optimal forces are defined as follows.

196   Chapter 6   Redundancy in Multirobot Coordination

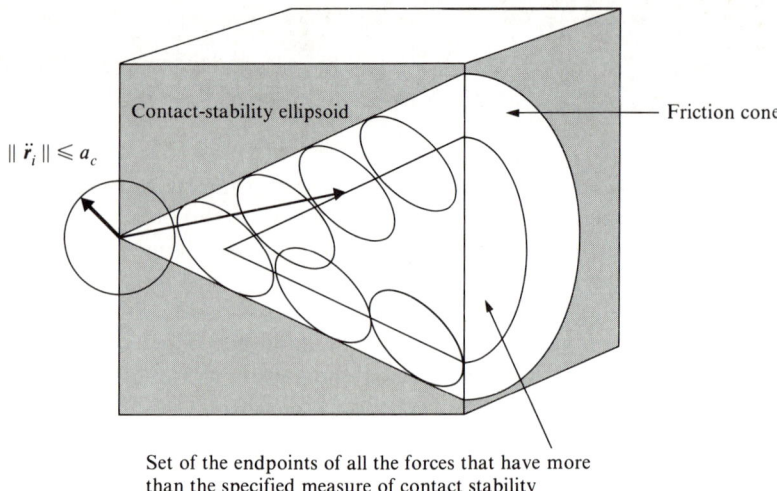

Figure 6.7 Finger force that has more than the specific contact stability.

Definition 6.4 (Optimal Force, Optimal Internal Forces)

The minimum norm force $\boldsymbol{F}$ among the forces that generate the specified resultant force $\boldsymbol{Q}$ under the contact-stability constraints of Eqs. 6.49 through 6.51 is called the *optimal force*, and the internal forces that yield the optimal force are called the *optimal internal forces*.

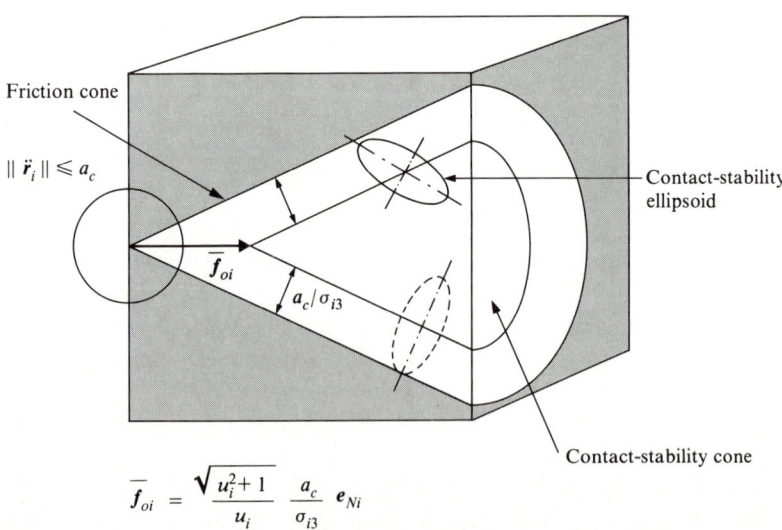

$$\overline{f}_{oi} = \frac{\sqrt{u_i^2+1}}{u_i} \frac{a_c}{\sigma_{i3}} \boldsymbol{e}_{Ni}$$

Figure 6.8 Contact-stability cone.

6.5.3 Computation of Optimal Forces

The problem of obtaining the optimal force and the optimal internal force is mathematically formulated as follows. For the given resultant force Q, the internal forces that satisfy

$$\min \|F\| \qquad 6.52$$

and net force F are solved under the constraints

$$Q = WF \qquad 6.53$$

$$e_{Ni}^T \overline{f}_i \geq \eta_i \|\overline{f}_i\| \qquad 6.54$$

If there exists an F that satisfies Eq. 6.53, then Eq. 6.53 is equivalently replaced by

$$\overline{F} = F_o - \overline{F}_o + Ay \qquad 6.55$$

$$\overline{F} = (\overline{f}_1^T \ \overline{f}_2^T \ \cdots \ \overline{f}_m^T)^T$$

$$\overline{F}_o = (\overline{f}_{o1}^T \ \overline{f}_{o2}^T \ \cdots \ \overline{f}_{om}^T)^T$$

This formulation is similar to the formulation of the minimum forces and the minimizing internal forces discussed in Sections 6.4.3 through 6.4.5. Therefore, we obtain the algorithms for the optimal forces and the optimal internal forces by slightly modifying Algorithms 6.1 and 6.2 and replacing Eqs. 6.27 and 6.38 by

$$\overline{g}_i(y) \leq 0 \qquad \text{for } i = 1, \ldots, 2m \qquad 6.56$$

$$\overline{g}_i(y) = \begin{cases} (f_{oi} - \overline{f}_{oi} + A_i y)^T B_i (f_{oi} - \overline{f}_{oi} + A_i y) \\ \qquad \qquad \text{for } i = 1, \ldots, m \\ -e_{Ni-m}^T (f_{oi-m} - \overline{f}_{oi-m} + A_{i-m} y) \\ \qquad \qquad \text{for } i = m+1, \ldots, 2m \end{cases}$$

$$y^o = -\{A^T (B_o + E_{3m}) A\}^{-1} A^T B_o (F_o - \overline{F}_o) \qquad 6.57$$

Assuming $\overline{f}_i \neq 0$ $(i = 1, \ldots, m)$, we obtain the optimal internal forces and the optimal forces according to Algorithm 6.1 by replacing $g_i(y^o)$, Eq. 6.27, and Eq. 6.38 with $\overline{g}_i(y^o)$, Eq. 6.56, and Eq. 6.57, respectively.

We compute the global solutions in the general case when $\overline{f}_i = 0$ is allowed using Algorithm 6.2 by replacing f_i and f_{oi} with $\overline{f}_i$ and $f_{oi} - \overline{f}_{oi}$, respectively.

198   Chapter 6   Redundancy in Multirobot Coordination

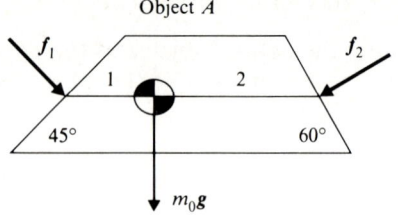

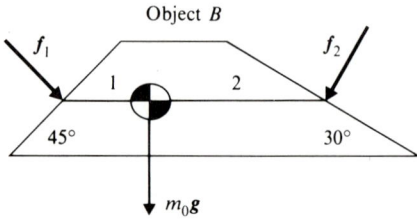

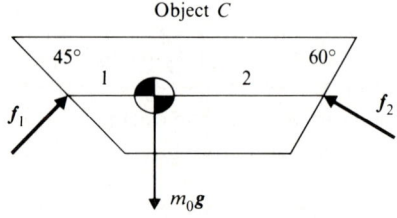

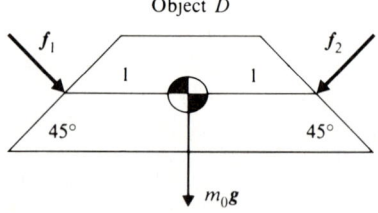

Figure 6.9 Numerical examples of minimum-force computation: Two-dimensional objects.

The computational procedure, complexity of the optimal forces, and the optimal internal forces are almost the same as those of the minimum forces and the minimizing internal forces, except for the computation of σ_{i3} to provide $\overline{\boldsymbol{f}}_{oi}$.

6.6 Examples of Minimum Force Computation

In this section, we give numerical examples of the computation of the minimizing internal forces. Since the computation of the optimal internal forces is almost the same, these examples will also show the complexity of the optimal internal-forces computation. The example tasks are to grasp four different kinds of two-dimensional objects, as shown in Fig. 6.9, applying the resultant forces to cope with the gravity force. The object coordinates are fixed at the mass center of each object, with the x axis in the right direction. The y axis is chosen to make a right-hand coordinate system. The contact points are on the x axis. The maximum static frictional coefficient is $\mu_i = \sqrt{3}$ ($i = 1, 2$), which means that the half-angle of the friction cone is 60°.

The minimizing internal forces are computed according to the procedure in Section 6.4.5. Table 6.1 shows the result, where α_i is the angle between $\boldsymbol{e}_{Ni}$ and $\boldsymbol{f}_i^o$. Table 6.1 indicates the following:

1. The minimizing internal forces for object A provide a grasp such that $\boldsymbol{f}_1$ is on the boundary of the friction cone and $\boldsymbol{f}_2$ is inside.

2. Since the minimizing internal forces do not exist for object B, it cannot be grasped with these finger placements.

Table 6.1. Computed minimum forces and minimizing internal forces.

	Object A	Object B	Object C	Object D
λ_1^0	$8(3+2\sqrt{3})/3$	no solution	0	$\lambda_1^0 + \lambda_2^0$
λ_2^0	0	no solution	0	$= 8(3+2\sqrt{3})/3$
$\boldsymbol{f}_1^o/m_0 g$	$4(1+\sqrt{3}/2)/3$	no solution	0	$1+\sqrt{3}/2$
	$2/3$		$2/3$	$1/2$
$\boldsymbol{f}_2^o/m_0 g$	$-4(1+\sqrt{3}/2)/3$	no solution	0	$-(1+\sqrt{3}/2)$
	$1/3$		$1/3$	$1/2$
$A_1 y^o/m_0 g$	$4(1+\sqrt{3}/2)/3$	no solution	0	$1+\sqrt{3}/2$
	0		0	0
$A_2 y^o/m_0 g$	$-4(1+\sqrt{3}/2)/3$	no solution	0	$-(1+\sqrt{3}/2)$
	0		0	0
α_1	60°	no solution	45°	60°
α_2	37.6°	no solution	60°	60°

3. Object C can be grasped with zero internal forces.
4. The minimizing internal forces for object D imply a grasp such that both $\boldsymbol{f}_1$ and $\boldsymbol{f}_2$ are on the boundaries of the friction cones.

For two-dimensional problems with two robotic mechanisms, $g_i(\boldsymbol{y}^o) = 0$ $(1 \leq i \leq m)$ becomes a quadratic equation. For three-dimensional problems, it becomes a $2b$th-order (b is the dimension of the null space of $\boldsymbol{W}$) algebraic equation. This means we have a sixth-order equation even for three robotic mechanisms, which would require numerical methods to solve.

Summary

We discussed the dynamics and stability aspects in manipulating an object coordinatively with multiple robot manipulators or a multifingered robot hand. The main results obtained are summarized as follows:

1. Coordinative manipulation is divided into two phases. One is the determination of the resultant force to control the object's trajectory and to specify the stiffness for the external forces. The other is the determination of the internal forces to cope with uncertainty of friction and unexpected distances.

2. We proposed a scheme to determine the resultant force by considering the dynamics of the object.

3. If a robotic mechanism is force controlled in the direction in which the end-effector works, it must also be position controlled such that it can follow the object motion that the force control suggests.

4. The robotic mechanisms can generate an arbitrary acceleration of the object if and only if $\boldsymbol{Q}^j$ and $-\boldsymbol{Q}^j$ $(j = 1, \ldots, 6)$ can be generated under the static frictional constraints, where $\boldsymbol{Q}^j$ $(j = 1, \ldots, 6)$ are linearly independent resultant forces.

5. We defined the minimizing internal forces as the internal forces that yield the minimum forces for the specified resultant force under the static frictional constraints, and established a computational procedure to obtain them. The minimizing internal forces are necessarily obtained by solving, at most, $\Sigma_{j=1}^{m} \binom{m}{j} \Sigma_{i=0}^{j} \binom{j}{i}$ (where m is the number of robotic mechanisms) sets of algebraic equations if they exist. The concept of minimum forces is important because it answers the fundamental question of whether the object can be grasped with the specified contact-point placement.

6. Contact stability was defined as the ability to maintain contact with an object without slipping for a class of unexpected disturbing forces, and the measure of contact stability was proposed to evaluate the contact stability of the end-effector forces.

7. We defined the optimal internal forces as the internal forces that yield the minimum forces that give the specified resultant force and have the specified measure of contact stability. We can compute the optimal internal forces by using algorithms similar to those for the minimizing internal forces.

8. Examples of computing the minimizing internal forces were given to show the effectiveness of the algorithms.

References

Asada, H. 1983. A geometrical representation of manipulator dynamics and its application to arm design. *Trans. ASME J. Dyn. Sys., Meas. Contr.* 105 (3):131–136.

Brady, M., Hollerbach, J. M., Johnson, T. L., Lozano-Perez, T., and Mason, M. T., 1982. *Robot motion.* Cambridge, MA: MIT Press.

Cutkosky, M. R. 1985. *Robotic grasping and fine manipulation.* Boston: Kluwer Academic.

Cutkosky, M. R., and Wright, P. K. 1986. Friction, stability and the design of robotic fingers. *Int. J. Robotics Res.* 5 (4): 20–37.

Fearing, R. S. 1986. Simplified grasping and manipulation with dextrous robot hands. *IEEE J. Robotics Automat.* 2 (4): 188–195.

Golub, G. H., and Van Loan, C. F. 1983. *Matrix computations.* Baltimore: Johns Hopkins University Press.

Hanafusa, H., and Asada, H. 1977 (Tokyo). Stable prehension by a robot hand with elastic fingers. *Proc. 7th Int. Symp. Industrial Robots*, pp. 311–368.

Hanafusa, H., Kobayashi, H., and Terasaki, N. 1983 (Tokyo). Fine control of the object with articulated multi-finger robot hands. *Proc. 1983 Int. Conf. Advanced Robotics*, pp. 245–251.

Hanafusa, H., Yoshikawa, T., Nakamura, Y., and Nagai, K. 1985 (Tokyo). Structural analysis and robust prehension of robotic hand-arm system. *Proc. 1985 Int. Conf. Advanced Robotics*, pp. 311–318.

Jameson, J. W., and Leifer, L. J. 1986 (San Francisco). Quasi-static analysis: A method for predicting grasp stability. *Proc. 1986 IEEE Int. Conf. Robotics and Automation*, pp. 876–883.

Kerr, J., and Roth, B. 1986. Analysis of multifingered hands. *Int. J. Robotics Res.* 4 (4): 3–17.

Khatib, O. 1985 (Tokyo). The operational space formulation in robot manipulator control. *Proc. 15th Int. Symp. Industrial Robots*, pp. 165–172.

Kobayashi, H. 1985. Control and geometrical consideration for an articulated robot hand. *Int. J. Robotics Res.* 4 (1): 3–12.

Kuhn, H. W., and Tucker, A. W. 1951. Nonlinear Programming. *Proc. Second Berkeley Symposium on Mathematical Statistics and Probability,* ed. J. Neyman, Berkeley: University of California Press, pp. 481–492.

Kumar, V., and Waldron, K. 1987 (Raleigh). Sub-optimal algorithms for force distribution in multifingered grippers. *Proc. 1987 IEEE Int. Conf. Robotics and Automation,* pp. 252–257.

Kurono, S. 1975. Coordinated control of a pair of artificial arms. In *Biomechanism 3,* pp.182–193.Tokyo: Tokyo University Press.

Luh, J. Y. S., Walker, M. W., and Paul, R. P. C. 1980. Resolved acceleration control of mechanical manipulators. *IEEE Trans. Automat. Contr.* 25 (3): 468–474.

Mason, M. T. 1981. Compliance and force control for computer controlled manipulators. *IEEE Trans. Syst., Man Cybernet.* SMC-11: 418–432.

Mine, H. 1966. *Operations research Vol. 1,* pp.122–130. Tokyo: Asakura.

Nakamura, Y. 1987 (Los Angeles). Force applicability of robotic mechanisms. *Proc. 26th IEEE Conf. Decision and Control,* pp. 570–575.

Nakamura, Y. 1988 (Minneapolis). Contact stability measure and optimal finger force control of multi-fingered robot hands. *Proc. 1988 USA–Japan Symposium on Flexible Automation,* pp. 523–528.

Nakamura, Y., Nagai, K., and Yoshikawa, T. 1986. Mechanics of coordinative manipulation by multiple robotic mechanisms. *J. Robotics Society of Japan* 4 (5): 489–498 *(in Japanese).*

Nakamura, Y., Nagai, K., and Yoshikawa, T. 1987 (Raleigh). Mechanics of coordinative manipulation by multiple robotic mechanisms. *Proc. 1987 IEEE Int. Conf. Robotics and Automation,* pp. 991–998.

Nakamura, Y., Nagai, K., and Yoshikawa, T. 1989. Dynamics and stability in coordination of multiple robotic mechanisms. *Int. J. Robotics Res.* 8 (2): 44–61.

Nakano, E., Ozaki, S., Ishida, T., and Kato, I. 1974 (Tokyo). Cooperational control of the anthropomorphous manipulator 'MELARM.' *Proc. 4th Int. Symp. Industrial Robots,* pp. 251–260.

Nguyen, V. 1986a. The synthesis of stable force-closure grasps. Technical report 905, Massachusetts Institute of Technology, Artificial Intelligence Laboratory.

Nguyen, V. 1986b (San Francisco). Constructing force-closure grasp. *Proc. 1986 IEEE Int. Conf. Robotics and Automation,* pp. 1368–1373.

Nguyen, V. 1986c (San Francisco). The synthesis of stable grasps in the plane. *Proc. 1986 IEEE Int. Conf. Robotics and Automation*, pp. 884–889.

Ohwovoriole, M. S. 1980. An extension of screw theory and its application to the automation of industrial assemblies. Ph.D. thesis, Stanford University, Dept. of Mechanical Engineering.

Raibert, M. H., and Craig, J. J. 1981. Hybrid position/force control of manipulators. *Trans. ASME J. Dyn. Sys., Meas. Contr.* 102 (2): 126–133.

Rao, C. R., and Mitra, S. K., 1971. *Generalized inverse of matrices and its applications.* New York: Wiley.

Salisbury, J. K. 1982. Kinematic and force analysis of articulated hands. Ph.D. thesis, Stanford University, Dept. of Mechanical Engineering.

Salisbury, J. K., and Craig, J. J. 1982. Articulated hands: Force control and kinematic issues. *Int. J. Robotics Res.* 1 (1): 4–17.

Takase, K. 1976. Generalized decomposition and control of a motion of a manipulator. *Trans. Society of Instrument and Control Engineers* 12 (3): 62–68 *(in Japanese)*.

Takase, K. 1985. Representation of constraint motion and dynamic control of manipulators under constraints. *Trans. Society of Instrument and Control Engineers* 21 (5): 508–513 *(in Japanese)*.

Trinkle, J. C., Abel, J. M., and Paul, R. P. 1985. An investigation of frictionless, enveloping grasping in the plane. Technical report MS-CIS-86-57 GRASP LAB 70, Dept. Computer and Information Science, University of Pennsylvania.

Uchiyama, M., Hakomori, K., and Shimizu, K. 1983 (Tokyo). A servo synthesis method for multi-arm cooperation. *Proc. 1st Annual Conf. Robotics Society of Japan*, pp. 101–102 *(in Japanese)*.

Yoshikawa, T. 1986 (San Francisco). Dynamic hybrid position/force control of robot manipulators: Description of hand constraints and calculation of joint driving force. *Proc. 1986 IEEE Int. Conf. Robotics and Automation*, pp. 1393–1398.

CHAPTER 7

Actuation Redundancy of Closed-Link Mechanisms

7.1 Introduction

Closed-link mechanisms have been widely used in industrial robots such as ASEA Robot, Cincinnati Milacron T^3, Bendix MA 510, FANUC Robot, and YASUKAWA Motoman. High mechanical stiffness of the mechanisms plays a crucial roll in attaining high trajectory accuracy in such applications as arc welding, cutting, and sealing. Recent developments in direct-drive robots have generated new interest in the applications of closed-link mechanisms, since the latter allow us to locate motors closer to the base. Closed-link mechanisms also have the potential for finding an efficient parallel computational scheme of inverse dynamics. In addition, they are suitable for making compact multiple-DOF mechanisms, such as parallel actuated wrists and dexterous fingers. Although closed-link mechanisms have various practical advantages, the theory of their dynamics has not been studied as extensively as has that of the open-link mechanisms.

This chapter is adapted from, by permission, Nakamura, Y., and Ghodoussi, M. 1988 (Philadelphia). "A computational scheme of closed link robot dynamics derived by the d'Alembert's principle," *Proc. 1988 IEEE International Conference on Robotics and Automation*, pp. 1354–1360; Nakamura, Y., and Ghodoussi, M. 1988 (Salo, Italy). "Closed link

One common method is to obtain the Lagrangian function and to compute the Lagrange equation analytically. The process for obtaining the Lagrangian function is driven by the peculiarity of the mechanism. Another defect of this method is the difficulty in finding recursive computational schemes, like those proposed for open-link mechanisms. The systematic dynamics computational scheme of closed-link mechanisms was studied by Smith (1973). He formulated this problem as the problem of computing the dynamics of an open-link tree-structure mechanism that is obtained by cutting a closed-link mechanism virtually, and is subjected to unknown reaction forces at the cut points. Paul (1975) had a similar approach to this problem involving the Lagrange multipliers. Although these were discussed for planer mechanisms, Featherstone (1987) extended them to spatial ones. For spatial mechanisms, Luh and Zheng (1985) proposed replacing the problem with computing the dynamics of the open-link tree-structure mechanisms subjected to unknown additional joint torques, and used the Lagrange multipliers to obtain the unknown torques. Kleinfinger and Khalil (1986) also developed a comparable computational scheme. Although these methods are general and systematic, there are many possible choices for the representation of mechanical constraints that become complicated when extended to multiloop closed-link mechanisms. On the other hand, Asada and Youcef-Toumi (1987) clarified the condition of mass distribution of links that makes the inertia matrix of closed-link mechanisms decoupled and configuration-invariant. The force distribution between multiple legs or arms that form a closed kinematic chain was discussed by Orin and Oh (1981), where the selection of optimization criteria and the numerical method of computation were considered.

In this chapter, we first develop a general and systematic computational scheme of the inverse dynamics of the closed-link mechanisms that works even for multiloop closed-link mechanisms without any modifications. This scheme is derived using the d'Alembert's principle, and this scheme is obtained without computing the Lagrange multipliers. Since the representation of constraints requires only the Jacobian matrix of the unactuated joint angles in terms of the actuated ones, the constraint is described uniquely even for complicated multiloop closed-link mechanisms, such as the Stewart Platform

---

manipulators with actuational redundancy," *Prep. NATO Advanced Research Workshop on Robots with Redundancy: Design, Sensing and Control* (Proceedings to be published from Springer-Varlag); Nakamura, Y. 1988 (Albuquerque). "Dynamics of closed link robots with actuational redundancy," *Robotics and Manufacturing: Recent Trends in Research, Education, and Applications*; eds. M. Jamshidi, J. Y. S. Luh, H. Seraji, and G. P. Starr (*Proc. Second International Symposium on Robotics and Manufacturing Research, Education and Application*) ASME Press, pp. 309–318; Nakamura, Y., and Ghodoussi, M. 1989. "Dynamics computation of closed link robot mechanisms with non-redundant and redundant actuators," *IEEE Journal of Robotics and Automation*, Vol. 5, No. 3, pp. 294–302 and; Nakamura, Y. 1989. "Dynamics computation of closed-link robots and optimization of actuation redundancy," *Trans. Society of Instrument and Control Engineers* 25(5): 600–607 (in Japanese).

with the conventional methods (Luh and Zheng 1985; Kleinfinger and Khalil 1986), since the computation concerning the Lagrange multipliers is not necessary.

Second, we discuss the inverse dynamics of closed-link mechanisms that contain redundant actuators, and examine the optimization of their redundancy. Closed-link mechanisms having redundant actuators are underdeterminate in computing inverse dynamics. For a redundant actuation system that contains N_r redundant actuators, the unactuated joint angles are represented by $N_r + 1$ independent ways as functions of actuated joints. Using their Jacobian matrices, we parameterize the actuation redundancy of closed-link mechanism by an N_r-dimensional arbitrary vector in a linear equation. For the case with one redundant actuator, the computational algorithm to minimize the joint torque taking account of the limits of actuator torques is presented. Numerical examples will be given to show the computational efficiency and practicality of closed-link manipulators with actuation redundancy.

7.2 Inverse Dynamics of Closed-Link Robots

7.2.1 Equivalent Tree-Structure Mechanism

The basic principle to compute the inverse dynamics of the closed-link mechanisms consists of the following three steps:

1. A closed-link mechanism is transformed into an open-loop tree-structure mechanism by virtually cutting loops at an unactuated joint of each loop. It is assumed that the unactuated joints of the tree-structure mechanism have virtual actuators except for those we cut.

2. The joint torques of the tree-structure mechanism with the virtual actuators are computed for exactly the same motion as is required of the closed-link mechanism. We assume that there is no force or torque interaction though the cut joints.

3. With the constraints taken into account, the joint torques of the actuated joints of the original closed-link mechanism are computed from the obtained joint torques of the tree-structure mechanism.

This principle is identical to the principle used in Luh and Zheng (1985) and Kleinfinger and Khalil (1986). There is an advantage in the fact that we can perform the second step by using the recursive computational schemes established for open-link mechanisms (Luh, Walker, and Paul 1980), which results in a highly efficient computation. Recursive computation for tree-structure mechanisms is inherently suitable for parallel computation. The previous works (Luh and Zheng 1985; Kleinfinger and Khalil 1986) used the Lagrange multipliers to carry out the third step. Although they are fairly general and systematic computational schemes, the choice of the representation

Lagrange multipliers to carry out the third step. Although they are fairly general and systematic computational schemes, the choice of the representation of mechanical constraints is dependent on the individual and will become complicated when applied to multiloop closed-link mechanisms. The efficient and systematic method of computing the third step will be discussed in Section 7.2.2.

Let the number of actuated joints and the number of unactuated joints included in the total robot mechanism be represented by N_{ta} and N_{tu}, respectively. By cutting each loop at one of the unactuated joints, we transform a closed-link mechanism into an open-link tree structure-mechanism, as shown in Fig. 7.1. Accordingly, if the number of closed loops is represented by N_{loop}, the number of unactuated joints contained in the tree structure mechanism becomes $N_u = N_{tu} - N_{\text{loop}}$.

The mobility of single closed-loop mechanisms was studied by Duffy (1980), who showed that the maximum number of unactuated joints in a single closed loop is determined by the structure of the closed loop; it is three for planer case and six for general spatial case. If the number of unactuated joints is greater than the maximum number, the closed loop is not *controllable*.[7.1] If the number of unactuated joints is less than the maximum number, the closed loop has redundant actuators and the joint torques become underdeterminate. We call this type of closed-link mechanism a *redundant actuation system*. A redundant actuation system has the potential to optimize joint torques for some purpose. In Section 7.3, we discuss the formulation of a redundant actuation system and its optimization method.

7.2.2 *Representation of Constraints Using the Jacobian Matrix*

Suppose that the joint angles and the joint torques of the open-loop tree-structure mechanism are represented, respectively, by

$$\boldsymbol{\theta} = \begin{pmatrix} \boldsymbol{\theta}_1 \\ \boldsymbol{\theta}_2 \end{pmatrix} \in R^{N_{ta}+N_u} \qquad 7.1$$

$$\boldsymbol{\tau} = \begin{pmatrix} \boldsymbol{\tau}_1 \\ \boldsymbol{\tau}_2 \end{pmatrix} \in R^{N_{ta}+N_u} \qquad 7.2$$

where $\boldsymbol{\theta}_1$ and $\boldsymbol{\tau}_1 \in R^{N_{ta}}$ correspond to the actuated joints, and $\boldsymbol{\theta}_2$ and $\boldsymbol{\tau}_2 \in R^{N_u}$ correspond to the unactuated ones. We assume that the closed-link mechanism is *controllable*. When the tree-structure mechanism makes the same motion as required for the original closed-link mechanism, $\boldsymbol{\theta}_2$ is uniquely determined as a function of $\boldsymbol{\theta}_1$ as follows:

$$\boldsymbol{\theta}_2 = \boldsymbol{\theta}_2(\boldsymbol{\theta}_1) \qquad 7.3$$

[7.1] *Controllable* means that the mechanism can find the finite actuator torques to maintain static equilibrium against any finite external force exerted on any point of the mechanism.

7.2 Inverse Dynamics of Closed-Link Robots

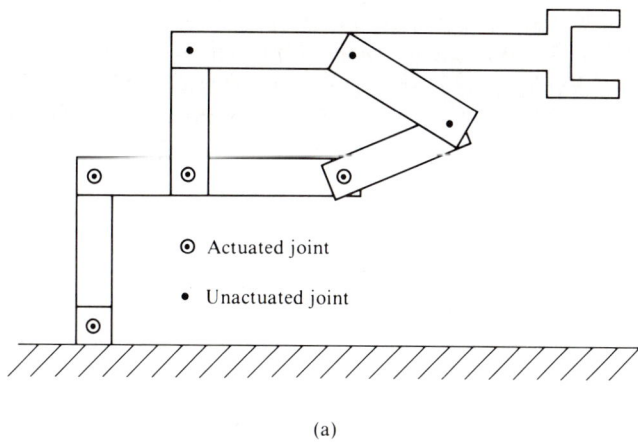

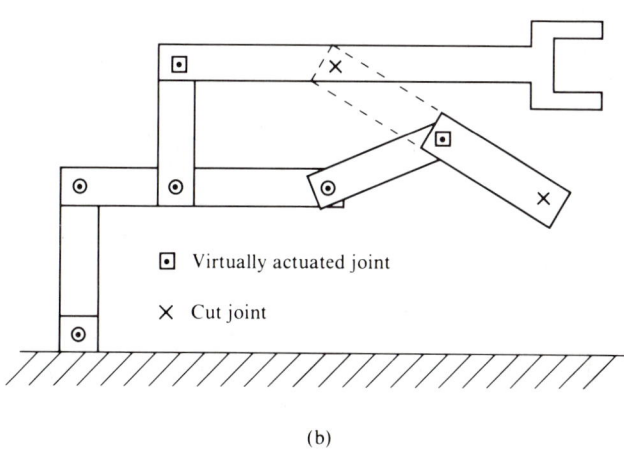

Figure 7.1 Equivalent transformation of (a) closed-link mechanisms to (b) open-link mechanisms.

In this case, the infinitesimal motion of the tree-structure mechanism is represented by

$$\delta\boldsymbol{\theta} = \boldsymbol{W}\delta\boldsymbol{\theta}_1 \qquad 7.4$$

$$\boldsymbol{W} = \begin{pmatrix} \boldsymbol{E} \\ \partial\boldsymbol{\theta}_2/\partial\boldsymbol{\theta}_1 \end{pmatrix} \in R^{(Nta+Nu)\times Nta}$$

where $\boldsymbol{E} \in R^{Nta \times Nta}$ is an identity matrix and $\partial\boldsymbol{\theta}_2/\partial\boldsymbol{\theta}_1$ is the Jacobian matrix of the unactuated joints with respect to the actuated ones. $\boldsymbol{W}$ is the Jacobian matrix of the total joints with respect to the actuated joints. Although we will use $\boldsymbol{W}$ to compute the joint torques of the closed-link mechanism, it is not necessary to obtain W analytically. For simple closed-link mechanisms W is obtained analytically, whereas for general complicated closed-link mechanisms we have to compute it numerically.

The joint torque τ of the tree structure mechanism is derived from the second step of the basic principles stated in Section 7.2.1. Now, let the torque of the actuated joints of the original closed-link mechanism be represented by $\tau_1^c \in R^{Nta}$. Then, τ_1^c can be computed from τ according to the following proposition. This proposition was derived by using the d'Alembert's principle and the principle of virtual work (Nakamura and Ghodoussi 1988). The proof is given in the next section.

Proposition 7.1

The joint torque τ_1^c of a closed-link mechanism for a given motion is computed by Eq. 7.5 from the joint torque τ of the corresponding open-link tree-structure mechanism required to make the same motion.

$$\tau_1^c = W^T \tau \qquad 7.5$$

Although Eq. 7.5 is physically equivalent to the equations in Luh and Zheng's paper (1985) (Eqs. 24 and 25), Eq. 7.5 is a simpler result with clearer physical meaning and has the following advantages:

1. Since W is uniquely represented, there is only one possible way to formulate or program the constraint.

2. Since the physical meaning of W is very clear, visual inspection can be used in computing W. We can compute W for many simple closed-link mechanisms, for some of which W becomes constant, as in the Example in Section 7.4.

3. Since the computation of the Lagrange multipliers is not necessary, Eq.7.5 is computationally more efficient.

We can compute the forward dynamics, which compute joint acceleration in response to given joint torque, by repeatedly using the method of inverse dynamics. The forward-dynamics computation for open-link mechanisms was discussed in detail by Walker and Orin (1982) and Featherstone (1987). The forward dynamics of closed-link mechanisms were also discussed by Orin and McGhee (1981) and Oh and Orin (1986). However, the closed-link mechanisms discussed are formed by the multiple legs or arms and do not include the general closed-link mechanism having unactuated joints in the middle of a chain. Using the inverse dynamics computation given in this section, we can solve the forward dynamics of general closed-link mechanisms by straightforwardly applying the *unit vector approach* developed for open-link mechanisms (Orin and McGhee 1981; Walker and Orin 1982; Featherstone 1987).

7.2.3 Proof of Proposition 7.1

Suppose that the Lagrangian functions of the original closed-link mechanism and the open-link tree-structure mechanism is described by L_c and L_o, respectively. That is,

$$L_c = L_c(\boldsymbol{\theta}_1, \dot{\boldsymbol{\theta}}_1) \qquad 7.6$$

$$L_o = L_o(\boldsymbol{\theta}_1, \boldsymbol{\theta}_2, \dot{\boldsymbol{\theta}}_1, \dot{\boldsymbol{\theta}}_2) \qquad 7.7$$

The dynamics of the tree-structure mechanism is represented by

$$\frac{d}{dt}\left(\frac{\partial L_o}{\partial \dot{\boldsymbol{\theta}}}\right) - \frac{\partial L_o}{\partial \boldsymbol{\theta}} = \boldsymbol{\tau}^T \qquad 7.8$$

Note that the partial derivative of a scalar function with respect to a column vector becomes a row vector. According to the d'Alembert's principle, the following equation is satisfied for virtual displacement $\delta\boldsymbol{\theta}$:

$$\left\{\frac{d}{dt}\left(\frac{\partial L_o}{\partial \dot{\boldsymbol{\theta}}}\right) - \frac{\partial L_o}{\partial \boldsymbol{\theta}}\right\}\delta\boldsymbol{\theta} - \boldsymbol{\tau}^T \delta\boldsymbol{\theta} = 0 \qquad 7.9$$

Since a virtual displacement $\delta\boldsymbol{\theta} \in R^{Nta+Nu}$ is defined as an infinitesimal displacement that is structually allowed for the mechanism, $\delta\boldsymbol{\theta}$ must be represented by Eq. 7.4 with an arbitrary $\delta\boldsymbol{\theta}_1$. Accordingly,

$$\left\{\frac{d}{dt}\left(\frac{\partial L_o}{\partial \dot{\boldsymbol{\theta}}}\right) - \frac{\partial L_o}{\partial \boldsymbol{\theta}}\right\}\boldsymbol{W}\delta\boldsymbol{\theta}_1 - \boldsymbol{\tau}^T \boldsymbol{W}\delta\boldsymbol{\theta}_1 = 0 \qquad 7.10$$

Now, the first term of Eq. 7.10 is computed as follows:

$$\frac{d}{dt}\left(\frac{\partial L_o}{\partial \dot{\boldsymbol{\theta}}}\right)\boldsymbol{W} = \frac{d}{dt}\left(\frac{\partial L_o}{\partial \dot{\boldsymbol{\theta}}_1}\right) + \frac{d}{dt}\left(\frac{\partial L_o}{\partial \dot{\boldsymbol{\theta}}_2}\right)\frac{\partial \dot{\boldsymbol{\theta}}_2}{\partial \dot{\boldsymbol{\theta}}_1} \qquad 7.11$$

$$\frac{\partial L_o}{\partial \boldsymbol{\theta}}\boldsymbol{W} = \frac{\partial L_o}{\partial \boldsymbol{\theta}_1} + \frac{\partial L_o}{\partial \boldsymbol{\theta}_2}\frac{\partial \boldsymbol{\theta}_2}{\partial \boldsymbol{\theta}_1} \qquad 7.12$$

Since L_c and L_o are physically equivalent though they are different in functional dependency, we have the following relationships:

$$\frac{d}{dt}\left(\frac{\partial L_c}{\partial \dot{\boldsymbol{\theta}}_1}\right) = \frac{d}{dt}\left(\frac{\partial L_o}{\partial \dot{\boldsymbol{\theta}}_1} + \frac{\partial L_o}{\partial \boldsymbol{\theta}_1}\frac{\partial \boldsymbol{\theta}_1}{\partial \dot{\boldsymbol{\theta}}_1} + \frac{\partial L_o}{\partial \boldsymbol{\theta}_2}\frac{\partial \boldsymbol{\theta}_2}{\partial \dot{\boldsymbol{\theta}}_1} + \frac{\partial L_o}{\partial \dot{\boldsymbol{\theta}}_2}\frac{\partial \dot{\boldsymbol{\theta}}_2}{\partial \dot{\boldsymbol{\theta}}_1}\right)$$

$$= \frac{d}{dt}\left(\frac{\partial L_o}{\partial \dot{\boldsymbol{\theta}}_1}\right) + \frac{d}{dt}\left(\frac{\partial L_o}{\partial \dot{\boldsymbol{\theta}}_2}\frac{\partial \dot{\boldsymbol{\theta}}_2}{\partial \dot{\boldsymbol{\theta}}_1}\right)$$

Chapter 7 Actuation Redundancy of Closed-Link Mechanisms

$$= \frac{d}{dt}(\frac{\partial L_o}{\partial \dot{\theta}_1}) + \frac{d}{dt}(\frac{\partial L_o}{\partial \dot{\theta}_2})\frac{\partial \theta_2}{\partial \theta_1} + \frac{\partial L_o}{\partial \dot{\theta}_2}\frac{d}{dt}(\frac{\partial \theta_2}{\partial \theta_1}) \qquad 7.13$$

$$\frac{\partial L_c}{\partial \theta_1} = \frac{\partial L_o}{\partial \theta_1} + \frac{\partial L_o}{\partial \dot{\theta}_1}\frac{\partial \dot{\theta}_1}{\partial \theta_1} + \frac{\partial L_o}{\partial \theta_2}\frac{\partial \theta_2}{\partial \theta_1} + \frac{\partial L_o}{\partial \dot{\theta}_2}\frac{\partial \dot{\theta}_2}{\partial \theta_1}$$

$$= \frac{\partial L_o}{\partial \theta_1} + \frac{\partial L_o}{\partial \theta_2}\frac{\partial \theta_2}{\partial \theta_1} + \frac{\partial L_o}{\partial \dot{\theta}_2}\frac{\partial \dot{\theta}_2}{\partial \theta_1} \qquad 7.14$$

The following relationships were used in calculating Eqs. 7.13 and 7.14. From the independence of θ_1 and $\dot{\theta}_1$, $\partial \theta_1/\partial \dot{\theta}_1$ and $\partial \dot{\theta}_1/\partial \theta_1$ are equal to zero. $\partial \theta_2/\partial \dot{\theta}_1$ is also equal to zero from Eq. 7.3. Moreover,

$$\frac{\partial \dot{\theta}_2}{\partial \dot{\theta}_1} = \frac{\partial}{\partial \dot{\theta}_1}(\frac{\partial \theta_2}{\partial \theta_1}\dot{\theta}_1) = \frac{\partial \theta_2}{\partial \theta_1} \qquad 7.15$$

Substituting Eqs. 7.13 and 7.14 into Eqs. 7.11 and 7.12, the following relations are obtained.

$$\frac{d}{dt}(\frac{\partial L_o}{\partial \dot{\theta}})\boldsymbol{W} = \frac{d}{dt}(\frac{\partial L_c}{\partial \dot{\theta}_1}) - \frac{\partial L_o}{\partial \dot{\theta}_2}\frac{d}{dt}(\frac{\partial \theta_2}{\partial \theta_1}) \qquad 7.16$$

$$\frac{\partial L_o}{\partial \theta}\boldsymbol{W} = \frac{\partial L_c}{\partial \theta_1} - \frac{\partial L_o}{\partial \dot{\theta}_2}\frac{\partial \dot{\theta}_2}{\partial \theta_1} \qquad 7.17$$

The second terms of Eqs. 7.16 and 7.17 are identical because

$$\{\frac{d}{dt}(\frac{\partial \theta_2}{\partial \theta_1})\}_{ij} = \sum_{k=1}^{Nta}(\frac{\partial^2 \theta_{2i}}{\partial \theta_{1j}\partial \theta_{1k}}\dot{\theta}_{1k}) = \{\frac{\partial \dot{\theta}_2}{\partial \theta_1}\}_{ij} \qquad 7.18$$

where $\{\boldsymbol{M}\}_{ij}$ means the (i, j) element of matrix $\boldsymbol{M}$ and θ_{1i} and θ_{2i} are the ith entries of $\boldsymbol{\theta}_1$ and $\boldsymbol{\theta}_2$, respectively.

Consequently, taking account of Eqs. 7.16, 7.17, and 7.18, Eq. 7.10 becomes:

$$\{\frac{d}{dt}(\frac{\partial L_c}{\partial \dot{\theta}_1}) - \frac{\partial L_c}{\partial \theta_1}\}\delta\theta_1 - \boldsymbol{\tau}^T\boldsymbol{W}\delta\theta_1 = 0 \qquad 7.19$$

Since Eq. 7.19 holds for arbitrary $\delta\theta_1$, the following relationship is obtained:

$$\frac{d}{dt}(\frac{\partial L_c}{\partial \dot{\theta}_1}) - \frac{\partial L_c}{\partial \theta_1} = \boldsymbol{\tau}^T\boldsymbol{W} \qquad 7.20$$

The dynamics of the original closed-link mechanism are given by

$$\frac{d}{dt}(\frac{\partial L_c}{\partial \dot{\theta}_1}) - \frac{\partial L_c}{\partial \theta_1} = \boldsymbol{\tau}_1^{cT} \qquad 7.21$$

Comparing Eq. 7.21 with Eq. 7.20 yields Eq. 7.5, which completes the proof of the proposition.†⁷·²

7.3 Redundant Actuation Systems

7.3.1 Parametric Representation of Redundancy

If a closed-link mechanism has less than the maximum number of unactuated joints, the closed-link mechanism becomes overconstrained and the joint torques are no longer determined uniquely. This system is called a *redundant actuation system*, because there are more actuators than necessary. Although this kind of system is not too attractive from the conventional design viewpoint because it has more complexity in mechanism and control, it has interesting advantages, such as

1. Redundancy in actuation can be used to increase the reliability of the system. That is, even if some of the actuators break down, the system can operate normally as long as the number of operating actuators is enough to keep the system *controllable*.

2. The joint torque is not determined uniquely. This characteristic can be utilized for optimization of some criterion. For example, the joint torque required for a given motion can be minimized. In other words, it increases the payload of a closed-link mechanism by actively controlling redundant actuators.

Closed kinematic chains formed by multiple legs, arms, or fingers also make redundant actuation systems (Orin and Oh 1981; Oh and Orin 1986; Kerr and Roth 1986; Nakamura, Nagai, and Yoshikawa 1987; Tarn, Bejczy, and Li 1988). Their redundancy is exhibited as one in choosing tip forces because it is guaranteed that any force can be generated by each limb unless that force breaks the joint torque limits. The redundant actuation systems we discuss in this section generally include unactuated joints in the middle of a chain. The redundancy of this class of closed-link mechanisms cannot be dealt with like that of multilimb systems, and must be parametrized first.

In our formulation, so that the loop can be cut, at least one unactuated joint per closed loop is essential. Moreover, for convenience of formulation, we assume an additional unactuated joint. That is, at least two unactuated joints

---

†⁷·² Similar proofs can be found in the papers by Volterra (1898) and Maggi (1901), which discuss the dynamics of systems of material points that have lower dimension of the generalized force than the dimension of coordinates that describe the systems. This similarity was pointed out by Professor Jorge Angeles, McGill University.

in a closed loop are assumed in the following discussion. The case where there is only one unactuated joint involves only a slight change in the formulation.

Let N_{oa}, N_r, and N_u indicate the number of the original actuated joints, the additional actuated joints with redundant actuators, and the unactuated joints. The unactuated joints are totally represented by the original actuated joints. The unactuated joints can be also represented by $N_{oa} - 1$ of the original actuated joints plus one of the additional actuated joints. It is also possible to describe the unactuated joints in terms of more than one of the additional actuated joints. However, since we only use the infinitesimal relationship, this description can be represented as a linear combination of cases where only one of the additional actuated joints is involved. Therefore, having N_r redundant actuators, we can represent the angles of the unactuated joints in $N_r + 1$ independent ways.[7.3] Accordingly, Eq. 7.3 is replaced by

$$\boldsymbol{\theta}_2 = \boldsymbol{\theta}_2^i(\boldsymbol{\theta}_1) \qquad (i = 1, \ldots, N_r + 1) \qquad 7.22$$

Using Eq. 7.22, we produce function $\boldsymbol{\theta}_2^0$ as follows:

$$\boldsymbol{\theta}_2 = \boldsymbol{\theta}_2^0(\boldsymbol{\theta}_1) = \sum_{i=1}^{Nr+1} \alpha_i \boldsymbol{\theta}_2^i(\boldsymbol{\theta}_1) \qquad 7.23$$

where the α_is are scalars and satisfy

$$\sum_{i=1}^{Nr+1} \alpha_i = 1 \qquad 7.24$$

Therefore, the joint torque of the closed-link mechanism is obtained as follows:

$$\boldsymbol{\tau}_1^c = \boldsymbol{W}^T \boldsymbol{\tau} \qquad 7.25$$

$$\boldsymbol{W} = \begin{pmatrix} \boldsymbol{E} \\ \partial \boldsymbol{\theta}_2^0 / \partial \boldsymbol{\theta}_1 \end{pmatrix} \in R^{(Nta+Nu)\times Nta}$$

$$= \begin{pmatrix} \boldsymbol{E} \\ \boldsymbol{O} \end{pmatrix} + \sum_{i=1}^{Nr+1} \alpha_i \begin{pmatrix} \boldsymbol{O} \\ \boldsymbol{J}^i \end{pmatrix} \qquad 7.26$$

$$\boldsymbol{J}^i \triangleq \frac{\partial \boldsymbol{\theta}_2^i}{\partial \boldsymbol{\theta}_1} \qquad 7.27$$

[7.3] Note that this discussion is for a general case. There are special structures and singular configurations of closed-link mechanisms where the discussion does not apply.

where all the α_is are not independent and are subject to Eq. 7.24. Eliminating α_{Nr+1} using Eq. 7.24 yields

$$W = \left(\frac{E}{J^{Nr+1}}\right) + \sum_{i=1}^{Nr} \alpha_i \left(\frac{O}{J^i} - \frac{O}{J^{Nr+1}}\right) \qquad 7.28$$

which includes independent α_is only. Accordingly, the joint torque of closed-link mechanism is computed by

$$\tau_1^c = \tau_1^0 + \Gamma_1 \alpha \qquad 7.29$$

$$\alpha = (\alpha_1 \ \cdots \ \alpha_{Nr})^T$$

$$\tau_1^0 = (E \ \ J^{Nr+1^T}) \tau$$

$$\tau_1^i = (O \ \ J^{i^T} - J^{Nr+1^T}) \tau$$

$$\Gamma_1 = (\tau_1^1 \ \cdots \ \tau_1^{Nr}) \in R^{Nta \times Nr}$$

Equation 7.29 shows that the joint torque of the closed-link mechanism is represented by a linear function of α_is. The α_is are the parameters that determine the dynamic load distribution between actuators in the closed loop. Once the joint torque of the tree-structure mechanism is computed, the joint torque of the closed-link mechanism can be optimized by an appropriate choice of α_i. From Eqs. 7.25 and 7.29, we can see that the joint torque of the closed-link mechanism forms a bilinear equation of the joint torque of the equivalent open-link tree mechanism and the arbitrary vector α.

7.3.2 *Optimal Load Distribution*

Actuators usually have the torque limits such as

$$|\tau_{1i}^c| \leq \tau_{1i\max}^c \quad \text{for } i = 1, \cdots, N_{ta} \qquad 7.30$$

$$\tau_{1i\max}^c > 0$$

where τ_{1i}^c is the ith component of τ_1^c, and $\tau_{1i\max}^c$ is constant. One of the possible criterian is to minimize the following quadratic function of τ_1^c:

$$\min \tau_1^{cT} A \tau_1^c \qquad 7.31$$

where $A \in R^{Nta \times Nta}$ is a symmetric positive definite matrix. The joint torque is generally proportional to the electric current for an electric motor. Therefore for example, if a diagonal matrix with appropriate entries is chosen as A, Eq. 7.31 minimizes the total output wattage of motors. This optimization problem formed by Eqs. 7.29, 7.30, and 7.31 is typical of *quadratic programming*. Several algorithms have been proposed for this sort of problem. Wolfe's method (1959), for example, reduces the problem to the *simplex method* of linear programming.

A practically significant case is that where a closed-link mechanism includes only one redundant actuator. In this case, we can find a simple method to solve this problem. Equation 7.29 becomes

$$\tau_1^c = \tau_1^0 + \tau_1^1 \alpha_1 \qquad 7.32$$

This is the equation of a line parameterized by α_1. Substituting Eq.7.32 into 7.31, we have

$$\min \tau_1^{cT} A \tau_1^c$$
$$= \min \{\alpha_1^2 (\tau_1^{1T} A \tau_1^1) + 2\alpha_1 (\tau_1^{1T} A \tau_1^0) + (\tau_1^{0T} A \tau_1^0)\} \qquad 7.33$$

Therefore, if $\tau_1^1 \neq O$, the objective function is a convex quadratic function of α_1 with the minimum value at $\alpha_1 = -(\tau_1^{1T} A \tau_1^0)/(\tau_1^{1T} A \tau_1^1)$. If $\tau_1^1 = O$, the solution becomes $\tau_1^c = \tau_1^0$ from Eq. 7.32. From Eq.7.30, we have $2N_{ta}$ boundary planes; that is,

$$\begin{aligned} \tau_{1i}^c &= \tau_{1i\max}^c \\ \tau_{1i}^c &= -\tau_{1i\max}^c \end{aligned} \qquad \text{for } i = 1, \cdots, N_{ta} \qquad 7.34$$

The computation algorithm is summarized in four steps as follows:

Algorithm 7.1

1. Compute the point of intersection between each boundary plane of Eq. 7.34 and the line of Eq. 7.32. The set of solutions collected for $2N_{ta}$ planes become

$$\alpha_1^1 \leq \alpha_1^2 \leq \cdots \leq \alpha_1^k, \qquad k \leq 2N_{ta} \qquad 7.35$$

   When the line of Eq. 7.32 is included in one of the planes, we do not need to obtain the solution for this plane.

2. For $i = 1, \cdots, k-1$, check whether a point on the line at $\alpha_1 = (\alpha_1^i + \alpha_1^{i+1})/2$ is included in the feasible space given by Eq. 7.30. If one is, the feasible span of α_1 is $\alpha_1^i \leq \alpha_1 \leq \alpha_1^{i+1}$. Note that the feasible span of α_1 is unique and is not separated into two or more spans because it corresponds to the intersection of the line of Eq. 7.32 and the convex polyhedron given by Eq. 7.30.

3. If $\alpha_1^i > -(\tau_1^{1T} A \tau_1^0)/(\tau_1^{1T} A \tau_1^1)$, then $\alpha_1^{opt} = \alpha_1^i$. If $\alpha_1^i \leq -(\tau_1^{1T} A \tau_1^0)/(\tau_1^{1T} A \tau_1^1) \leq \alpha_1^{i+1}$, then $\alpha_1^{opt} = -(\tau_1^{1T} A \tau_1^0)/(\tau_1^{1T} A \tau_1^1)$. If $\alpha_1^i < -(\tau_1^{1T} A \tau_1^0)/(\tau_1^{1T} A \tau_1^1)$, then $\alpha_1^{opt} = \alpha_1^{i+1}$.

4. The solution τ_1^{copt} is yielded by substitution of α_1^{opt} into Eq. 7.32.

7.4 Examples

7.4.1 Nonredundant Actuator System

Figure 7.2 shows the UCSB HAnd developed in the Center for Robotic Systems in Microelectronics, University of California at Santa Barbara. The robot hand has three 3-DOF fingers driven by nine DC servo motors. This hand was developed for research on coordinative manipulation by multiple robotic mechanisms (Nakamura, Nagai, and Yoshikawa 1987) described in Chapter 6. Since precise force control of each finger is required, a simple gear-transmission system with a low reduction ratio of 1 : 16 was adopted. By using the five-bar closed-link mechanism and a special transmission, the developers simplified the finger structure as shown in Fig. 7.3, and all the motors were located in the wrist portion. The transmission is shown in Fig. 7.4. The use of a closed-link mechanism for robotic finger was also discussed by Youcef-Toumi and Yahiaoui (1987) from the viewpoint of decoupled dynamics.

The dynamics of the closed-link finger mechanism was computed using the computational scheme proposed in Section 7.2. The coordinate frames and the corresponding open-link tree-structure mechanism are shown in Fig. 7.5. Although we chose the Denavit–Hartenberg convention (Denavit and Hartenberg 1955) to define the link coordinates, the choice is not essential to our formulation and you may choose the modified Denavit–Hartenberg convention by Craig (1989). The θ_i indicates the rotational angle along the z axis of the

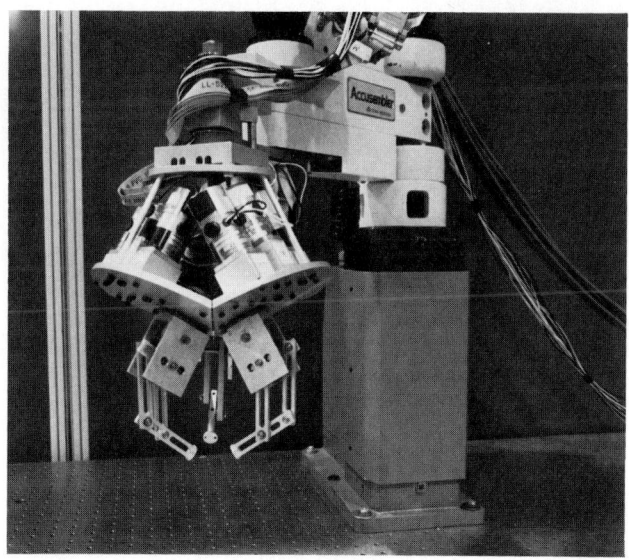

Figure 7.2 UCSB HAND: Three 3-DOF closed-link fingers are driven by nine DC motors.

$(i-1)$th coordinate frame. The θ_1, θ_2, and θ_3 axes are the actuated joints. Accordingly,

$$\boldsymbol{\theta}_1 = (\theta_1 \ \theta_2 \ \theta_3)^T \qquad 7.36$$

$$\boldsymbol{\theta}_2 = (\theta_4 \ \theta_5)^T \qquad 7.37$$

From the visual inspection, θ_4 and θ_5 are represented as the function of θ_1, θ_2, and θ_3 as follows:

$$\theta_4 = \theta_2 - \theta_3 \qquad 7.38$$

$$\theta_5 = -\pi - \theta_2 + \theta_3 \qquad 7.39$$

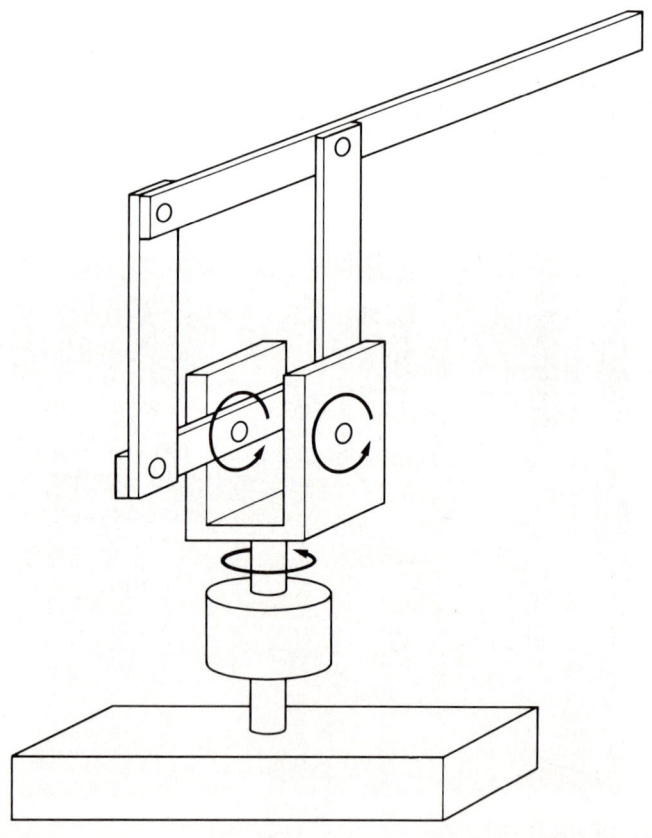

Figure 7.3 Finger structure of UCSB HAND: Five-bar closed-link mechanism.

7.4 Examples 219

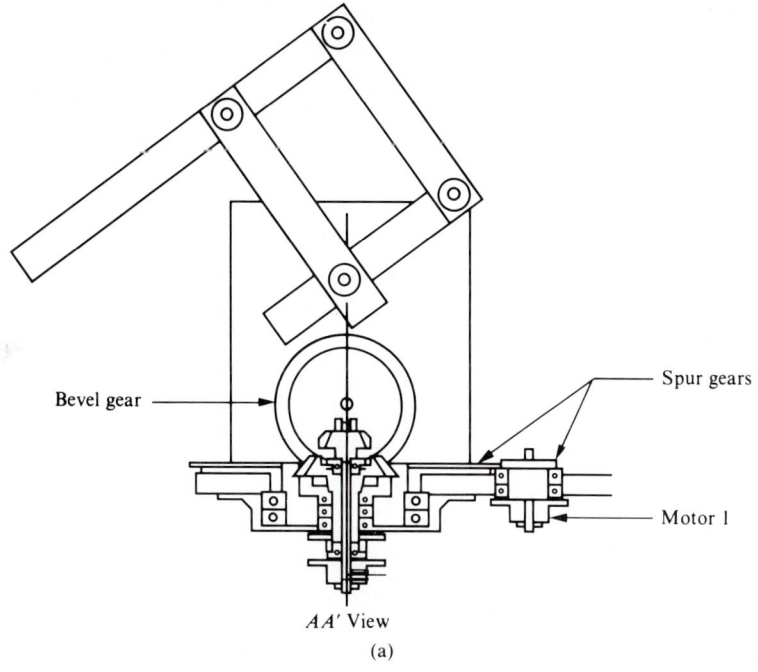

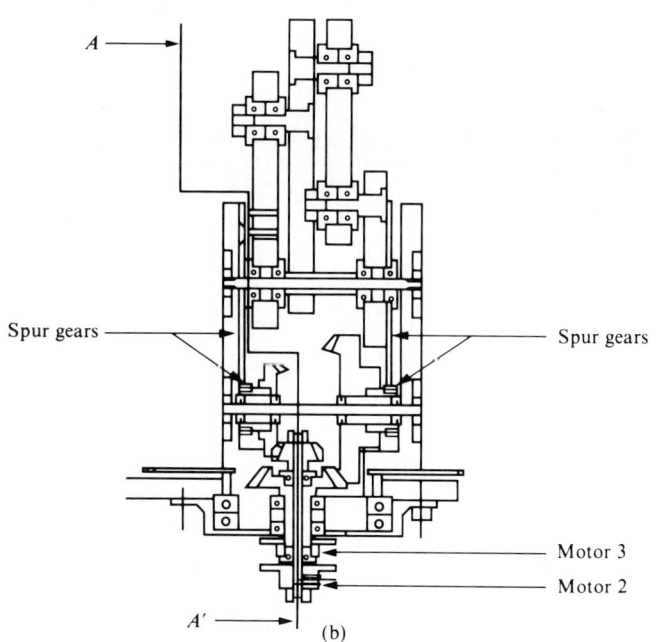

Figure 7.4 Gear reduction and transmission of UCSB HAND.

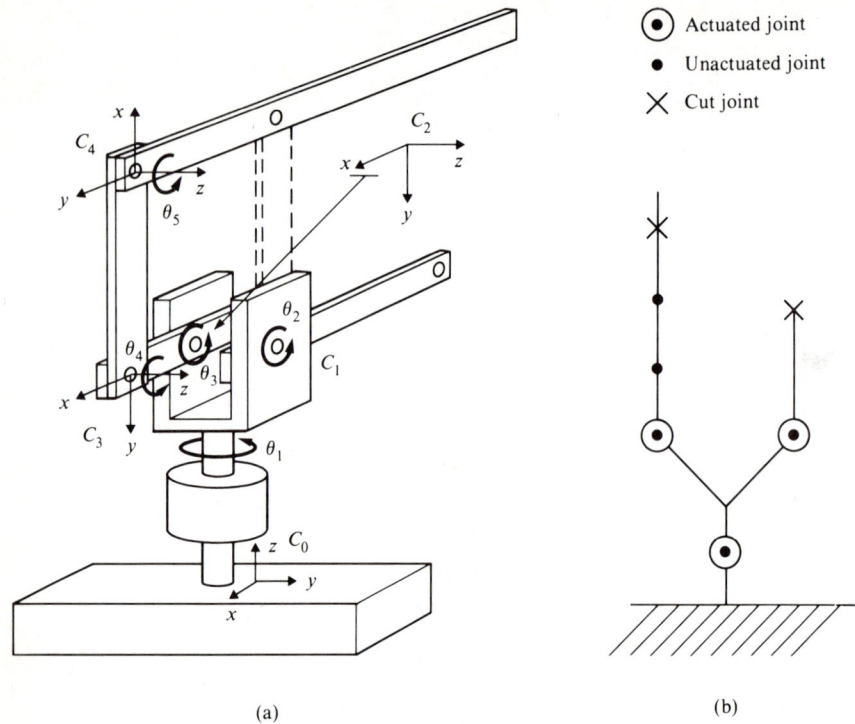

Figure 7.5 Corresponding open-link tree-structure mechanism. (a) Coordinate frames. (b) Tree structure.

Equations 7.38 and 7.39 correspond to Eq. 7.3. Therefore, matrix W becomes a constant matrix as follows:

$$W = \begin{pmatrix} 1 & 0 & 0 \\ 0 & 1 & 0 \\ 0 & 0 & 1 \\ 0 & 1 & -1 \\ 0 & -1 & 1 \end{pmatrix} \qquad 7.40$$

To compute direct kinematics, inverse kinematics, and inverse dynamics for controlling the fingers, we applied the unified recursive computational scheme (Nakamura, Yokokohji, Hanafusa, and Yoshikawa 1986), which was developed to reduce the total computation. Therefore, we cannot show separately the amount of computation required for only the inverse dynamics to which the proposed scheme contributes. The whole computation included 1493 arithmetic operations (798 multiplications, 510 additions, 185 subtractions) and took 2.0 ms on a SUN3/160 with a Floating Point Accelerator. Multiplications or additions of zero and multiplications by one, which are expected from the structure of the coordinate transformation matrices, were eliminated carefully. We wrote all the operations explicitly in FORTRAN without loops or matrix representations, to eliminate the overhead.

Since the open-link tree-structure mechanism has five joints and five links and the additional computation to evaluate Eq. 7.5 is only two additions and two subtractions, from Eq. 7.40, the computational complexity of the closed-link mechanism can be compared with that of a 5-DOF open-link robot mechanism. According to the recursive Newton–Euler scheme (Luh, Walker, and Paul 1980), the total number of arithmetic operations becomes 1309 ($150n-48$ multiplications, $131n-48$ additions (Hollerbach 1980), and $n=5$) for inverse dynamics computation only.

7.4.2 Redundant Actuator System

In this subsection, we show a simple numerical example of the redundant actuator system. Suppose that unactuated joint θ_4 of the same closed-link mechanism as that shown in Fig. 7.5(a) has been replaced with an actuated joint as indicated in Fig. 7.6. That is,

$$\boldsymbol{\theta}_1 = (\theta_1 \quad \theta_2 \quad \theta_3 \quad \theta_4)^T \qquad 7.41$$

$$\boldsymbol{\theta}_2 = \theta_5 \qquad 7.42$$

Accordingly, $\boldsymbol{\theta}_2$ is represented as a function of $\boldsymbol{\theta}_1$ in two different ways:

$$\boldsymbol{\theta}_2^1 = \theta_5 = -\pi - \theta_2 + \theta_3 \qquad 7.43$$

$$\boldsymbol{\theta}_2^2 = \theta_5 = -\pi - \theta_4 \qquad 7.44$$

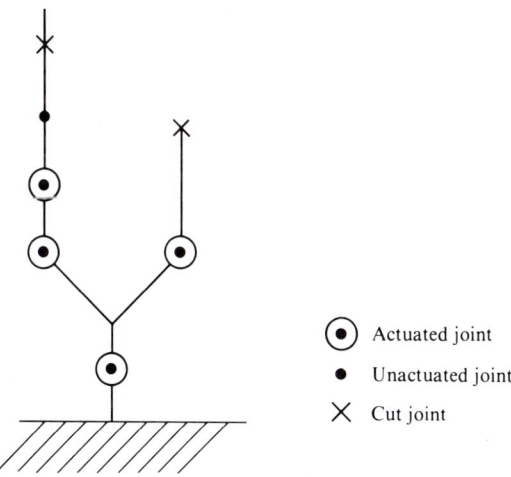

Figure 7.6 Tree structure of redundant actuation system.

From Eq.7.27, J^1 and J^2 become

$$J^1 = \frac{\partial \theta_2^1}{\partial \theta_1} = (0 \quad -1 \quad 1 \quad 0)$$

$$J^2 = \frac{\partial \theta_2^2}{\partial \theta_1} = (0 \quad 0 \quad 0 \quad -1)$$

7.45

Therefore, W can be computed from Eqs. 7.28 and 7.45 by

$$W = \begin{pmatrix} 1 & 0 & 0 & 0 \\ 0 & 1 & 0 & 0 \\ 0 & 0 & 1 & 0 \\ 0 & 0 & 0 & 1 \\ 0 & 0 & 0 & -1 \end{pmatrix} + \alpha_1 \begin{pmatrix} 0 & 0 & 0 & 0 \\ 0 & 0 & 0 & 0 \\ 0 & 0 & 0 & 0 \\ 0 & 0 & 0 & 0 \\ 0 & -1 & 1 & 1 \end{pmatrix}$$

7.46

Now, suppose that the robot finger has to maintain static equilibrium against a force exerted on the end link as shown in Fig. 7.7, where we neglect the effect of gravity to simplify the discussion. In this case, the joint torque of the corresponding open-link tree-structure mechanism becomes

$$\tau = (0 \quad 0 \quad 0.640 \quad 0.890 \quad 0.890)^T \qquad \text{(N/m)} \qquad 7.47$$

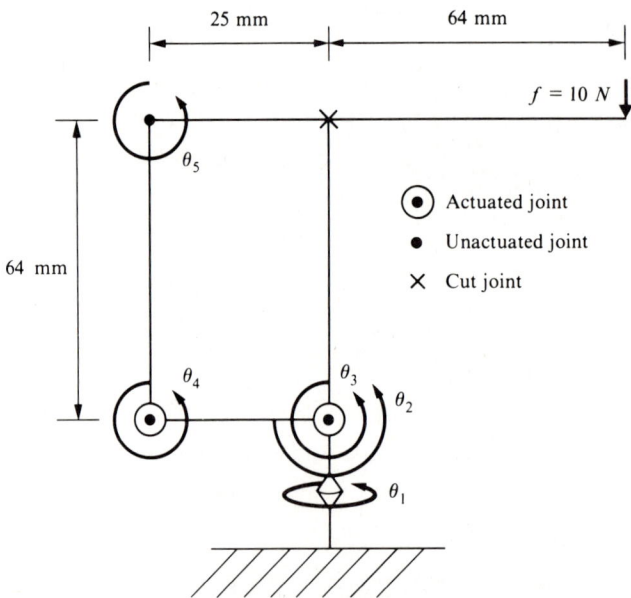

Figure 7.7 Example of redundant actuation system: Statical equilibrium against a vertical load at the tip.

From Eqs. 7.5, 7.40, and 7.47, the joint torque of the closed-link mechanism with the nonredundant actuator system is computed:

$$\boldsymbol{\tau}_1^c = (0 \quad 0 \quad 0.640)^T \quad \text{(N/m)} \quad 7.48$$

From Eqs. 7.29, 7.45, and 7.47, the joint torque of redundant actuation system becomes

$$\boldsymbol{\tau}_1^c = \boldsymbol{\tau}_1^0 + \boldsymbol{\tau}_1^1 \alpha_1 \quad 7.49$$
$$\boldsymbol{\tau}_1^0 = (0 \quad 0 \quad 0.640 \quad 0)^T \quad \text{(N/m)}$$
$$\boldsymbol{\tau}_1^1 = (0 \quad -0.890 \quad 0.890 \quad 0.890)^T \quad \text{(N/m)}$$

We assume that $\boldsymbol{A} = \boldsymbol{E}$ and $\tau_{1i\text{max}}^c = 0.640$ (Nm) for all i. Since τ_{11}^c has nothing to do with optimization, we can graphically represent the polyhedron of admissible joint torques for τ_{1i}^c ($i = 2, 3, 4$) and the line of Eq. 7.49 as shown in Fig. 7.8.

Now, we apply the optimization method proposed in Section 7.3.2. The feasible span of α_1 is obtained as

$$-0.719 \leq \alpha_1 \leq 0 \quad 7.50$$

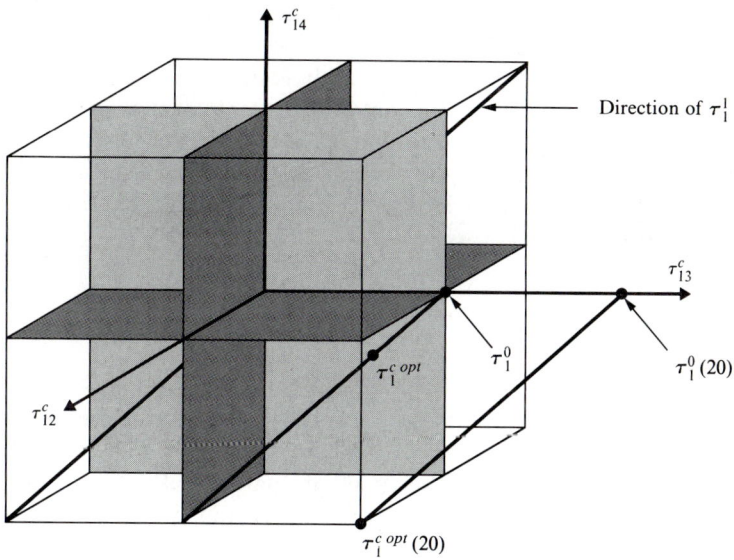

τ_1^0 : τ_1^0 for load 10 (N)
$\tau_1^{c\ opt}$: optimal torque for load 10 (N)
$\tau_1^0 (20)$: τ_1^0 for load 20 (N)
$\tau_1^{c\ opt}(20)$: optimal torque for load 20 (N)

Figure 7.8 Polyhedron of admissible joint torque and the line of Eq. 7.49.

The α_1 that yields the extreme value becomes

$$-\frac{\tau_1^{1T} A \tau_1^0}{\tau_1^{1T} A \tau_1^1} = -0.240 \qquad 7.51$$

Since $-(\tau_1^{1T} A \tau_1^0)/(\tau_1^{1T} A \tau_1^1)$ is included in the feasible span of Eq. 7.50, the optimal value becomes $\alpha_1^{opt} = -0.240$. Substituting α_1^{opt} into Eq. 7.49, we obtain the optimal joint torque as follows:

$$\tau_1^{copt} = (0 \quad 0.213 \quad 0.427 \quad -0.213)^T \qquad (N/m) \qquad 7.52$$

When the magnitude of load increases in Fig. 7.7, the maximum magnitude that can be supported by the redundant actuation system can be computed in the same way, which turns out to be 20 (N), whereas the maximum load for the nonredundant actuation system is 10 (N). The joint torque, in this case, becomes

$$\tau_1^c = (0 \quad 0.640 \quad 0.640 \quad -0.640)^T \qquad (N/m) \qquad 7.53$$

Note that, although this example is a static case, our discussion is general enough to deal with the dynamic case. The advantage of having redundant actuators depends on the direction of vector τ_1^0. Only when τ_1^1 is perpendicular to τ_1^0 if $A = E$—more generally, only when $\tau_1^{1T} A \tau_1^0 = 0$—the actuation redundancy cannot contribute to reduce the joint torque and, therefore, the maximum load is the same as that of nonredundant system. Practically speaking, the comparison between redundant and nonredundant actuation systems must be done taking account of the increase of mass and inertia due to inclusion of redundant actuators. However, it is clear that redundant actuation systems have the potential for increasing the payload and improving the dynamic response of manipulators.

Summary

A new computational scheme for the inverse dynamics of closed-link mechanisms was derived by using the d'Alembert's principle. The constraint was taken into consideration using the Jacobian matrix of the unactuated joint angles in terms of the actuated ones. This is a simpler scheme with clearer physical meaning than the previously known results, and provides new insight into the problem. This scheme is computationally more efficient than are the conventional ones.

The redundancy in inverse dynamics computation for closed-link mechanisms with redundant actuators was discussed. Using the Jacobian matrices of the unactuated joint angles with respect to the actuated joints, we formulated the actuation redundancy as a bilinear equation of the joint torque of the equivalent open-tree mechanism and an arbitrary vector.

The optimization of actuation redundancy was also discussed as the problem of minimizing a quadratic function of the joint torques subjected by the torque limit constraints. This problem is a typical quadratic-programming problem.

A simple algorithm of optimal load distribution was presented for the case of one redundant actuator that is of practical significance.

Numerical examples clearly show the potential advantage in increasing the payload and improving the dynamic response of manipulators. Closed-link mechanisms with actuation redundancy will be especially advantageous for making direct-drive manipulators with kinematic redundancy.

References

Asada, H., and Youcef-Toumi, K. 1987. *Direct-drive robots.* Cambridge, MA: MIT Press.

Craig, J. J. 1989. *Introduction to robotics: Mechanics & control*, second edition. Reading, MA: Addison-Wesley.

Denavit, J., and Hartenberg, R.S. 1955. A kinematic notation for lower-pair mechanisms based on matrices. *ASME Trans. J. Appl. Mech.* 77 (2): 215–221.

Duffy, J. 1980. *Analysis of mechanisms and robot manipulators.* London: Edward Arnold.

Featherstone, R. 1987. *Robot dynamics algorithms.* Norwell: Kluwer Academic.

Hollerbach, J. M. 1980. A recursive Lagrangian formulation of manipulator dynamics and a comparative study of dynamics formulation complexity. *IEEE Trans. Syst., Man Cybernet.* 10 (11): 730–736.

Kerr, J. R., and Roth, B. 1986. Analysis of multi-fingered hands. *Int. J. Robotics Res.* 4 (4): 3–17.

Kleinfinger, J. F., and Khalil, W. 1986 (Brussels, Belgium). Dynamic modeling of closed-loop robots. *Proc. 16th Int. Symp. Industrial Robots*, pp. 401–412.

Luh, J. Y. S., Walker, M. W., and Paul, R. P. C. 1980. On-line computational scheme for mechanical manipulators. *ASME J. Dyn. Sys., Meas. Contr.* 102 (2): 468–474.

Luh, J. Y. S., and Zheng, Y. 1985. Computation of input generalized forces for robots with closed kinematic chain mechanisms. *IEEE J. Robotics and Automat.* 1 (2): 95–103.

Maggi, G. A. 1901. Di alcune nuove forme delle equazioni della dinamica applicabili ai sistemi anolonomi. *Atti Accad. Naz. Lincei Rend. Cl. Fis. Mat. Nat.* 5 (10): 287–291.

Nakamura, Y. 1988. Dynamics of closed link robots with actuational redundancy. In *Robotics and Manufacturing: Recent Trends in Research, Education, and Applications*, eds: M. Jamshidi, J. Y. S. Luh, H. Seraji, and G. P. Starr. (*Proc. 2nd Int. Symp. Robotics and Manufacturing Res., Educ. Appl.*), pp. 309–318, ASME Press.

Nakamura, Y. 1989. Dynamics computation of closed link robots and optimization of actuation redundancy. *Trans. Society of Instruments and Control Engineers*, 25 (5): 600–607 (*In Japanese*).

Nakamura, Y., and Ghodoussi, M. 1988a (Philadelphia). A computational scheme of closed link robot dynamics derived by d'Alembert principle. *Proc. 1988 IEEE Int. Conf. Robotics and Automation*, pp. 1354–1360.

Nakamura, Y., and Ghodoussi, M. 1988b (Salo, Italy). Closed link manipulators with actuational redundancy. *Prep. NATO Advanced Research Workshop on Robots with Redundancy: Design, Sensing and Control*, (Proceedings to be published from Springer-Verlag).

Nakamura, Y., and Ghodoussi, M. 1989. Dynamics computation of closed link robot mechanisms with nonredundant and redundant actuators. *IEEE J. Robotics and Automat.* 5 (3): 294–302.

Nakamura, Y., Nagai, K., and Yoshikawa, T. 1987 (Raleigh). Mechanics of coordinative manipulation by multiple robotic mechanisms. *Proc. 1987 IEEE Int. Conf. Robotics and Automation*, pp. 991–998.

Nakamura, Y., Yokokohji, Y., Hanafusa, H., and Yoshikawa, T. 1986 (Osaka). Unified recursive formulation of kinematics and dynamics of robot manipulators. *Proc. Japan–USA Symp. Flexible Automation*, pp. 53–60.

Oh, S. Y., and Orin, D. 1986 (San Francisco). Dynamic computer simulation of multiple closed-chain robotic mechanisms. *Proc. 1986 IEEE Int. Conf. Robotics and Automation*, pp. 15–20.

Orin, D. E., and McGhee, R. B. 1981. Dynamic computer simulation of robotic mechanisms. *Proc. 4th Symp. Theory and Practice of Robots and Manipulators*, eds. A. Morecki, G. Bianchi, and K. Kedzior, pp. 286–296. Warsaw: Polish Scientific Publications.

Orin, D. E., and Oh, S. Y. 1981. Control of force distribution in robotic mechanisms containing closed kinematic chains. *ASME J. Dyn. Sys., Meas. Contr.* 102 (2): 134–141.

Paul, B. 1975. Analytical dynamics of mechanisms—a computer oriented overview. *Mechanism and Machine Theory* 10 (6): 481–507.

Smith, D. A. 1973. Reaction force analysis in generalized machine systems. *ASME J. Engineering for Industry* 95 (2): 617–623.

Stewart, D. 1965. A platform with six degrees of freedom. *Proc. Inst. Mech. Eng.*, 180 (1): 371–386.

Tarn, T. J., Bejczy, A. K., and Li, Z. 1988 (Atlanta). Dynamic workspace analysis of two cooperating robot arms. *Proc. 1988 Amer. Contr. Conf.*, pp. 489–498.

Volterra, V. 1898. Sopra una classe di equazione dinamiche. *Atti Accad. Sci. Trino* 33 (Feb.): 451–475.

Walker, M. W., and Orin, D. E. 1982. Efficient dynamic computer simulation of robot manipulators. *ASME J. Dyn. Sys., Meas. Contr.* 104 (3): 205–211.

Wolfe, P. 1959. The simplex method for quadratic programming. *Econometrica* 27: 382–398.

Youcef-Toumi, K., and Yahiaoui, M. 1987 (Boston). The design of a mini direct-drive finger with decoupled dynamics. *Proc. ASME Winter Annual Meeting: Modeling and Control of Robotic Manipulators and Manufacturing Processes*, DSC-Vol. 6: pp. 411–416.

CHAPTER 8

A Manipulator with Kinematic and Actuation Redundancy

8.1 Introduction

Kinematic redundancy of robot manipulators is useful because it allows variety in access to the object in the workspace and avoids singularity (Ligeois 1977; Nakamura and Hanafusa 1987; Nakamura, Hanafusa, and Yoshikawa 1987). In Chapters 4 and 5, we studied the advantages of kinematic redundancy and the techniques to utilize it. Direct-drive actuation (Asada, Kanade, and Takeyama 1983; Asada and Youcef-Toumi 1984, 1987) has the advantage of improving the quality of force control and the speed of motion. Although both concepts promise wide use in the future for advanced robot mechanisms, they are somewhat contradictory, since kinematic redundancy tends to have more actuators than necessary, which makes direct-drive actuation rather bulky and less practical. Asada and Youcef-Toumi (1984) used closed linkages to realize practical nonredundant direct-drive manipulators.

This chapter is adapted from, by permission, Nakamura, Y., and Ropponen, T. 1989 (San Francisco). "A closed link manipulator with kinematic and actuationdancies," *Robotics Research–1989, Proc. 1989 ASME Winter Annual Meeting*, eds. K. Youcef-Toumi and H. Kazerooni, pp. 203–210; Ropponen, T., and Nakamura, Y. 1990 (Cincinnati). "Parameterization and performance analysis of actuation redundancy," *Proc. 1990 IEEE International Conference on Robotics and Automation*, pp. 806–811; Nakamura, Y., and Ropponen, T.

Low payload due to motor torque limitation is another disadvantage of direct-drive mechanisms. To overcome this limitation, in Chapter 7 we have proposed incorporating additional actuators within closed linkages. This concept is called *actuation redundancy*. Redundant actuators in closed loops can be coordinated dynamically so that they maximize the payload, as we saw in Chapter 7.

Due to the lack of torque, some developed direct-drive manipulators have two motors to actuate a single joint (for example, the second and fourth joints of CMU DD Arm Robot model I (Asada, Kanade, and Takeyama 1983)). The actuation redundancy can be considered an extension of this multimotor versus single-joint relationship to the multimotor versus multijoint relationship.

In this chapter, we propose a closed-link mechanism for kinematically redundant direct-drive manipulators, and we discuss its redundant actuation. The mechanism is kinematically equivalent to four joints of the shoulder through the elbow of an anthropomorphic 7-DOF manipulator. The mechanism uses five direct-drive motors to control the 3-DOF position of the end-effector. One DOF is used for kinematic redundancy; another DOF is for actuation redundancy.

We also show that the parameterization of actuation redundancy developed in Chapter 7 cannot take full advantage of actuation redundancy in some situations due to the singularity of the parameterization. We propose a new parameterization that allows constant full utilization of actuation redundancy. The advantages of actuation redundancy and the differences due to the parameterizations are investigated numerically.

8.2 A Closed-Link Mechanism with Kinematic Redundancy

Figure 8.1 shows a closed-link mechanism with kinematic redundancy. If we appropriately allocate actuators among joint 0 through joint 8, it behaves equivalently to four joints of the shoulder through the elbow of an anthropomorphic manipulator shown in Fig. 8.2. Note that, if joints 2, 6, and 7 are fixed and joints 0, 1, and 5 are actuated, the mechanism coincides with the nonredundant mechanism of MIT DD Arm II (Asada and Youcef-Toumi 1987). If a 3-DOF wrist mechanism is mounted, the mechanism forms a kinematic equivalent of a 7-DOF anthropomorphic manipulator.

In addition to the one at joint 0, we have to allocate three actuators within the closed kinematic chain. Among several possibilities, the two shown in Fig.

1990 (San Diego). "Actuation redundancy of a closed link manipulator," *Proc. 1990 American Control Conference*, pp. 2294–2299; and Nakamura, Y., and Ropponen, T. 1990 (Kyoto). "Dynamic payload maximization of a closed link robot with actuation redundancy," *Proc. 1990 Japan-USA Symposium on Flexible Automation*, pp. 321–327.

8.2 A Closed-Link Mechanism with Kinematic Redundancy

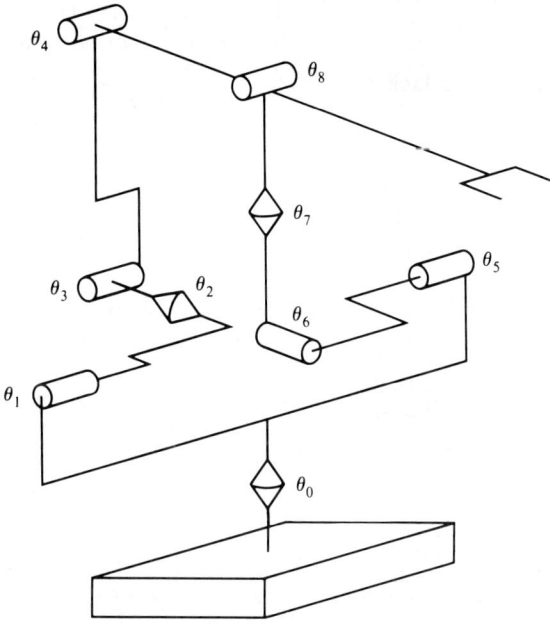

Figure 8.1 A closed-link mechanism with kinematic redundancy.

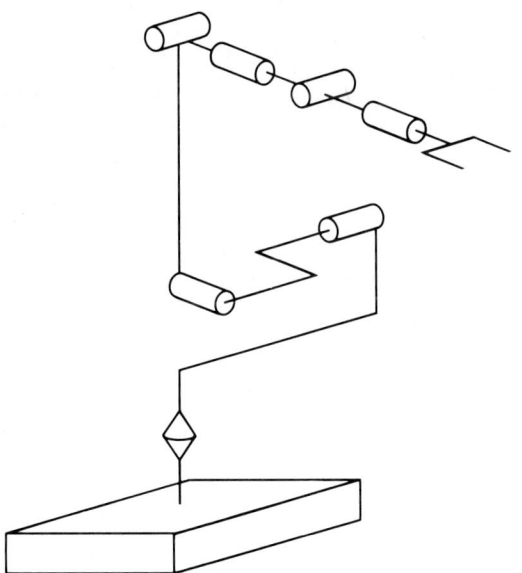

Figure 8.2 A 7-DOF anthropomorphic manipulator.

8.3(a) and (b) would be practical choices, since it is desirable to locate the actuators closer to the base in order to reduce the gravity load.

8.3 Redundant Actuation of the Closed-Link Mechanism

To compensate for the lack of motor torque, we sometimes use two motors to actuate a single joint (for example, the second and fourth joints of CMU DD Arm Robot model I (Asada, Kanade, and Takeyama 1983)). Redundant actuation is to include one or more additional actuators in a closed-kinematic chain to compensate for the lack of torque of the actuators. By choosing appropriate locations for additional actuators, we can compensate for the most demanding actuators among all the actuators in the closed kinematic chain depending on the instantaneous load and motion of the chain. We developed the basic discussion of actuation redundancy in Chapter 7 taking the total dynamics into consideration.

Let us include an additional actuator in the closed-link mechanism proposed in Section 8.2. At the arm configuration shown in Fig. 8.1, note that joints 2 and 6 have similar contribution to the mechanism and that joints 1, 3, 4, 5, and 8 are coupled to one another. Statically speaking, the latter joints are related to the end-effector load in the plane that contains the closed-link mechanism. The former joints are related to the end-effector load in the perpendicular direction.

In practical situations, major portions of static loads, including the gravity load, are involved in the plane defined by joints 1, 3, 4, 5, and 8. Hence, we propose including an additional actuator among these joints. Taking into account the practical consideration that actuators should be located closer to the base, we choose the redundant actuation as shown in Fig. 8.4, which we obtain by locating the fourth actuator at joint 3 of Fig. 8.3(b). A similar allocation is possible for Fig. 8.3(a), although it tends to have more motors in one side of the branches than in the other.

Note that, when joint 6 rotates from the configuration of Fig. 8.1 in either the positive or negative direction, all joints become coupled, which implies that the additional joint now can contribute to compensate for the torques of all other actuators in the closed kinematic chain.

8.4 Singularity-Free Parameterization of Actuation Redundancy

8.4.1 Quick Review of Parameterization

We overview the parameterization developed in Section 7.3.1 by applying it to our specific mechanism of Fig. 8.4. The computation proceeds following the three steps we studied in Section 7.2.1. Although the method can consistently

8.4 Singularity-Free Parameterization of Actuation Redundancy

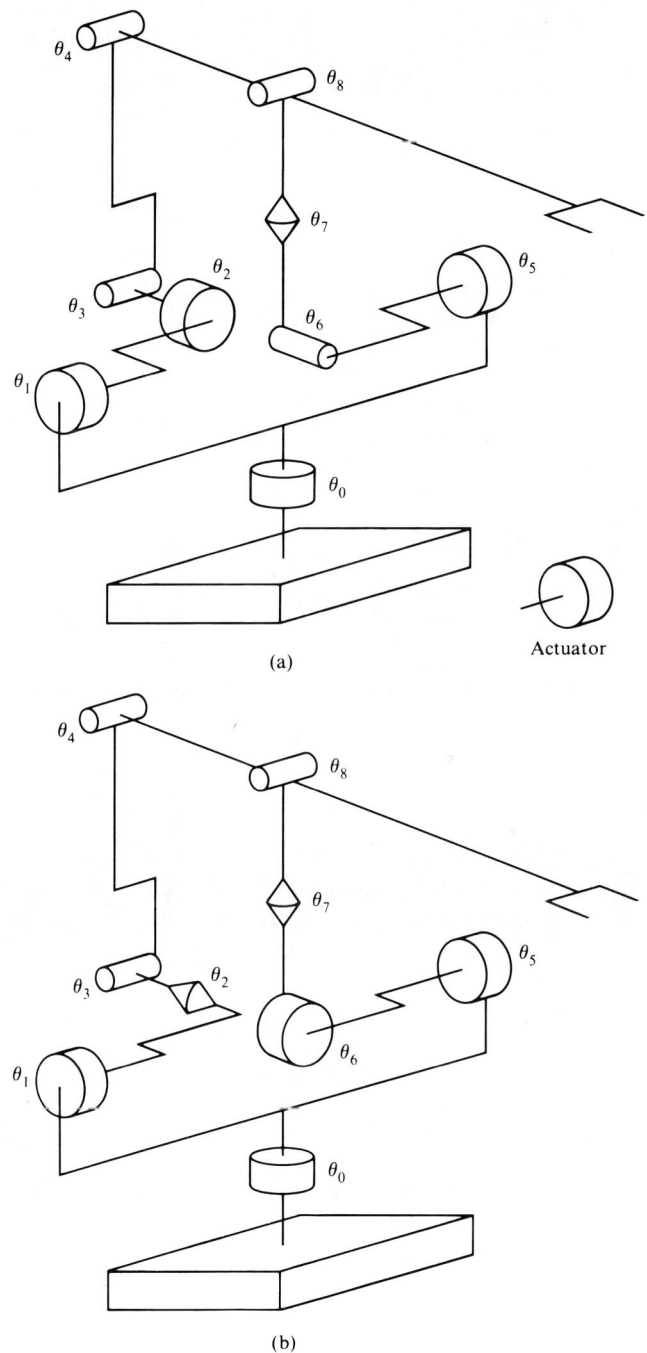

Figure 8.3 Actuator location of the kinematically redundant closed-link mechanism.

234   Chapter 8   A Manipulator with Kinematic and Actuation Redundancy

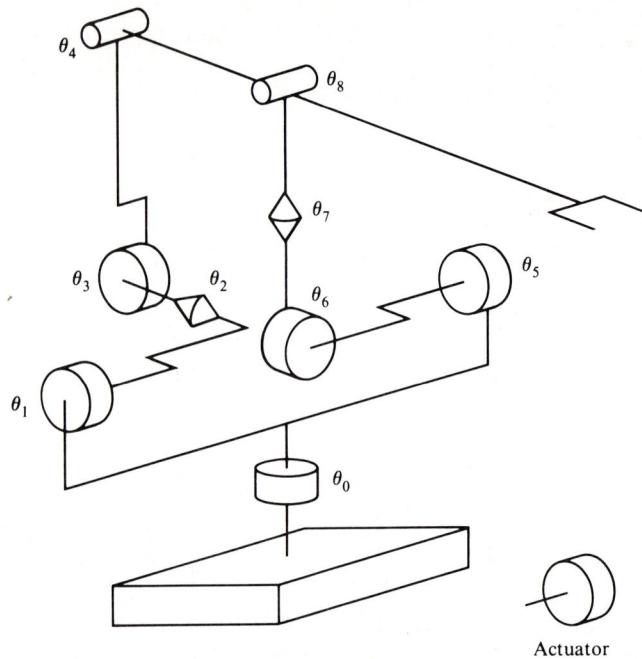

Figure 8.4 Redundant actuation of the kinematically redundant closed-link mechanism.

include joint 0, which is outside the closed loop, we disregard it to simplify the discussion. As step 1, we cut the closed linkage of Fig. 8.4 at joint 8. The link coordinates are assigned as shown in Fig. 8.5. The Denavit–Hartenberg parameters (Denavit and Hartenberg 1955) are shown in Table 8.1. Joints 1 through 7 are classified into the actuated joints and the unactuated joints. The former is represented by vector $\boldsymbol{\theta}_1$, the latter by $\boldsymbol{\theta}_2$. Namely,

$$\boldsymbol{\theta}_1 = (\theta_1 \ \theta_3 \ \theta_5 \ \theta_6)^T \qquad 8.1$$

$$\boldsymbol{\theta}_2 = (\theta_2 \ \theta_4 \ \theta_7)^T \qquad 8.2$$

We represent $\boldsymbol{\theta}_2$ in terms of $\boldsymbol{\theta}_1$ in two different ways:

$$\boldsymbol{\theta}_2 = \begin{cases} \boldsymbol{\theta}_2^1(\boldsymbol{\theta}_1) \\ \boldsymbol{\theta}_2^2(\boldsymbol{\theta}_1) \end{cases} \qquad 8.3$$

Step 2 computes the joint torque, $\boldsymbol{\tau}$, of the tree-structure open-link mechanism. As step 3, the joint torque, $\boldsymbol{\tau}$, is now transformed to the net closed-link actuator torque, $\boldsymbol{\tau}_1^c$, using the Jacobian matrices of both representations of Eq. 8.3:

8.4 Singularity-Free Parameterization of Actuation Redundancy

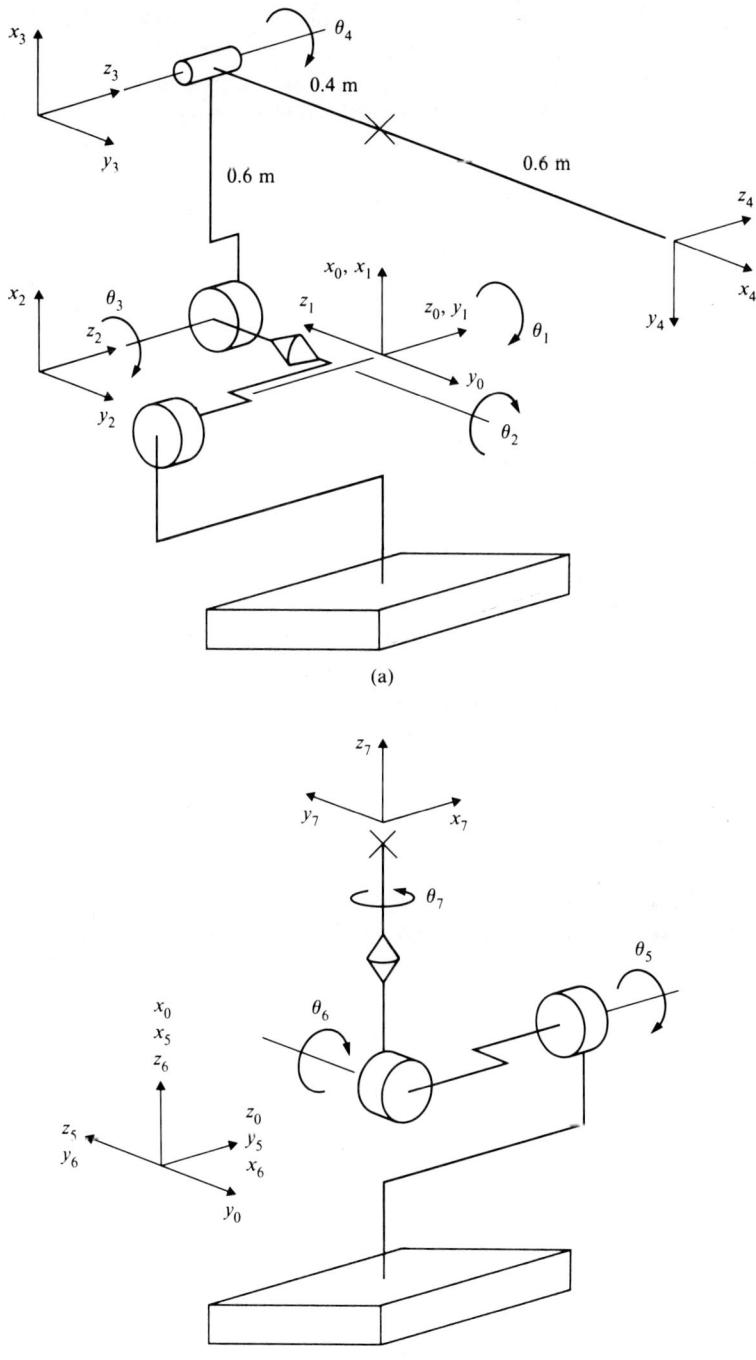

Figure 8.5 Open tree structure and link coordinates.

Table 8.1. Denavit–Hartenberg parameters. The i indicates DH parameters between the $(i-1)$th and ith frames—except for $i = 5$, where they are between the zeroth and fifth frames.

i	a_i	α_i	d_i	θ_i
1	0	90°	0	θ_1
2	0	−90°	l_2	θ_2
3	l_3	0	0	θ_3
4	l_4	0	0	θ_4
5	0	90°	0	θ_5
6	0	90°	0	θ_6
7	0	0	l_3	θ_7

$(l_2, l_3, l_4) = (0.4, 0.6, 1.0) 9m)$

$$\tau_1^c = W^T \tau \qquad 8.4$$

where matrix W is computed as follows:

$$W = \begin{pmatrix} E \\ J^2 \end{pmatrix} + \alpha \begin{pmatrix} J^1 \overset{O}{-} J^2 \end{pmatrix} \qquad 8.5$$

where J^1 and J^2 are $\partial \theta_2^1 / \partial \theta_1$ and $\partial \theta_2^2 / \partial \theta_1$, respectively, and E is an identity matrix. The physical meaning of W in Eq. 8.5 is that W is the Jacobian matrix of the whole joints—namely, $(\theta_1^T\ \theta_2^T)^T$—with respect to the actuated joints, θ_1.

In Eqs. 8.4 and 8.5, α is the parameter we can adjust to control the actuation redundancy. In Section 7.3, our discussion concerned general closed-link systems with multiple additional actuators. A noteworthy feature is that the closed-link actuator torque is represented as a bilinear equation of the tree-structure torque and parameter α (α becomes a vector when more than one additional actuators are involved).

We now repeat Eqs. 7.28 and 7.29 to summarize the parameterization for a general case with N_r redundant actuators:

$$W = \begin{pmatrix} E \\ J^{Nr+1} \end{pmatrix} + \sum_{i=1}^{Nr} \alpha_i \begin{pmatrix} J^i \overset{O}{-} J^{Nr+1} \end{pmatrix} \qquad 8.6$$

The net closed-link actuator torque is given by

$$\tau_1^c = \tau_1^0 + \Gamma_1 \alpha \qquad 8.7$$

$$\alpha = (\alpha_1 \ \cdots \ \alpha_{N_r})^T \qquad 8.8$$

$$\tau_1^0 = (E \ \ J^{N_r+1^T}) \tau \qquad 8.9$$

8.4 Singularity-Free Parameterization of Actuation Redundancy 237

$$\tau_1^i = \begin{pmatrix} O & J^{iT} & -J^{N_r+1^T} \end{pmatrix} \tau \qquad 8.10$$

$$\Gamma_1 = (\tau_1^1 \quad \cdots \quad \tau_1^{N_r}) \in R^{N_{ta} \times N_r} \qquad 8.11$$

where N_{ta} is the total number of actuated joints.

8.4.2 Singularity of Parameterization

Based on the numerical analysis we shall show in Section 8.5, it turns out that the parameterization in the preceding subsection has singularity at some situation.

In Eq. 8.10, the coefficient matrix is $N_{ta} \times (N_{ta} + N_u)$, where N_u is the number of unactuated joints of the tree-structure mechanism. Therefore, it has the null space with at least max (N_{ta}, N_u) dimension. When the null space includes τ, the corresponding α_i becomes useless since, whatever value α_i may take, the contribution of α_i on τ_1^c is always zero.

The occurance of this singularity depends on τ that is dynamically determined by the load, velocity, and acceleration as well as the kinematic configuration. However, whether or not additional actuators can contribute in torque distribution should be structually determined and depend only on the kinematic configuration. Therefore, we have to conclude that the above singularity is due to the method of parameterization of actuation redundancy. Namely, it is an algorithmic singularity.

In the case of the mechanism of Fig. 8.4, the second part of matrix W in Eq. 8.5 has at most rank 3. Therefore, the dimension of the null space is at least four. For the tree-structure torque τ that is included in this null space, the parameterization degenerates although the actuation redundancy still may be present.

8.4.3 New Singularity-Free Parameterization

We obtain Eq. 8.4 by equating the virtual work done by the joints of the tree-structure mechanism with that done by the real actuators of the closed-link mechanism. This is a very intuitive and static explanation of the derivation. The rigorous proof was given in Section 7.2.3, taking the full dynamics into consideration. The derivation is summarized by the following equation:

$$\tau_1^{cT} \delta\theta_1 = \tau^T \begin{pmatrix} \delta\theta_1 \\ \delta\theta_2 \end{pmatrix}$$

$$= \tau^T W \delta\theta_1 \qquad 8.12$$

where $\delta\theta_1$ and $\delta\theta_2$ are virtual displacements.

Although we contained the arbitrary variable α_i in W with the different representation of functional dependency of θ_2 or θ_1 in the previous parameterization, here we use just one representation of θ_2; namely,

$$W = W_0 \triangleq \begin{pmatrix} E \\ J^1 \end{pmatrix} \qquad 8.13$$

With nonredundant actuation of closed-link mechanisms, we can conclude Eq. 8.4 as a necessary and sufficient condition for Eq. 8.12, since $\boldsymbol{\theta}_1$ is independent in determining the motion of the closed-link mechanisms. However, it is merely a sufficient condition for redundant actuation systems, since $\boldsymbol{\theta}_1$ is no longer fully independent. For instance, only three out of four actuators of the redundant actuation system in Fig. 8.4 are independent.

Generally, let $\boldsymbol{\theta}_{10} \in R^{N_{in}}$ be the collection of independent actuators among $\boldsymbol{\theta}_1 \in R^{N_{ta}}$, where $N_{in} < N_{ta}$ is the number of independent actuators. Then, Eq. 8.12 becomes

$$\boldsymbol{\tau}_1^{cT} \boldsymbol{S} \delta \boldsymbol{\theta}_{10} = \boldsymbol{\tau}^T \boldsymbol{W}_0 \boldsymbol{S} \delta \boldsymbol{\theta}_{10} \qquad 8.14$$

where we substituted the following equation into Eq. 8.12:

$$\delta \boldsymbol{\theta}_1 = \boldsymbol{S} \delta \boldsymbol{\theta}_{10} \qquad 8.15$$

where $\boldsymbol{S} \in R^{N_{ta} \times N_{ia}}$ means the instantaneous coefficient that relates $\boldsymbol{\theta}_1$ and $\boldsymbol{\theta}_{10}$.

Since $\delta \boldsymbol{\theta}_{10}$ is totally independent in Eq. 8.14, the following equation is obtained as a necessary and sufficient condition for Eq. 8.14:

$$\boldsymbol{S}^T \boldsymbol{\tau}_1^c = \boldsymbol{S}^T \boldsymbol{W}_0^T \boldsymbol{\tau} \qquad 8.16$$

Accordingly, the general solution of closed-link actuator torque is represented using Theorem 2.5 as follows:

$$\boldsymbol{\tau}_1^c = \boldsymbol{S} \boldsymbol{S}^\# \boldsymbol{W}_0^T \boldsymbol{\tau} + (\boldsymbol{E} - \boldsymbol{S} \boldsymbol{S}^\#) \boldsymbol{y} \qquad 8.17$$

where $\boldsymbol{S}^\#$ is the pseudoinverse of $\boldsymbol{S}$, and $\boldsymbol{y} \in R^{N_{ta}}$ is the arbitrary vector that represents the actuation redundancy. An equality of pseudoinverse, $(\boldsymbol{S} \boldsymbol{S}^\#)^T = \boldsymbol{S} \boldsymbol{S}^\#$, was used in obtaining Eq. 8.17. Assuming the full rankness of $\boldsymbol{S}$, the rank of the coefficient matrix of $\boldsymbol{y}$ is $N_{ta} - N_{ia}$. Therefore, the actuation redundancy spans an $(N_{ta} - N_{ia})$-dimensional space.

The first term of Eq. 8.17 implies the minimum norm solution of $\boldsymbol{\tau}_1^c$. Although it is meaningful to use Eq. 8.17, we now derive the similar general solution to Eq. 8.7. Choosing the arbitrary vector as

$$\boldsymbol{y} = \boldsymbol{W}_0^T \boldsymbol{\tau} + \boldsymbol{z} \qquad 8.18$$

Eq. 8.17 becomes

$$\boldsymbol{\tau}_1^c = \boldsymbol{W}_0^T \boldsymbol{\tau} + (\boldsymbol{E} - \boldsymbol{S} \boldsymbol{S}^\#) \boldsymbol{z} \qquad 8.19$$

Making matrix $\boldsymbol{\Gamma} \in R^{N_{ta} \times (N_{ta} - N_{ia})}$ that contains the independent column vectors of $\boldsymbol{E} - \boldsymbol{S} \boldsymbol{S}^\#$, Eq. 8.19 finally forms

$$\boldsymbol{\tau}_1^c = \boldsymbol{W}_0^T \boldsymbol{\tau} + \boldsymbol{\Gamma} \boldsymbol{\alpha} \qquad 8.20$$

where $\boldsymbol{\alpha} \in R^{N_{ta}-N_{ia}}$ is the independent parameter of actuation redundancy. Comparing Eq. 8.20 with Eq. 8.7, we observe the following features of the new parameterization:

1. The closed-link actuator torque is linear to the parameter $\boldsymbol{\alpha}$.

2. The first term of Eq. 8.20 is basically equivalent to that of Eq. 8.7.

3. This parameterization is free of the algorithmic singularity we discussed in Section 8.4.2, since $\boldsymbol{\Gamma}$ in Eq. 8.20 is totally kinematically determined and has nothing to do with $\boldsymbol{\tau}$, whereas $\boldsymbol{\Gamma}_1$ in Eq. 8.7 is linear to it.

8.5 Numerical Analysis

8.5.1 Kinematics of the Closed-Link Mechanism

Visual inspection of the mechanism of Fig. 8.4 suggests the following relationships:

$$\begin{aligned}
\boldsymbol{e}_{z4} &= \boldsymbol{e}_{x7} \\
\boldsymbol{e}_{x3} &= \boldsymbol{e}_{z6} \\
\boldsymbol{e}_{x4} &= -\boldsymbol{e}_{z1} \\
\boldsymbol{e}_{y3} &= -\boldsymbol{e}_{y7}
\end{aligned} \qquad 8.21$$

where $\boldsymbol{e}_{xi}$, $\boldsymbol{e}_{yi}$, and $\boldsymbol{e}_{zi}$ ($i = 1, \cdots, 7$) are the unit vectors in the directions of x, y, and z axes of the ith link coordinates represented in terms of the zeroth link coordinates.

By comparing the unit vectors obtained from the corresponding transformation matrices, the unactuated joints θ_2, θ_4, and θ_7 can be represented in terms of the actuated joints θ_1, θ_3, θ_5, θ_6 in two different ways:

$$\begin{aligned}
\sin \theta_2 &= -\cos \theta_6 / \cos \theta_3 \\
\sin \theta_4 &= \cos \theta_3 \\
\sin \theta_7 &= \tan \theta_3 \cot \theta_6
\end{aligned} \qquad 8.22$$

$$\begin{aligned}
\sin \theta_2 &= -\frac{\cos \theta_6}{\sqrt{1 - (\sin \theta_{15} \sin \theta_6)^2}} \\
\sin \theta_4 &= \sqrt{1 - (\sin \theta_{15} \sin \theta_6)^2} \\
\sin \theta_7 &= -\frac{\sin \theta_{15} \cos \theta_6}{\sqrt{1 - (\sin \theta_{15} \sin \theta_6)^2}}
\end{aligned} \qquad 8.23$$

where $\sin \theta_{15} = \sin(\theta_1 - \theta_5)$ and $\cos \theta_{15} = \cos(\theta_1 - \theta_5)$. Equation 8.22 represents the unactuated joints using only θ_3 and θ_6. Equation 8.23 represents them without using θ_3 and with using θ_1, θ_5, and θ_6. Equations 8.22 and 8.23 provide the two representations of Eq. 8.3.

The Jacobian matrices $\bm{J}^1$ and $\bm{J}^2$ are computed from Eqs. 8.22 and 8.23 as follows:

$$\bm{J}^1 = \begin{pmatrix} 0 & -k_1 S_3 C_6/C_3 & 0 & k_1 S_6 \\ 0 & -1 & 0 & 0 \\ 0 & k_2 C_6/C_3 & 0 & -k_2 S_3/S_6 \end{pmatrix} \qquad 8.24$$

$$k_1 = 1/\sqrt{C_3^2 - C_6^2}$$

$$k_2 = 1/\sqrt{(C_3 S_6)^2 - (S_3 C_6)^2}$$

$$\bm{J}^2 = \begin{pmatrix} -k_3 S_6 C_6 S_{15} & 0 & k_3 S_6 C_6 S_{15} & k_3 C_{15} \\ -k_4 S_6 C_{15} & 0 & k_4 S_6 C_{15} & -k_4 C_6 S_{15} \\ -k_3 C_6 & 0 & k_3 C_6 & k_3 S_{15} C_{15} S_6 \end{pmatrix} \qquad 8.25$$

$$k_3 = 1/(1 - (S_{15} S_6)^2)$$

$$k_4 = 1/\sqrt{1 - (S_{15} S_6)^2}$$

where S and C indicate sin and cos, respectively.

We choose the independent actuators as follows:

$$\bm{\theta}_{10} = (\theta_1 \; \theta_5 \; \theta_6)^T \qquad 8.26$$

Comparing the second equalities of Eqs. 8.22 and 8.23 provides

$$\sin\theta_3 = -\sin(\theta_1 - \theta_5)\sin\theta_6 \qquad 8.27$$

where the minus sign is determined from Fig. 8.5. From Eqs. 8.26 and 8.27, the instantaneous coefficient matrix $\bm{S}$ in Eq. 8.15 becomes

$$\bm{S} = \begin{pmatrix} 1 & 0 & 0 \\ -C_{15} S_6/C_3 & C_{15} S_6/C_3 & -S_{15} C_6/C_3 \\ 0 & 1 & 0 \\ 0 & 0 & 1 \end{pmatrix} \qquad 8.28$$

For the sake of comparison, we call the structure of Fig. 8.3(b) the *nonredundant* case. The actuated joints of the nonredundant case coincide with $\bm{\theta}_{10}$ of Eq. 8.26. Note that the denominators of Eqs. 8.22 through 8.25 and 8.28 become zero only when the closed-link mechanism collapses, which should be considered outside of the workspace.

8.5.2 Computation of Static Joint Torque

The control of actuation redundancy in Section 8.4 works for any dynamic motion of closed-link mechanisms. In Section 8.5.3, we shall show the results of numerical analysis. To understand physical meaning easily, we shall merely consider the torque distribution for static loads at the end-effector. Here we explain the computational method of the equivalent joint torque.

From Fig. 8.5(a), the position of the end-effector is represented in terms of θ_1 through θ_4 as follows:

$$\boldsymbol{x} = \begin{pmatrix} l_4(C_1C_2C_{34} - S_1S_{34}) + l_3(C_1C_2C_3 - S_1S_3) + l_2S_1 \\ l_4(S_1C_2C_{34} + C_1S_{34}) + l_3(S_1C_2C_3 + C_1S_3) - l_2C_1 \\ l_4S_2C_{34} + l_3S_2C_3 \end{pmatrix} \quad 8.29$$

where $S_{34} = \sin(\theta_3 + \theta_4)$ and $C_{34} = \cos(\theta_3 + \theta_4)$. The 3×4 Jacobian matrix $\boldsymbol{J}$ with respect to θ_1, θ_2, θ_3, and θ_4 is directly computed from Eq. 8.29. The (i, j) entry of $\boldsymbol{J}$ is represented by J_{ij} and is computed as follows:

$$\begin{aligned}
J_{11} &= -l_4(S_1C_2C_{34} + C_1S_{34}) - l_3(S_1C_2C_3 + C_1S_3) + l_2C_1 \\
J_{12} &= -l_4C_1S_2C_{34} - l_3C_1S_2C_3 \\
J_{13} &= -l_4(C_1C_2S_{34} + S_1C_{34}) - l_3(C_1C_2S_3 + S_1C_3) \\
J_{14} &= -l_4(C_1C_2S_{34} + S_1C_{34}) \\
J_{21} &= l_4(C_1C_2C_{34} - S_1S_{34}) + l_3(C_1C_2C_3 - S_1S_3) + l_2S_1 \\
J_{22} &= -l_4(S_1S_2C_{34}) - l_3S_1S_2C_3 \\
J_{23} &= l_4(C_1C_{34} - S_1C_2S_{34}) + l_3(C_1C_3 - S_1C_2S_3) \\
J_{24} &= l_4(C_1C_{34} - S_1C_2S_{34}) \\
J_{31} &= 0 \\
J_{32} &= l_4C_2C_{34} + l_3C_2C_3 \\
J_{33} &= -l_4S_2S_{34} - l_3S_2S_3 \\
J_{34} &= -l_4S_2S_{34}
\end{aligned} \quad 8.30$$

The torque necessary to maintain static equilibrium against force $\boldsymbol{F} \in R^3$ exerted on the end-effector is computed by

$$\begin{aligned}
\boldsymbol{\tau}_0 &\triangleq (\tau_1 \ \tau_2 \ \tau_3 \ \tau_4)^T \\
&= -\boldsymbol{J}^T \boldsymbol{F}
\end{aligned} \quad 8.31$$

Since the end of the other branch has no external force, the joint torques of the branch are all zero and the joint torques of the whole tree structure used in computing Eqs. 8.7 and 8.20 become

$$\boldsymbol{\tau} = (\tau_1 \ \tau_3 \ 0 \ 0 \ \tau_2 \ \tau_4 \ 0)^T \quad 8.32$$

Note that the first four components correspond to the actuated joints, and the rest correspond to the unactuated ones.

8.5.3 Results of Numerical Analysis

The main aim of utilizing actuation redundancy is to increase the static and dynamic payload of robotic mechanisms. Joint torques are usually subject to their limits

$$|\tau_{1i}^c| \leq \tau_{1i\max}^c \quad \text{for } i = 1, \cdots, N_{ta} \quad \quad 8.33$$
$$\tau_{1i\max}^c > 0$$

where τ_{1i}^c is the ith component of $\boldsymbol{\tau}_1^c$, and $\tau_{1i\max}^c$ is the maximum magnitude. A problem where the following quadratic function is to be minimized

$$\min \boldsymbol{\tau}_1^{cT} \boldsymbol{A} \boldsymbol{\tau}_1^c \quad \quad 8.34$$

is called a *quadratic-programming* problem (Luenberger 1984). This problem is easy to solve when only one additional actuator is involved in a closed-link mechanism, as we discussed in Section 7.3.2 and Algorithm 7.1.

In this subsection, we see the advantage of actuation redundancy and the improvement of its new singularity-free parameterization. To compare the redundant actuation system (Fig. 8.4) and the nonredundant actuation system (Fig. 8.3b), we modify the performance index of Eq. 8.34 as follows:

$$\min \sqrt{\frac{1}{N_{ta}} \sum_{i=1}^{N_{ta}} (\tau_{1i}^c / \tau_{1i\max}^c)^2} \quad \quad 8.35$$

Equation 8.35 is equivalent to minimizing a quadratic function of $\boldsymbol{\tau}_1^c$ and can be considered as a variation of Eq. 8.34. Equation 8.35 tends to evaluate

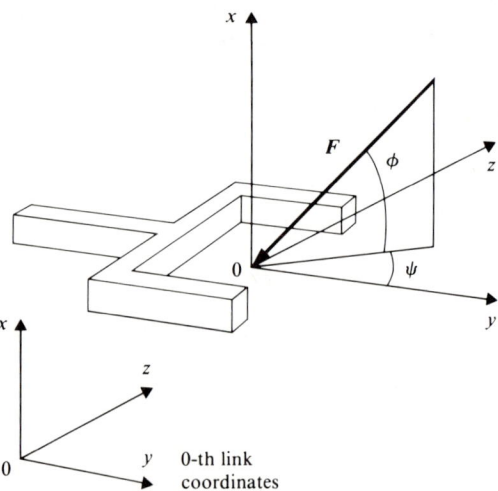

Figure 8.6 Spherical coordinates to describe the direction of the end-effector force.

a mean load of joints, and becomes one when all the actuators need their maximum torques.

Performance Index Analysis. We first investigate the change of joint torque as the change of the direction of end-effector force $\boldsymbol{F}$. The direction of $\boldsymbol{F}$ is indicated by two spherical coordinates ϕ and ψ, as shown in Fig. 8.6.

Figure 8.7 shows the change of the performance index for the nonredundant system of Fig. 8.3(b). We assumed $\|\boldsymbol{F}\| = 10$ N and $\tau^c_{1i\,\text{max}} = 6$ N/m. These values are picked up simply for the purpose of the numerical analysis.

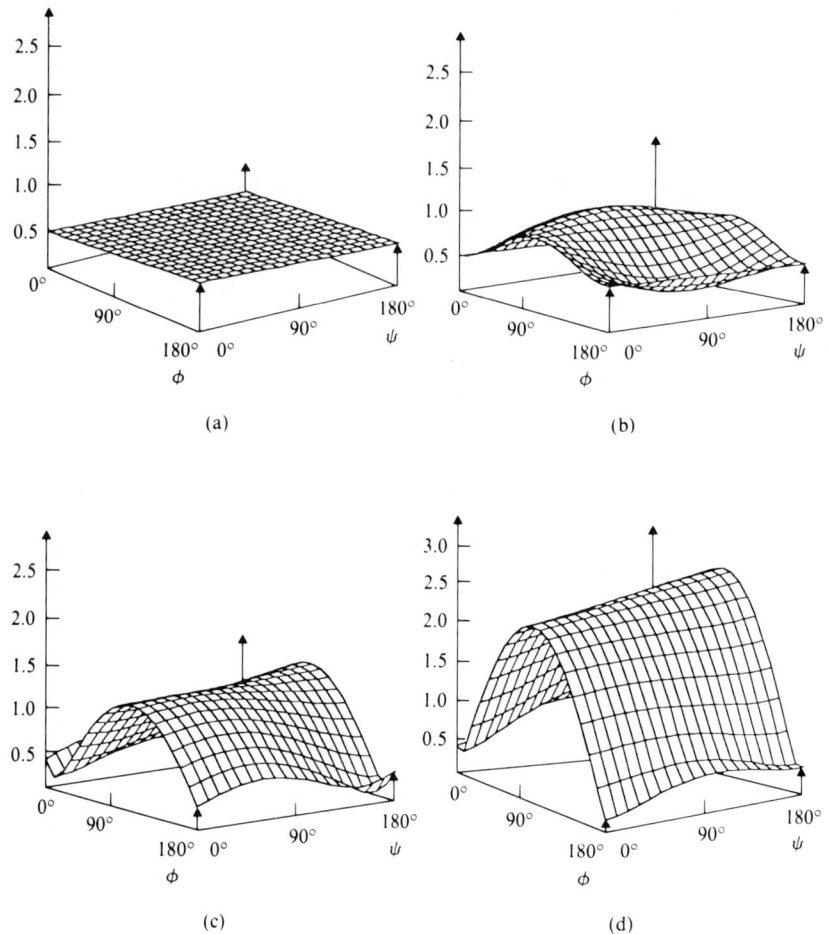

Figure 8.7 Change of performance index of Eq. 8.35 in response to the change of external force direction at the end-effector. $\|\boldsymbol{F}\| = 10\,\text{N}$, $\tau^c_{1i\,\text{max}} = 6\,\text{N/m}$: Nonredundant case (Fig. 8.3b). (a) $(\theta_1, \theta_5, \theta_6) = (0°, 0°, 90°)$. (b) $(\theta_1, \theta_5, \theta_6) = (0°, 0°, 60°)$. (c) $(\theta_1, \theta_5, \theta_6) = (0°, 60°, 60°)$. (d) $(\theta_1, \theta_5, \theta_6) = (0°, 60°, 90°)$.

244 Chapter 8 A Manipulator with Kinematic and Actuation Redundancy

Figure 8.8 shows the change of the minimum value of the performance index obtained by adjusting actuation redundancy. Figure 8.8 is computed using the previous parameterization of Eq. 8.7 developed in Chapter 7 for the redundant system of Fig. 8.4. The same assumption was made as in the nonredundant case. In Fig. 8.8(a) and (b), discontinuous large changes appear for some directions of $\boldsymbol{F}$. These directions are those included in the horizontal plane ($\phi = 0$ or $180°$), which is the null space discussed in Section 8.4.2. The discontinuity was caused when the minimization of Eq. 8.35, using the second

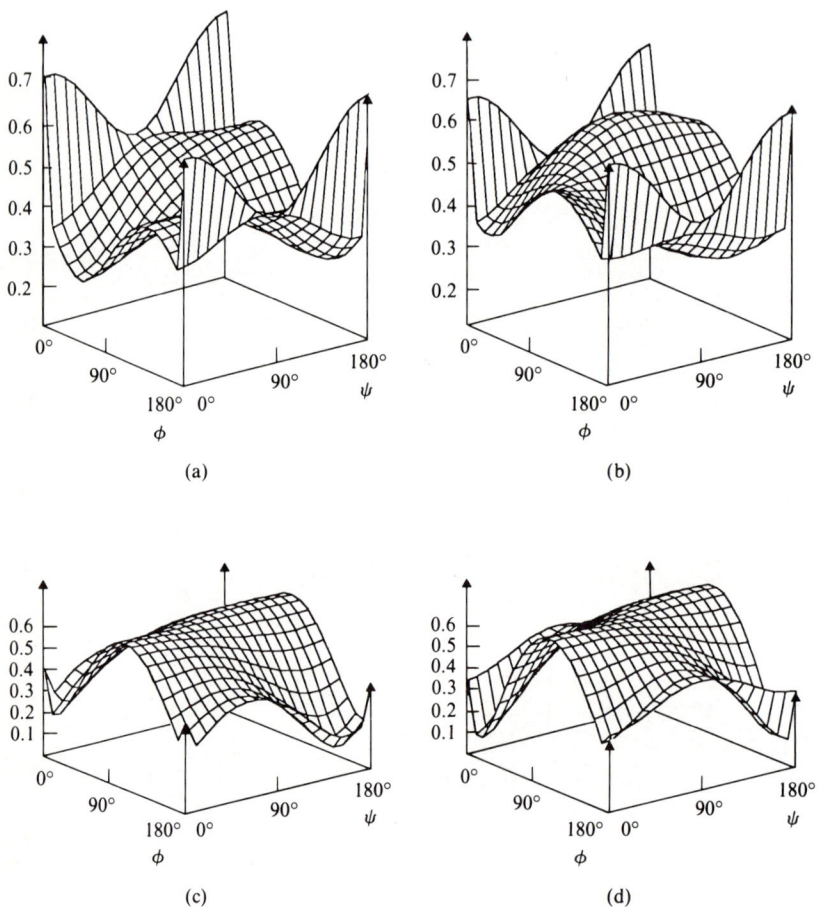

Figure 8.8 Change of performance index of Eq. 8.35 in response to the change of external force direction at the end-effector. $\|\boldsymbol{F}\| = 10\,\text{N}$, $\tau^c_{1i\,\text{max}} = 6\,\text{N/m}$: Redundant case (Fig. 8.4) by the parameterization developed in Chapter 7. (a) $(\theta_1, \theta_5, \theta_6) = (0°, 0°, 90°)$. (b) $(\theta_1, \theta_5, \theta_6) = (0°, 0°, 60°)$. (c) $(\theta_1, \theta_5, \theta_6) = (0°, 60°, 60°)$. (d) $(\theta_1, \theta_5, \theta_6) = (0°, 60°, 90°)$.

term of Eq. 8.7, became impossible due to the sudden encounter of the algorithmic singularity. The large values at the discontinuous points correspond to the first term of Eq. 8.7.

Figure 8.9 shows the change of the minimum value of the performance index obtained using the new parameterization of Eq. 8.20 for the redundant system. By comparing Fig. 8.9 with Fig. 8.8, we can see that the discontinuity of Fig. 8.8 disappeared. The values of the performance index, except for those at the singular situation, are equivalent to those of Fig. 8.8.

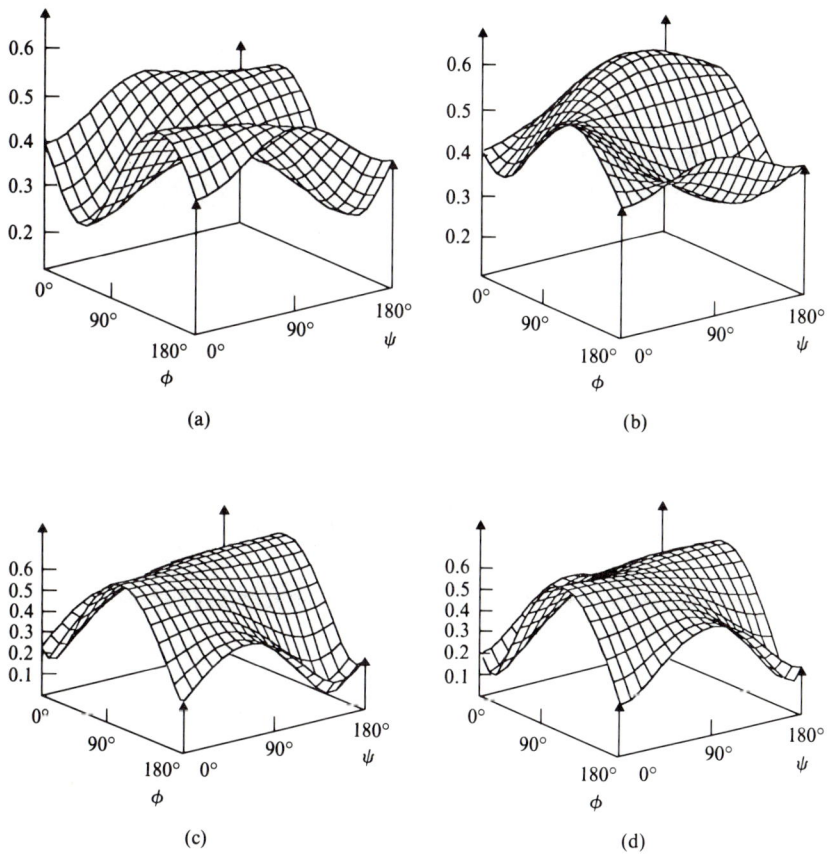

Figure 8.9 Change of performance index of Eq. 8.35 in response to the change of external force direction at the end-effector. $\| F \| = 10\,\text{N}$, $\tau^c_{1i\,\text{max}} = 6\,\text{N/m}$: Redundant case (Fig. 8.4) by the new parameterization. (a) $(\theta_1, \theta_5, \theta_6) = (0°, 0°, 90°)$. (b) $(\theta_1, \theta_5, \theta_6) = (0°, 0°, 60°)$. (c) $(\theta_1, \theta_5, \theta_6) = (0°, 60°, 60°)$. (d) $(\theta_1, \theta_5, \theta_6) = (0°, 60°, 90°)$.

The following observations are obtained from the comparison of Figs. 8.7 and 8.9:

1. The actuation redundancy significantly reduces the values of the performance index, compared with the nonredundant case.

2. The surface becomes wavier as it proceeds from (a) to (d) in Fig. 8.7. In contrast, the surface tends to become less wavy in Fig. 8.9. This is because all the actuated joints become more coupled in (d) than in (a), and the actuation redundancy can contribute to the most demanding joints.

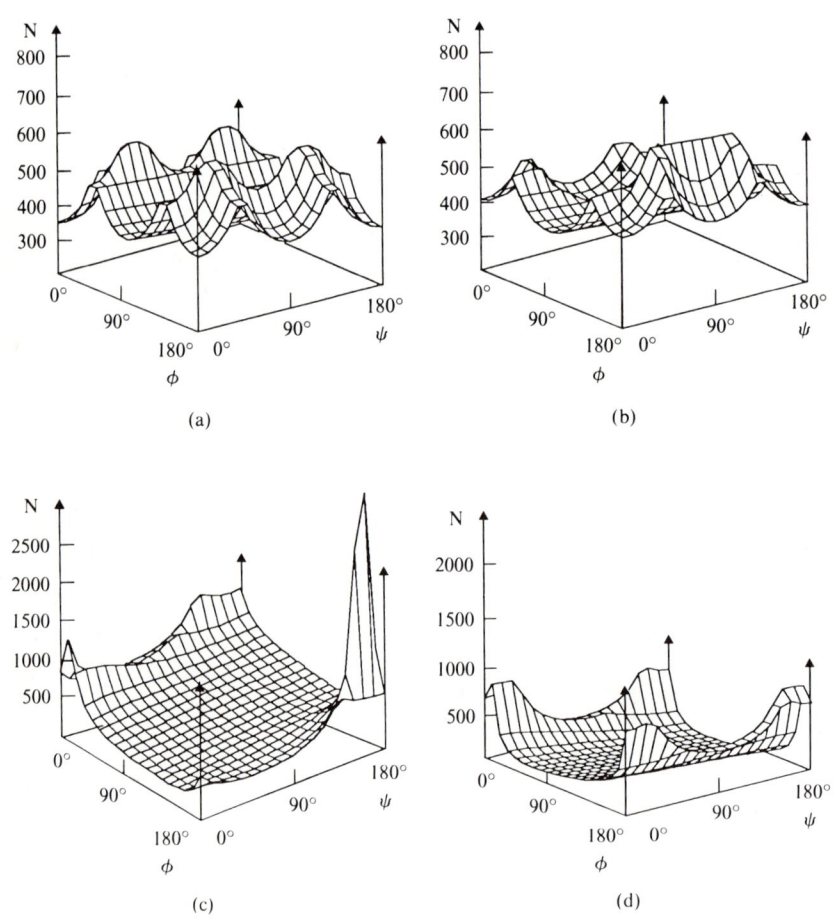

Figure 8.10 Maximum supportable external force at the end-effector: Nonredundant case (Fig. 8.3b). $\tau^c_{1i\,\max} = 220\,\text{N/m}$ ($i = 1, 2, 3$). (a) (θ_1, θ_5, θ_6) = ($0°$, $0°$, $90°$). (b) (θ_1, θ_5, θ_6) = ($0°$, $0°$, $60°$). (c) (θ_1, θ_5, θ_6) = ($0°$, $60°$, $60°$). (d) (θ_1, θ_5, θ_6) = ($0°$, $60°$, $90°$).

8.5 Numerical Analysis 247

Maximum Payload Analysis. Next, we investigate the change of the maximum magnitude of supportable end-effector force in response to the change of the direction of end-effector force. The direction is again indicated by ϕ and ψ, shown in Fig. 8.6. The maximum motor torque was assumed 220 Nm.

Figure 8.10 shows the maximum supportable end-effector force for the nonredundant system of Fig. 8.3(b). The surfaces of Fig. 8.10(a) and (b) are somewhat wavy. This might be because at these configurations ((a) $(\theta_1, \theta_5, \theta_6) = (0°, 0°, 90°)$, and (b) $(\theta_1, \theta_5, \theta_6) = (0°, 0°, 60°)$), the contributions of three actuators are decoupled in three orthogonal directions of the end-effector force and, therefore, the maximum magnitude is sensitive in direction. At the configuration of (c) and (d), the maximum magnitude surfaces are less wavy.

Figure 8.11 shows the effect of putting two motors at θ_1 (four motors total). As compared with Fig. 8.10(a) and (b), the middle parts (near $\phi = 90°$) of the surfaces of Fig. 8.11(a) and (b) are lifted; however, the boundary parts (near $\phi = 0$ and $180°$) remain almost unchanged. The surfaces of Fig. 8.11(c) and (d) are very similar to those in Fig. 8.10(c) and (d).

Figure 8.12 corresponds to the redundant system of Fig. 8.4. The shapes of surfaces of Fig 8.12(a) and (b) are quite different from those in Figs. 8.10(a) and (b) and 8.11(a) and (b). Note that, as well as the middle part, the boundary parts near corners are significantly lifted. The improvement in Fig 8.12(c) and (d) is remarkable. The entire surfaces are raised.

Table 8.2 shows the minimum and maximum values of each surface of Figs. 8.10 through 8.12. The minimum vales of Figs. 8.10 and 8.11 are the same. The increase of maximum values from Fig. 8.10 to Fig. 8.11 is marginal. The minimum values of Fig. 8.12(a) and (b) are also the same as those of Fig. 8.10(a) and (b) and Fig. 8.11(a) and (b). The improvements near four corners are between the maximum and minimum values and, therefore, are invisible in the table. The minimum values of Fig. 8.12(c) and (d) are 2.46 and 4.22 times larger than those of Fig. 8.10(c) and (d) and Fig. 8.11(c) and (d), respectively.

Table 8.2. Minimum and maximum supportable forces at each configuration of Figs. 8.10, 8.11, and 8.12.

	Fig. 8.10 min	Fig. 8.10 max	Fig. 8.11 min	Fig. 8.11 max	Fig. 8.12 min	Fig. 8.12 max
(a)	367	553	367	847	367	1019
(b)	367	734	367	1141	367	1130
(c)	166	4927	166	5457	408	2538
(d)	87	905	87	1549	367	2145

248 Chapter 8 A Manipulator with Kinematic and Actuation Redundancy

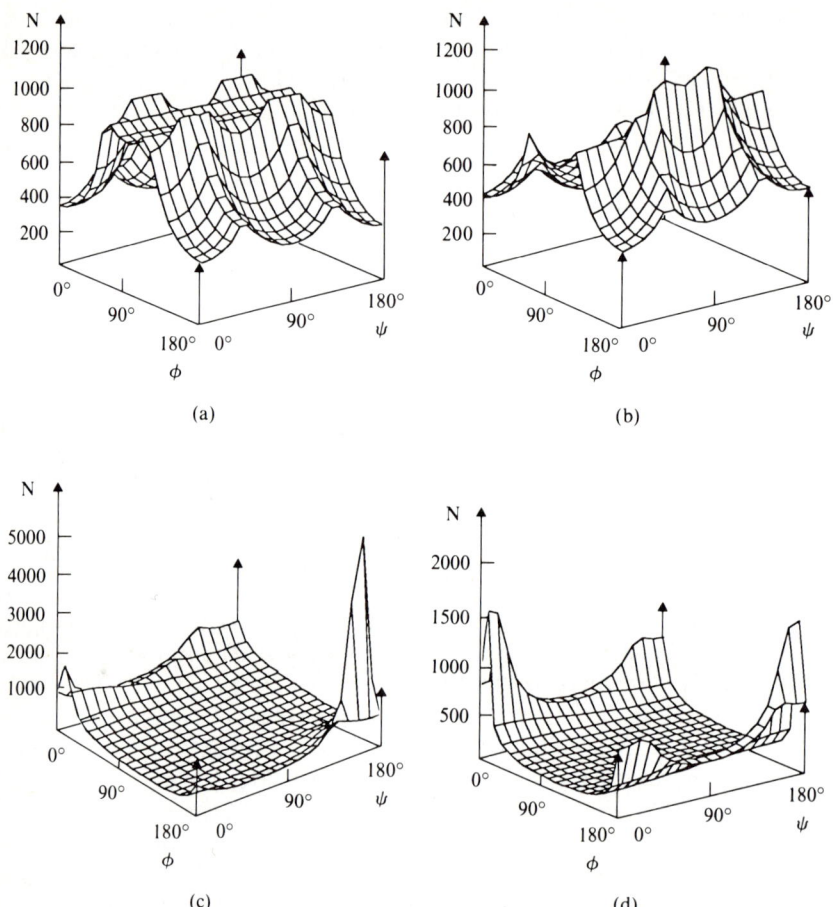

Figure 8.11 Maximum supportable external force at the end-effector: Nonredundant case (Fig. 8.3b). $\tau^c_{11\,\text{max}} = 440\,\text{N/m}$, $\tau^c_{1i\,\text{max}} = 220\,\text{N/m}$ ($i = 2, 3$). (a) $(\theta_1, \theta_5, \theta_6) = (0°, 0°, 90°)$. (b) $(\theta_1, \theta_5, \theta_6) = (0°, 0°, 60°)$. (c) $(\theta_1, \theta_5, \theta_6) = (0°, 60°, 60°)$. (d) $(\theta_1, \theta_5, \theta_6) = (0°, 60°, 90°)$.

Summary

A closed-link mechanism for kinematically redundant direct-drive manipulators was proposed, and its redundant actuation was discussed. The mechanism is kinematically equivalent to the base four joints of an anthropomorphic 7-DOF manipulator. Five actuators were used to control the 3-DOF position of the end-effector.

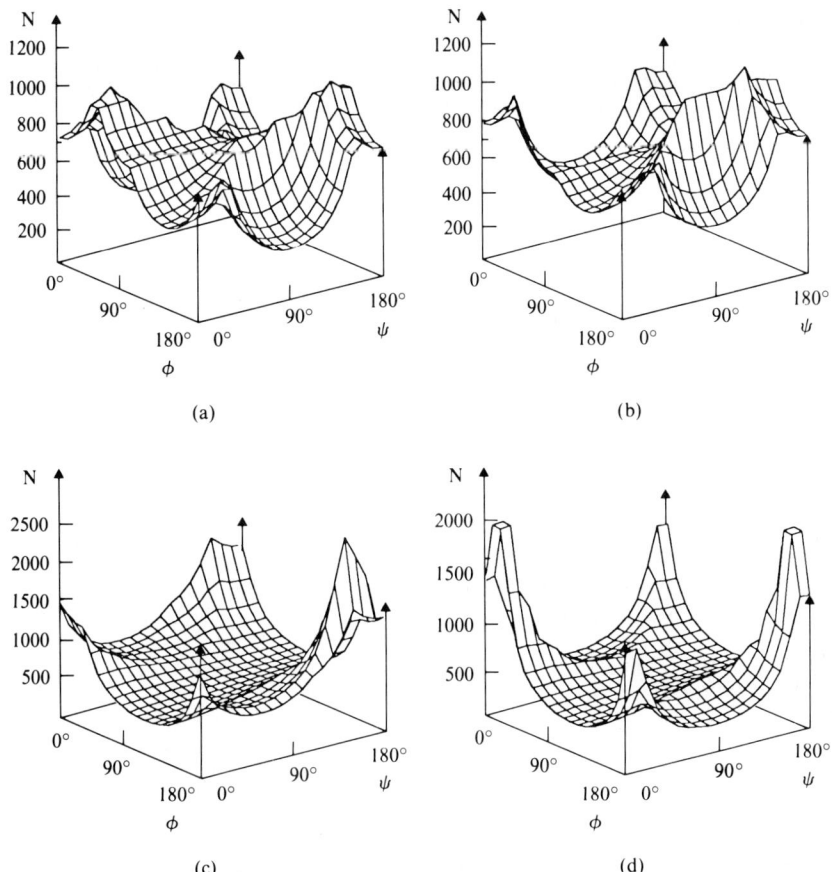

Figure 8.12 Maximum supportable external force at the end-effector: Redundant case (Fig. 8.4). $\tau^c_{1i\,\max} = 220\,\mathrm{N/m}$ ($i = 1, 2, 3, 4$). (a) $(\theta_1, \theta_5, \theta_6) = (0°, 0°, 90°)$. (b) $(\theta_1, \theta_5, \theta_6) = (0°, 0°, 60°)$. (c) $(\theta_1, \theta_5, \theta_6) = (0°, 60°, 60°)$. (d) $(\theta_1, \theta_5, \theta_6) = (0°, 60°, 90°)$.

The singularity of the parameterization of actuation redundancy developed in Chapter 7 (Nakamura and Ghodoussi 1989) was pointed out, and its physical meaning was clarified. We then proposed a new singularity-free parameterization that allows constant full utilization of actuation redundancy. The improvement due to the new parameterization was demonstrated numerically.

The advantage of actuation redundancy was investigated in the light of performance index minimization and maximum payload. The results of optimization of actuation redundancy were remarkable for both purposes.

References

Asada, H., Kanade, T., and Takeyama, I. 1983. Control of a direct-drive arm. *ASME J. Dyn. Sys., Meas. Contr.* 105 (3): 136–142.

Asada, H., and Youcef-Toumi, K. 1984. Analysis and design of a direct-drive arm with a five-bar-link parallel drive mechanism. *ASME J. Dyn. Sys., Meas. Contr.* 106 (3): 225–230.

Asada, H., and Youcef-Toumi, K. 1987. *Direct-drive robots.* Cambridge, MA: MIT Press.

Denavit, J., and Hartenberg, R. S. 1955. A kinematic notation for lower-pair mechanisms based on matrices. *ASME Trans. J. Appl. Mech.* 77 (2): 215–221.

Ligeois, A. 1977. Automatic supervisory control of the configuration and behavior of multibody mechanisms. *IEEE Trans. Syst., Man Cybernet.* 7 (12): 868–871.

Luenberger, D. G. 1984. *Linear and non-linear programming*, second edition. Reading, MA: Addison-Wesley.

Nakamura, Y., and Hanafusa, H. 1987. Optimal redundancy control of robot manipulators. *Int. J. Robotics Res.* 6 (1): 32–42.

Nakamura, Y., Hanafusa, H., and Yoshikawa, T. 1987. Task-priority based redundancy control of robot manipulators. *Int. J. Robotics Res.* 6 (2): 3–15.

Nakamura, Y., and Ghodoussi, M. 1988 (Philadelphia). A computational scheme of closed link robot dynamics derived by d'Alembert principle. *Proc. 1988 IEEE Int. Conf. Robotics and Automation*, pp. 1354–1360.

Nakamura, Y., and Ghodoussi, M. 1989. Dynamics computation of closed link robot mechanisms with non-redundant and redundant actuators. *IEEE J. Robotics and Automat.* 5 (3): 294–302.

Nakamura, Y., and Ropponen, T. 1989 (San Francisco). A closed link manipulator with kinematic and actuation redundancies. *Robotics Research—1989*, eds. K. Youcef-Toumi and H. Kazerooni, 1989 ASME Winter Annual Meeting, ASME, pp. 203–210.

Nakamura, Y., and Ropponen, T. 1990 (San Diego). Actuation redundancy of a closed link manipulator. *Proc. 1990 American Control Conference*, pp. 2294–2299.

Nakamura, Y., and Ropponen, T. 1990 (Kyoto). Dynamic payload maximization of a closed link robot with actuation redundancy. *Proc. 1990 Japan-USA Symposium on Flexible Automation*, pp. 321–327.

Rao, C. R., and Mitra, S. K. 1971. *Generalized inverse of matrices and its applications.* New York: Wiley.

Ropponen, T., and Nakamura, Y. 1990 (Cincinnati). Singularity-free parameterization and performance analysis of actuation redundancy. *Proc. 1990 IEEE Int. Conf. Robotics and Automation,* pp. 806–811.

CHAPTER 9

Singularity-Robust Inverse of Jacobian Matrix

9.1 Introduction

Inverse kinematics is one of the functions indispensable to robot manipulator control systems. The position and orientation, r, of the end-effector is described as a nonlinear function of the joint variable $\theta \in R^n$ as follows:

$$r = f(\theta) \qquad 9.1$$

The differential relationships of Eqs. 3.9 and 3.15 can be used to relate the velocities or accelerations of r and θ. As we discussed in Section 3.3.1, both cases are commonly described by the following linear equation with the Jacobian matrix, $J(\theta) \triangleq \partial f / \partial \theta$:

This chapter is adapted from, by permission, Nakamura, Y., and Hanafusa, H. 1984. "Singularity low-sensitive motion resolution of articulated robot arms," *Trans. Society of Instrument and Control Engineers*, Vol. 20, No. 5 pp. 453–459 (*in Japanese*); Nakamura, Y., and Hanafusa, H. 1985. "Inverse kinematic solutions with singularity robustness for robot manipulator control," *Robotics and Manufacturing Automation-PED-*, Vol. 15 (Book No. G00321) (presented at 1985 ASME Winter Annual Meeting, Miami Beach), eds. M. Donath and M. Leu, ASME, pp. 193–204; and Nakamura, Y., and Hanafusa, H. 1986. "Inverse kinematic solutions with singularity robustness for robot manipulator control," *Journal of Dynamic Systems, Measurement, and Control*, Vol. 108, pp. 163–171.

$$\delta r = J(\theta)\delta\theta \qquad 9.2$$

To calculate the inverse kinematics, we solve either Eq. 9.1 for θ or Eq. 9.2 for $\delta\theta$. When r and θ have the same dimension, calculating the inverse kinematics is considered as finding the inverse function of either the nonlinear function of Eq. 9.1 or the linear function of Eq. 9.2. For both cases, the condition for the existence of inverse function is det $J(\theta) \neq 0$ (Lang 1968). Hence, θ is called a *singular point* if det $J(\theta) = 0$. For kinematically redundant manipulators, a singular point is defined as θ that makes $J(\theta)$ not full rank.

Since singular points are found not only in the boundary of the workspace, but also in the middle of it, the singular points limit the end-effector trajectories that a manipulator should follow in solving inverse kinematics. However, the most serious problem of singularity is not at singular points, which have zero volume in the workspace, but rather in the neighborhood of singular points, which has finite volume. In the neighborhood of singular points, even for a small change of r, an enormous change of θ is often required. This causes a large error in the end-effector motion, since the actuator torques and velocities are restricted. The singularity problem is an inherent problem in controlling articulated manipulators.

For 6-DOF manipulators, Pieper (1969) discussed the solution of Eq. 9.1, and clarified the kinematic structures that have analytical solutions. Equation 9.1 has plural solutions in general. Uchiyama (1979) pointed out that singular points are the points where multiple joint trajectories that have the same end-effector motion should meet one another. He also suggested that, in the neighborhood of singular points, the control mode should be switched from operational space control using inverse kinematics to joint control without using inverse kinematics. Whitney (1972) proposed calculating an approximate solution at singular points by disregarding the degenerated block of the Jacobian matrix. This approach would be generalized into a scheme using pseudoinverses, as we studied in Chapter 2. Waldron, Wang, and Bolin (1984) discussed the Jacobian matrices of serial manipulators. Orin and Schrader (1984) proposed an efficient computation algorithm of the Jacobian matrix.

In Uchiyama's mode switching, we modify the desired trajectory in the neighborhood of singular points. The modification, if it is necessary, should be done based on a reasonable criterion. Moreover, the joint trajectory in joint control mode cannot be readily obtained if the desired end-effector motion is generated from sensory information. Although use of Whitney's approximation or of the pseudoinverse of the Jacobian matrix provides a reasonable approximation at singular points, it would not help in the neighborhood of singular points. Moreover, the sudden switching from the exact solution to an approximation would need further research to ensure the stability and accuracy of control systems that include inverse kinematics in their feedback loops.

In Chapters 4 and 5, we learned that kinematic redundancy may allow us to avoid singularities while moving the end-effector along the desired trajectories. Hollerbach (1985) discussed where to locate the seventh joint of a PUMA-type 6-DOF manipulator in order to reduce the internal singularities. Yoshikawa (1985) developed a 4-DOF wrist mechanism and discussed the utilization of redundancy to avoid wrist singularities. Although kinematic redundancy is certainly effective in reducing the risk that a robot manipulator will fall into singularities, it cannot essentially do away with singularities. Therefore, careful consideration of unavoidable singularities must be done from a control point of view.

In this chapter, we study a method to solve inverse kinematics while simultaneously evaluating the feasibility of joint motion. Since the method provides a feasible joint motion even at or in the neighborhood of singular points, it has robustness in terms of kinematic singularity. As an alternative to the pseudoinverse, the *singularity-robust inverse* (SR-inverse) is introduced. The SR-inverse of the Jacobian matrix offers a feasible motion close to the desired trajectory of the end effector, even when the inverse kinematic solution by the inverse or the pseudoinverse is not feasible at or in the neighborhood of singular points. We clarify the properties of the SR-inverse by comparing with the inverse and the pseudoinverse. The computational complexity of the SR-inverse is also discussed. Simulation results are to be given to illustrate the singularity problem and the effectiveness of an inverse kinematic solution with singularity robustness.

9.2 Singularity and Pseudoinverse

When a robot manipulator is controlled by velocity control, we calculate inverse kinematics by solving the following equation for $\dot{\boldsymbol{\theta}}$:

$$\dot{\boldsymbol{r}} = \boldsymbol{J}(\boldsymbol{\theta})\dot{\boldsymbol{\theta}} \qquad 9.3$$

When a robot manipulator is controlled by acceleration control, we calculate the inverse kinematics by solving the following equation for $\ddot{\boldsymbol{\theta}}$:

$$\ddot{\boldsymbol{r}} - \dot{\boldsymbol{J}}(\boldsymbol{\theta})\dot{\boldsymbol{\theta}} = \boldsymbol{J}(\boldsymbol{\theta})\ddot{\boldsymbol{\theta}} \qquad 9.4$$

Since the left-hand side of Eq. 9.4 is determined by the desired acceleration $\ddot{\boldsymbol{r}}$ and the current $\boldsymbol{\theta}$ and $\dot{\boldsymbol{\theta}}$, the necessary computation for Eq. 9.4 is to solve the same linear equation as Eq. 9.3. Therefore, the differential kinematics is generally represented by Eq. 9.2, whether velocity control or acceleration control is chosen. Note that the inverse kinematic solutions obtained from Eqs. 9.1 and 9.2 are equivalent if the continuity of $\boldsymbol{\theta}$ is taken into account and the initial values of $\boldsymbol{\theta}$ are the same.

Let $\boldsymbol{r} \in R^m$, $m \leq n$. If $m = n$ and $\det \boldsymbol{J}(\boldsymbol{\theta}) \neq 0$, the differential joint motion, $\delta\boldsymbol{\theta}$, for the desired differential motion of the end effector, $\delta\boldsymbol{r}$, can be solved by the inverse of the Jacobian matrix as follows:

$$\delta\boldsymbol{\theta} = \boldsymbol{J}^{-1}(\boldsymbol{\theta})\delta\boldsymbol{r} \qquad 9.5$$

At singular points, Eq. 9.5 cannot be used, since $\boldsymbol{J}^{-1}(\boldsymbol{\theta})$ is not defined.

The pseudoinverse of the Jacobian matrix has been proposed as a remedy for this problem since it is defined even at singular points. Using the pseudoinverse of the Jacobian matrix, $\boldsymbol{J}^{\#}(\boldsymbol{\theta}) \in R^{n \times m}$, Eq. 9.2 is now solved as follows:

$$\delta\boldsymbol{\theta} = \boldsymbol{J}^{\#}(\boldsymbol{\theta})\delta\boldsymbol{r} \qquad 9.6$$

As we saw in Theorem 2.5 in Section 2.4.3, Eq. 9.6 offers a least-squares solution with a minimum norm for Eq. 9.2; namely, $\delta\boldsymbol{\theta}$ of Eq. 9.6 satisfies

$$\min \ \| \ \delta\boldsymbol{\theta} \ \| \qquad 9.7$$

among all of the $\delta\boldsymbol{\theta}$ that fulfill

$$\min \ \| \ \delta\boldsymbol{r} - \boldsymbol{J}(\boldsymbol{\theta})\delta\boldsymbol{\theta} \ \| \qquad 9.8$$

where $\| * \|$ denotes the Euclidean norm of $*$.

When $m = n$, the difference between $\boldsymbol{J}^{-1}(\boldsymbol{\theta})$ and $\boldsymbol{J}^{\#}(\boldsymbol{\theta})$ is that $\boldsymbol{J}^{-1}(\boldsymbol{\theta})$ is not defined at singular points, whereas $\boldsymbol{J}^{\#}(\boldsymbol{\theta})$ is defined and provides an approximation in the sense of Eqs. 9.7 and 9.8. This is an advantage of using Eq. 9.6 over using Eq. 9.5. However, from the control point of view, we have to pay attention to the fact that the solution by Eq. 9.6 is switched discontinuously from an exact solution to an approximation at singular points. Moreover, in the neighborhood of singular points, the pseudoinverse of the Jacobian matrix as well as the inverse is forced to find an exact solution even if $\delta\boldsymbol{r}$ is in the almost-degenerated direction, which often results in prohibitively large $\delta\boldsymbol{\theta}$. This problem can be considered to be a result of the rigid hierarchical structure of Eqs. 9.7 and 9.8—Eq. 9.8 has priority higher than that of Eq. 9.7. In other words, the exactness of the solution is paid more attention than is the feasibility of the solution.

Figure 9.1 shows the block diagram of a typical robot control system. Since the block of inverse kinematics is inside the closed loop, the discontinuity or the prohibitively large solution by $\boldsymbol{J}^{\#}(\boldsymbol{\theta})$ would cause unstable or fatally disturbed behavior of the system. This tendency of the pseudoinverse is maintained even for redundant manipulators. Indeed, the author sometimes experienced unstable oscillations in the neighborhood of singular points in controlling UJIBOT with the pseudoinverse, which was used for the experiment in Chapter 4.

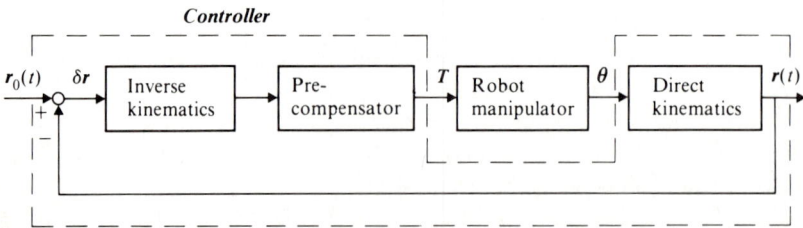

Figure 9.1 Block diagram of a typical robot control system.

We have two different approaches to overcoming this inherent singularity of robot manipulators. In Chapter 5, we discussed a method for planning a trajectory that is far from singular points. In this chapter, we establish a method to pass through singular points and their neighborhood.

9.3 SR-Inverse of the Jacobian Matrix

In this section, we introduce the *singularity-robust inverse* (SR-inverse), where singularity robustness is defined as a property of inverse kinematics that allows us to provide continuous and feasible solutions even at or in the neighborhood of singular points.

In place of the hierarchical structure of Eqs. 9.7 and 9.8, we propose using the following equation as an evaluation criterion for the solution of Eq. 9.2:

$$\min \| \delta e \|_W \qquad 9.9$$

$$\delta e \triangleq \begin{pmatrix} \delta r - J(\theta)\delta\theta \\ \delta\theta \end{pmatrix}$$

where $\| * \|_W$ implies the weighted norm that we discussed in Section 2.4.4, and is defined by Eq. 2.169. Now we choose the weighting matrix as follows:

$$W = \begin{pmatrix} W_1 & O \\ O & W_2 \end{pmatrix} \qquad 9.10$$

where $W \in R^{(m+n)\times(m+n)}$, $W_1 \in R^{m\times m}$, and $W_2 \in R^{n\times n}$. Here, W_1 and W_2 are symmetric positive definite matrices. Then, $\| \delta e \|_W$ satisfies

$$\| \delta e \|_W^2 = \| \delta r - J(\theta)\delta\theta \|_{W_1}^2 + \| \delta\theta \|_{W_2}^2 \qquad 9.11$$

Equations 9.9 and 9.11 imply that we determine the solution by simultaneously evaluating the exactness and the feasibility.

From Eqs. 2.169 and 9.10, Eq. 9.11 is now computed as follows:

$$\| \delta e \|_W^2 = (\delta\theta - \delta\theta_s)^T(J^T W_1 J + W_2)(\delta\theta - \delta\theta_s) + \delta r^T W_r \delta r \qquad 9.12$$

$$\delta\theta_s \triangleq (J^T W_1 J + W_2)^{-1} J^T W_1 \delta r$$

$$W_r \triangleq W_1 - W_1 J (J^T W_1 J + W_2)^{-1} J^T W_1$$

Note that $J^T W_1 J + W_2$ is always positive definite and, therefore, nonsingular. Since Eq. 9.12 is a quadratic function of $\delta\theta$, $\delta\theta_s$ is the unique solution for the evaluation criterion of Eq. 9.9. Consequently, the solution of Eq. 9.2 for the evaluation criterion of Eq. 9.9 becomes

$$\delta\theta = J^*(\theta)\delta r \qquad 9.13$$

$$J^*(\theta) \triangleq (J^T W_1 J + W_2)^{-1} J^T W_1 \qquad 9.14$$

Equation 9.14 is the formal definition of the *SR-inverse* of the Jacobian matrix.

W_1 and W_2 are generally represented as follows:

$$W_1 = N_r^T N_r \qquad 9.15$$
$$W_2 = k N_\theta^T N_\theta \qquad 9.16$$

where N_r and N_θ are symmetric positive definite matrices, and $k > 0$.

We choose N_r and N_θ as the normalizing matrices; namely,

$$\delta \hat{r} \triangleq N_r \delta r \qquad 9.17$$
$$\delta \hat{\theta} \triangleq N_\theta \delta \theta \qquad 9.18$$
$$\hat{J}(\theta) \triangleq N_r J(\theta) N_\theta^{-1} \qquad 9.19$$

Then, Eq. 9.2 is replaced by

$$\delta \hat{r} = \hat{J}(\theta) \delta \hat{\theta} \qquad 9.20$$

N_r and N_θ should be determined depending on the task of end effector, the physical limits of joints, the dynamic characteristics of joints, and so on. Scalar k determines the weighting between the exactness and the feasibility. Using Eqs. 9.17 through 9.19, the solution by Eqs. 9.13 and 9.14 is equivalently represented by

$$\delta \hat{\theta} = (\hat{J}^T \hat{J} + k E)^{-1} \hat{J}^T \delta \hat{r} \qquad 9.21$$

where E is an identity matrix of appropriate dimension. Accordingly, if we use the normalized form of Eqs. 9.17 through 9.20, the computation of the SR-inverse is the same as Eq. 9.14 when N_r and N_θ are identity matrices. Hence, without losing generality, we can replace Eq. 9.14 by the following equation:

$$J^*(\theta) = (J^T J + k E)^{-1} J^T \qquad 9.22$$

It is readily shown that, for any $J(\theta)$, Eq. 9.22 is equivalent to

$$J^*(\theta) = J^T (J J^T + k E)^{-1} \qquad 9.23$$

Note that the computation for Eq. 9.23 is less expensive than is that for Eq. 9.22, since $(J^T J + k E)^{-1} \in R^{n \times n}$, $(J J^T + k E)^{-1} \in R^{m \times m}$, and $m \leq n$. Also note that, from $\mathcal{R}(J^T) = \mathcal{N}(J)^\perp$ and Eqs. 2.192 and 9.23, we have

$$\mathcal{R}(J^*) = \mathcal{R}(J^\#) = \mathcal{N}(J)^\perp \qquad 9.24$$

which implies that the solution by the SR-inverse belongs to the orthogonal complement of the redundant space as well as the solution by the pseudoinverse.

An approach similar to Eqs. 9.13, 9.22, and 9.23 has been used in signal processing and regularization of experimental data (Tihonov 1964; Twomey 1965). Atkerson, An, and Hollerbach (1985) used it to regularize the experimental data for the identification of dynamic parameters of a robot manipulator.

9.4 Properties of the SR-Inverse

9.4.1 Singular Value Decomposition of the Jacobian Matrix and Its Inverses

The singular value decomposition (SVD; Section 2.3) of the Jacobian matrix, $J(\theta) \in R^{m \times n}$ ($m \leq n$ and rank $J(\theta) = l$), is represented by

$$J(\theta) = U_J \Sigma_J V_J^T \qquad 9.25$$

where $U_J \in R^{m \times m}$, $V_J \in R^{n \times n}$ are orthogonal matrices and $\Sigma_J \in R^{m \times n}$ is given by

$$\Sigma_J \triangleq \begin{pmatrix} \Sigma_{J0} & O \\ O & O \end{pmatrix} \qquad 9.26$$

$$\Sigma_{J0} \triangleq \mathrm{diag}(\sigma_i) \in R^{l \times l}$$

$$\sigma_1 \geq \sigma_2 \geq \cdots \geq \sigma_l > 0$$

where σ_i ($i = 1, \ldots, l$) are the singular values of $J(\theta)$ and are uniquely determined.

From Eq. 2.78, the pseudoinverse of the same Jacobian matrix is now represented using U_J, V_J, and σ_i as follows:

$$J^\#(\theta) = V_J \Sigma_J^\# U_J^T \qquad 9.27$$

$$\Sigma_J^\# = \begin{pmatrix} \Sigma_{J1} & O \\ O & O \end{pmatrix} \in R^{n \times m} \qquad 9.28$$

$$\Sigma_{J1} \triangleq \mathrm{diag}\left(\frac{1}{\sigma_i}\right) \in R^{l \times l}$$

If $l = m = n$, Eqs. 9.27 and 9.28 apply to the inverse of the Jacobian matrix, $J^{-1}(\theta)$.

On the other hand, from Eqs. 9.23, 9.25, and 9.26, the SR-inverse of the Jacobian matrix is represented by

$$J^*(\theta) = V_J \Sigma_J^* U_J^T \qquad 9.29$$

Chapter 9 Singularity-Robust Inverse of Jacobian Matrix

$$\Sigma_J^* = \begin{pmatrix} \Sigma_{J2} & O \\ O & O \end{pmatrix} \in R^{n \times m} \qquad 9.30$$

$$\Sigma_{J2} \triangleq \operatorname{diag}\left(\frac{\sigma_i}{\sigma_i^2 + k}\right) \in R^{l \times l}$$

Now, we apply the following orthogonal transformations to $\delta\boldsymbol{\theta}$ and $\delta\boldsymbol{r}$, respectively:

$$\delta\boldsymbol{\theta}^0 = \boldsymbol{V}_J^T \delta\boldsymbol{\theta} \in R^n \qquad 9.31$$

$$\delta\boldsymbol{r}^0 = \boldsymbol{U}_J^T \delta\boldsymbol{r} \in R^m \qquad 9.32$$

Then, the transformed forms of the solution of Eq. 9.2 in terms of the pseudoinverse and the SR-inverse are respectively obtained using Eqs. 9.27 and 9.29 as follows:

$$\delta\boldsymbol{\theta}^0{}_\# = \boldsymbol{\Sigma}_J^\# \delta\boldsymbol{r}^0 \qquad 9.33$$

$$\delta\boldsymbol{\theta}^0{}_* = \boldsymbol{\Sigma}_J^* \delta\boldsymbol{r}^0 \qquad 9.34$$

If we pick up the lth components of Eqs. 9.33 and 9.34 that are the closest components to the singularity, we have the following scalar equations:

$$\delta\theta^0{}_{\#l} = \frac{1}{\sigma_l} \delta r^0{}_l \qquad 9.35$$

$$\delta\theta^0{}_{*l} = \frac{\sigma_l}{\sigma_l^2 + k} \delta r^0{}_l \qquad 9.36$$

where $\delta\theta^0{}_{\#l}$, $\delta\theta^0{}_{*l}$, and $\delta r^0{}_l$ denote the lth components of $\delta\boldsymbol{\theta}^0{}_\#$, $\delta\boldsymbol{\theta}^0{}_*$, and $\delta\boldsymbol{r}^0$, respectively. Equations 9.35 and 9.36 clearly display the difference between the pseudoinverse and the SR-inverse. Figure 9.2 shows the curves of $\delta\theta^0{}_{\#l}$ and $\delta\theta^0{}_{*l}$ versus σ_l for $\delta r^0{}_l = 1$. From the figure, we observe that the closer σ_l approaches the singular point ($\sigma_l = 0$), the more $\delta\theta^0{}_{\#l}$ approaches infinity. Note that $\delta\theta^0{}_{\#l}$ becomes zero when $\sigma_l = 0$, from Eqs. 9.26 and 9.28. This indicates the discontinuity at $\sigma_l = 0$ and the infeasibility near $\sigma_l = 0$. On the other hand, $\delta\theta^0{}_{*l}$ tends to interpolate smoothly the discontinuity of $\delta\theta^0{}_{\#l}$ at and in the neighborhood of $\sigma_l = 0$. The feature of interpolation is determined by the scale factor k.

9.4.2 Error Analysis of the SR-Inverse Solution

For a kinematically redundant manipulator, as we studied in Section 2.4.3 and Theorem 2.5, the general form of the solution of Eq. 9.2 is provided using the pseudoinverse as follows:

$$\delta\boldsymbol{\theta} = \delta\boldsymbol{\theta}^1_\# + \delta\boldsymbol{\theta}^2_\# \qquad 9.37$$

$$\delta\boldsymbol{\theta}^1_\# \triangleq \boldsymbol{J}^\#(\boldsymbol{\theta})\delta\boldsymbol{r}$$

$$\delta\boldsymbol{\theta}^2_\# \triangleq \{\boldsymbol{E} - \boldsymbol{J}^\#(\boldsymbol{\theta})\boldsymbol{J}(\boldsymbol{\theta})\}\boldsymbol{y}$$

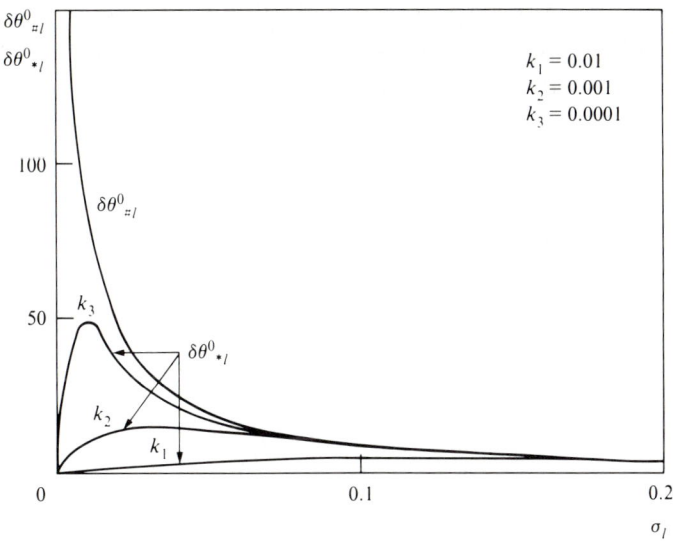

Figure 9.2 Difference between the pseudoinverse and the SR-inverse of the Jacobian matrix.

where $y \in R^n$ is an arbitrary vector. Replacing the pseudoinverse with the SR-inverse of Eq. 9.22 or 9.23, Eq. 9.37 is now represented by

$$\delta\theta = \delta\theta_*^1 + \delta\theta_*^2 \qquad 9.38$$
$$\delta\theta_*^1 \triangleq J^*(\theta)\delta r$$
$$\delta\theta_*^2 \triangleq \{E - J^*(\theta)J(\theta)\}y$$

In Eqs. 9.37 and 9.38, $\delta\theta_\#^1$ and $\delta\theta_*^1$ are for realizing δr, and $\delta\theta_\#^2$ and $\delta\theta_*^2$ are for utilizing kinematic redundancy. In this section, we analyze the errors involved in using Eq. 9.38 in place of Eq. 9.37.

We define the following four error vectors:

$$e_{11} \triangleq J(\theta)(\delta\theta_\#^1 - \delta\theta_*^1) \qquad 9.39$$
$$e_{12} \triangleq \{E - J^\#(\theta)J(\theta)\}\delta\theta_*^1 \qquad 9.40$$
$$e_{21} \triangleq J(\theta)\delta\theta_*^2 \qquad 9.41$$
$$e_{22} \triangleq \delta\theta_\#^2 - \delta\theta_*^2 \qquad 9.42$$

where e_{11} denotes the end-effector motion error caused by the difference of $\delta\theta_\#^1$ and $\delta\theta_*^1$; e_{12} is the disturbance of $\delta\theta_*^1$ to the redundant space; e_{21} is the disturbance of $\delta\theta_*^2$ to the end-effector motion; and e_{22} is the difference of $\delta\theta_\#^2$ and $\delta\theta_*^2$.

We again use SVD. We compute the SVD of the four error vectors using Eqs. 9.25, 9.27, 9.29, and 9.39 through 9.42 as follows:

$$e_{11} = U_J \Sigma_1 U_J^T \delta r \qquad 9.43$$

$$\Sigma_1 \triangleq \text{diag}(a_i) \in R^{m \times m}$$

$$a_i \triangleq \begin{cases} k/(\sigma_i^2 + k) & \text{for } i = 1, \cdots, l \\ 0 & \text{otherwise} \end{cases}$$

$$e_{12} = O \qquad 9.44$$

$$e_{21} = U_J \Sigma_2 V_J^T y \qquad 9.45$$

$$\Sigma_2 \triangleq \text{diag}(b_i) \in R^{m \times n}$$

$$b_i \triangleq \begin{cases} k\sigma_i/(\sigma_i^2 + k) & \text{for } i = 1, \cdots, l \\ 0 & \text{otherwise} \end{cases}$$

$$e_{22} = V_J \Sigma_3 V_J^T y \qquad 9.46$$

$$\Sigma_3 \triangleq \text{diag}(c_i) \in R^{n \times n}$$

$$c_i \triangleq \begin{cases} -k/(\sigma_i^2 + k) & \text{for } i = 1, \cdots, l \\ 0 & \text{otherwise} \end{cases}$$

Equation 9.44 is obtained from Eqs. 9.23 and 9.40 and $(E - M^\# M)M^T = O$. Equation 9.44 implies that $\delta\theta_*^1$ does not disturb the redundant space at all.

In addition to Eqs. 9.31 and 9.32, we introduce four more orthogonal transformations:

$$e_{11}^0 = U_J^T e_{11} \qquad 9.47$$

$$e_{21}^0 = U_J^T e_{21} \qquad 9.48$$

$$e_{22}^0 = V_J^T e_{22} \qquad 9.49$$

$$y^0 = V_J^T y \qquad 9.50$$

Then, the lth components of e_{11}^0, e_{21}^0, and e_{22}^0, which are the closest ones to singularity, are given by

$$e_{11}^0{}_l = \frac{k}{\sigma_l^2 + k} \delta r^0{}_l \qquad 9.51$$

$$e_{21}^0{}_l = \frac{k\sigma_l}{\sigma_l^2 + k} y^0{}_l \qquad 9.52$$

$$e_{22}^0{}_l = -\frac{k}{\sigma_l^2 + k} y^0{}_l \qquad 9.53$$

where the subscript l denotes the lth components. The curves of $e_{11}^0{}_l$, $e_{21}^0{}_l$, and $e_{22}^0{}_l$ for $\delta r^0{}_l = y^0{}_l = 1$ are shown versus σ_l in Fig. 9.3.

9.4 Properties of the SR-Inverse

The $e_{11}{}^0{}_l$ and $e_{22}{}^0{}_l$ take the maximum errors at the singular point ($\sigma_l = 0$). The magnitudes of errors decrease monotonically as σ_l increases. On the other hand, $e_{21}{}^0{}_l$ takes the maximum error, $\sqrt{k/2}$, at $\sigma_l = \sqrt{k}$, and becomes zero at the singular point. Figure 9.3(a) and (b) implies that the smaller value

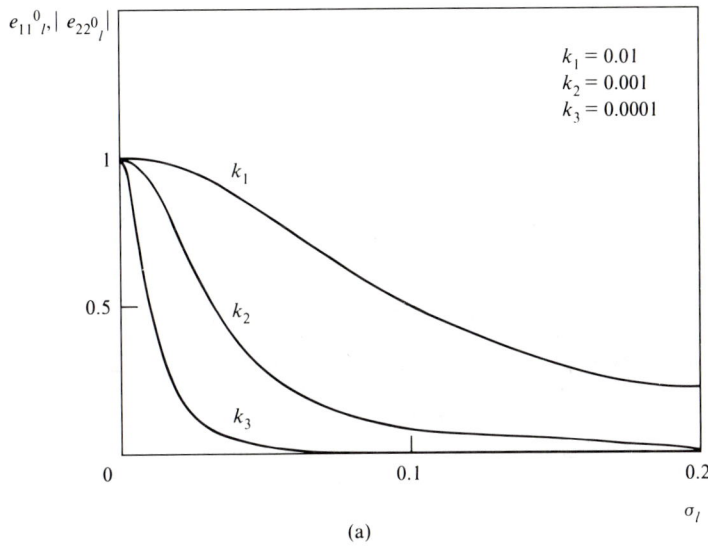

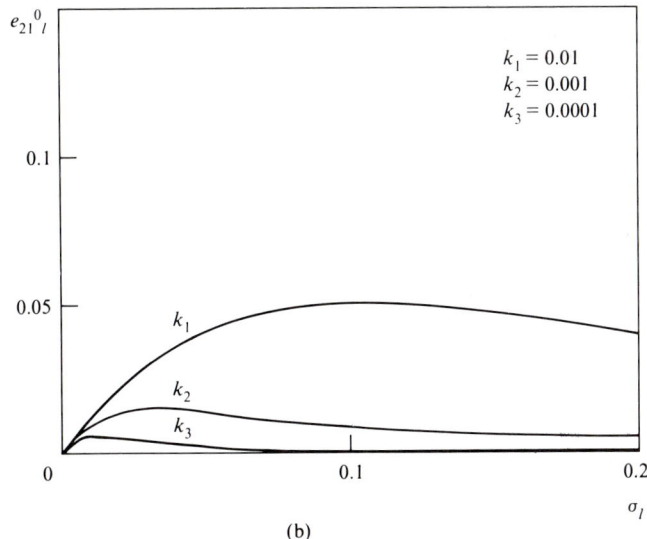

Figure 9.3 Error analysis of the solutions by the SR-inverse. (a) $e_{11}{}^0{}_l, |e_{22}{}^0{}_l|$. (b) $e_{21}{}^0{}_l$.

of k results in the lower amount of errors. Recall that a small value of k permits large magnitude of $\delta\theta$, as we saw in Fig. 9.2. Therefore, the feasibility and the exactness are traded off, and the selection of the scale factor k determines the weighting between them. In Section 9.6, we shall discuss how to change the scale factor, k, adaptively.

9.5 Computational Complexity of the SR-Inverse

In this section, we compare the computational complexities of the following five different types of solutions of inverse kinematics:

$$\text{Solution 1}: \quad \delta\theta = J^{-1}(\theta)\delta r \qquad 9.54$$

$$\text{Solution 2}: \quad \delta\theta = J^{\#}(\theta)\delta r \qquad 9.55$$

$$\text{Solution 3}: \quad \delta\theta = J^{*}(\theta)\delta r \qquad 9.56$$

$$\text{Solution 4}: \quad \delta\theta = J^{\#}(\theta)\delta r + \{E - J^{\#}(\theta)J(\theta)\}y \qquad 9.57$$

$$\text{Solution 5}: \quad \delta\theta = J^{*}(\theta)\delta r + \{E - J^{*}(\theta)J(\theta)\}y \qquad 9.58$$

Solution 1 can be applied to nonredundant manipulators only when they are not at singular points. Solutions 2 and 3 are used for both redundant and nonredundant manipulators even at singular points, but cannot actively utilize redundancy. Solutions 4 and 5 are applicable to utilization of redundancy of redundant manipulators.

Linear equation $Mq = p$, $M \in R^{m \times m}$ and $p, q \in R^m$, can be numerically solved using the Gaussian elimination (Golub and Van Loan 1983), which is known as an efficient algorithm. The computational amount for the Gaussian elimination becomes

$$N_{GE} = \left(\frac{m^3}{3} + m^2 - \frac{m}{3}\right) \text{mul} + m \text{ div} + \left(\frac{m^3}{3} + \frac{m^2}{2} - \frac{5m}{6}\right) \text{sub} \qquad 9.59$$

where mul, div, and sub denote multiplication, division, and subtraction, respectively. The pivoting operation was not considered in evaluating Eq. 9.59. Therefore, the computational complexity of solution 1 is

$$N_1 = N_{GE} \qquad 9.60$$

As we studied in Section 2.4, the computation of the pseudoinverse is expensive for a general matrix. However, for a full rank matrix, Algorithm 2.2 allows an efficient computation. Using Eq. 2.203, the following computation provides solution 2:

$$\{J(\theta)J^T(\theta)\}\alpha = \delta r \qquad 9.61$$

$$\delta\theta = J^T(\theta)\alpha \qquad 9.62$$

9.5 Computational Complexity of the SR-Inverse

where $\boldsymbol{\alpha} \in R^m$ is solved by the Gaussian elimination. The computational amount for solution 2 becomes

$$N_2 = N_{GE} + N_a + N_b \qquad 9.63$$

$$N_a = \left(\frac{nm^2}{2} + \frac{nm}{2}\right)\text{mul} + \left\{\frac{(n-1)m^2}{2} + \frac{(n-1)m}{2}\right\}\text{add} \qquad 9.64$$

$$N_b = nm\,\text{mul} + (nm - n)\,\text{add} \qquad 9.65$$

where N_a and N_b indicate the computations for $\boldsymbol{J}(\boldsymbol{\theta})\boldsymbol{J}^T(\boldsymbol{\theta})$ and $\boldsymbol{J}^T(\boldsymbol{\theta})\boldsymbol{\alpha}$, and add denotes addition. The symmetry of $\boldsymbol{J}(\boldsymbol{\theta})\boldsymbol{J}^T(\boldsymbol{\theta})$ was taken into account in counting N_a.

Similarly to Eqs. 9.61 and 9.62, the following computation scheme can be used for solution 3:

$$\{\boldsymbol{J}(\boldsymbol{\theta})\boldsymbol{J}^T(\boldsymbol{\theta}) + k\boldsymbol{E}\}\boldsymbol{\alpha} = \delta\boldsymbol{r} \qquad 9.66$$

$$\delta\boldsymbol{\theta} = \boldsymbol{J}^T(\boldsymbol{\theta})\boldsymbol{\alpha} \qquad 9.67$$

which are based on Eq. 9.23, and where $\boldsymbol{\alpha}$ is to be solved by the Gaussian elimination. Note that this computation can be executed even for nonfull rank $\boldsymbol{J}(\boldsymbol{\theta})$, whereas Eq. 9.61 cannot be, because $\boldsymbol{J}(\boldsymbol{\theta})\boldsymbol{J}(\boldsymbol{\theta})^T + k\boldsymbol{E}$ is always nonsingular for any $\boldsymbol{J}(\boldsymbol{\theta})$. The computation required for solution 3 is thus

$$N_3 = N_{GE} + N_a + N_b + N_c \qquad 9.68$$

$$N_c = m\,\text{add} \qquad 9.69$$

where N_c denotes the computation to add $k\boldsymbol{E}$ to $\boldsymbol{J}(\boldsymbol{\theta})\boldsymbol{J}(\boldsymbol{\theta})^T$.

For solution 4, Klein and Huang (1983) proposed the following efficient computation (Algorithm 2.3):

$$\{\boldsymbol{J}(\boldsymbol{\theta})\boldsymbol{J}^T(\boldsymbol{\theta})\}\boldsymbol{\alpha} = \delta\boldsymbol{r} - \boldsymbol{J}(\boldsymbol{\theta})\boldsymbol{y} \qquad 9.70$$

$$\delta\boldsymbol{\theta} = \boldsymbol{J}^T(\boldsymbol{\theta})\boldsymbol{\alpha} + \boldsymbol{y} \qquad 9.71$$

where $\boldsymbol{\alpha}$ is again solved by the Gaussian elimination. The computation for solution 4 is thus

$$N_4 = N_{GE} + N_a + N_b + N_d + N_e \qquad 9.72$$

$$N_d = nm\,\text{mul} + (n-1)m\,\text{add} + m\,\text{sub} \qquad 9.73$$

$$N_e = n\,\text{add} \qquad 9.74$$

where N_d shows the requirement to compute $J(\theta)y$ and to subtract it from δr, and N_e is the computation to add y to $J^T(\theta)\alpha$.

Similarly to Eqs. 9.70 and 9.71, the following computation is proposed for solution 5:

$$\{J(\theta)J^T(\theta) + kE\}\alpha = \delta r - J(\theta)y \qquad 9.75$$

$$\delta\theta = J^T(\theta)\alpha + y \qquad 9.76$$

where we use the Gaussian elimination to solve α. Consequently, the following is the computational requirement for solution 5:

$$N_5 = N_{GE} + N_a + N_b + N_c + N_d + N_e \qquad 9.77$$

The computational complexity for the inverse kinematic solutions discussed here is summarized in Table 9.1. Table 9.2 shows the specific computational amount for $m = 6$ and $n = 6, 7, 8$. The computational time was estimated assuming 64-bit floating-point arithmetic operations on the Intel 8087 (5 MHz), where time for data transfer was disregarded. From the tables, the following conclusions are derived:

1. $J^*(\theta)$ needs little more computation, compared with $J^\#(\theta)$. Note that Eqs. 9.61 and 9.70 cannot be applied for singular Jacobian matrices. In this case, we have to use one of the general algorithms, such as the one using SVD and the Greville's algorithm in Section 2.4.6. However, Eqs. 9.66 and 9.75 can be computed even for singular Jacobian matrices.

2. A short extra computation time is required to utilize kinematic redundancy, no matter which of $J^\#(\theta)$ or $J^*(\theta)$ is used (16 percent for $n = 7$, and 17 percent for $n = 8$).

3. The computational requirements for solutions 2 through 5 are represented by linear equations of n. When the number of joints increases from seven to eight, N_2 through N_5 increase at the rate of approximately 10 percent.

Table 9.1. Computational requirements for inverse kinematic solutions.

	Multiply	Divide	Add	Subtract
N_1	$\frac{1}{3}m^3 + m^2 - \frac{1}{3}m$	m	0	$\frac{1}{3}m^3 + \frac{1}{2}m^2 - \frac{5}{6}m$
N_2	$\frac{1}{3}m^3 + \frac{(n+2)}{2}m^2 + \frac{(9n-2)}{6}m$	m	$\frac{(n-1)}{2}m^2 + \frac{(3n-1)}{2}m - n$	$\frac{1}{3}m^3 + \frac{1}{2}m^2 - \frac{5}{6}m$
N_3	$\frac{1}{3}m^3 + \frac{(n+2)}{2}m^2 + \frac{(9n-2)}{6}m$	m	$\frac{(n-1)}{2}m^2 + \frac{(3n+1)}{2}m - n$	$\frac{1}{3}m^3 + \frac{1}{2}m^2 - \frac{5}{6}m$
N_4	$\frac{1}{3}m^3 + \frac{(n+2)}{2}m^2 + \frac{(15n-2)}{6}m$	m	$\frac{(n-1)}{2}m^2 + \frac{(5n-3)}{2}m$	$\frac{1}{3}m^3 + \frac{1}{2}m^2 + \frac{1}{6}m$
N_5	$\frac{1}{3}m^3 + \frac{(n+2)}{2}m^2 + \frac{(15n-2)}{6}m$	m	$\frac{(n-1)}{2}m^2 + \frac{(5n-1)}{2}m$	$\frac{1}{3}m^3 + \frac{1}{2}m^2 + \frac{1}{6}m$

Table 9.2. Computational requirements for inverse kinematic solutions ($m = 6$ and $n = 6, 7, 8$).

		mul	div	add	sub	time*(msec)
N_1	6	106	6	0	85	5.1568
	7	—	—	—	—	—
	8	—	—	—	—	—
N_2	6	268	6	135	85	12.7276
	7	295	6	161	85	14.0748
	8	322	6	187	85	15.422
N_3	6	268	6	141	85	12.874
	7	295	6	167	85	14.2212
	8	322	6	193	85	15.5684
N_4	6	—	—	—	—	—
	7	337	6	204	91	16.3792
	8	370	6	237	91	18.0556
N_5	6	—	—	—	—	—
	7	337	6	210	91	16.5256
	8	370	6	243	91	18.202

* Suppose that mul = 26.4, div = 47.4, add = 24.4 and sub = 24.4 μsec (64 bits floating point arithmetic operations by 8087 (5MHz)).

Hollerbach (1980) reported that $(150n - 48)$ mul + $(131n - 48)$ add are required for inverse dynamics using the Newton–Euler method. For the Intel 8087, it is estimated that 40.50 msec, 47.66 msec, and 54.81 msec are necessary for $n = 6, 7$, and 8, respectively. Accordingly, the computation of inverse kinematics is approximately one-third of that for inverse dynamics, even if kinematic redundancy is utilized. In Table 9.2, N_5 for $n = 7$ implies that the computation could be performed at the sampling rate of 60 Hz.

9.6 Variable Scale Factor

9.6.1 Scale Factor as a Function of the Measure of Manipulability

As we saw in Section 9.4.2, the feasibility and the exactness of solutions are contradictory requirements for the scale factor, k. In this section, we propose a method to change k adaptively depending on the configuration of a robot manipulator, such that k takes larger values in the neighborhood of singular points, and small values or zero outside the neighborhood.

Yoshikawa (1985) introduced the measure of manipulability to evaluate quantitatively the kinematic ability of robot manipulators (Section 3.4). The measure of manipulability, w, is defined as follows:

$$w \triangleq \sqrt{\det\{\boldsymbol{J}(\boldsymbol{\theta})\boldsymbol{J}^T(\boldsymbol{\theta})\}} \qquad 9.78$$

The measure of manipulability becomes zero at singular points, and takes positive values elsewhere. Since the measure of manipulability changes continuously as a robot manipulator moves, it can be considered as a kind of distance from the singular points.

We change the scale factor based on the following equation:

$$k = \begin{cases} k_0\left(1 - w/w_0\right)^2 & \text{for } w < w_0 \\ 0 & \text{for } w \geq w_0 \end{cases} \qquad 9.79$$

where k_0 is the scale factor at singular points, and w_0 is the measure of manipulability on the boundary of the neighborhood of singular points. In other words, w_0 defines the boundary of the neighborhood of singular points. Equations 9.23 and 9.79 imply that the SR-inverse coincides with the pseudoinverse outside the neighborhood of singular points. This guarantees the exactness of the solution outside of the neighborhood. Note that, although k is originally defined positive in Eq. 9.16, Eq. 9.23 holds even for $k = 0$ if $w \geq w_0$, since this means full rankness of the Jacobian matrix. Also note that Eqs. 9.14 and 9.22 are not defined for $k = 0$.

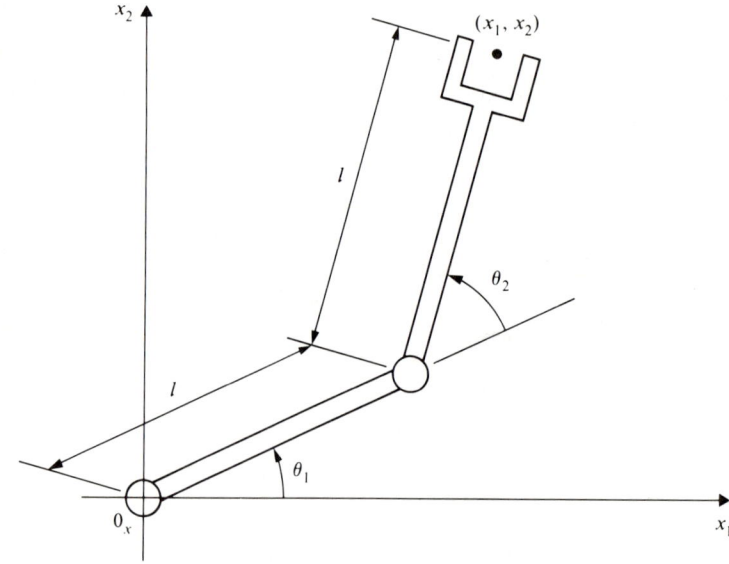

Figure 9.4 A 2-DOF manipulator.

9.6.2 Numerical Analysis

Properties of the SR-inverse with the variable scale factor are examined using a 2-DOF manipulator as shown in Fig. 9.4. Choosing $\boldsymbol{r} = (x_1 \; x_2)^T$, the Jacobian matrix becomes

$$\boldsymbol{J}(\boldsymbol{\theta}) = \begin{pmatrix} -l\{\sin\theta_1 + \sin(\theta_1+\theta_2)\} & -l\sin(\theta_1+\theta_2) \\ l\{\cos\theta_1 + \cos(\theta_1+\theta_2)\} & l\cos(\theta_1+\theta_2) \end{pmatrix} \quad 9.80$$

where l is the length of the both links. With weighting matrices, $\boldsymbol{N_r} = (1/l)\boldsymbol{E}$ and $\boldsymbol{N_\theta} = \boldsymbol{E}$, the measure of manipulability for the normalized Jacobian matrix becomes

$$\hat{w} = \hat{W}(\theta_2) = |\sin\theta_2| \quad 9.81$$

Figure 9.5 shows the $\hat{w}/\hat{w}_0$ and k/k_0 versus θ_2. We choose $\hat{w}_0 = \hat{w}(30°)$ as the threshold.

Figures 9.6 and 9.7 show the effect of the variable scale factor. In Fig. 9.6, the magnitude of solution, $\|\delta\hat{\boldsymbol{\theta}}\|$, and the magnitude of error, $\|\delta\hat{\boldsymbol{r}} - \delta\hat{\boldsymbol{r}}^*\|$, for the constant scale factors are plotted against $|\theta_2|$, where $\theta_1 = 0°$ and $\delta\hat{\boldsymbol{r}} = (1 \; 0)^T$ are assumed, and $\delta\hat{\boldsymbol{r}}^* \triangleq \hat{\boldsymbol{J}}(\boldsymbol{\theta})\hat{\boldsymbol{J}}^*(\boldsymbol{\theta})\delta\hat{\boldsymbol{r}}$ is used. Here, $\delta\hat{\boldsymbol{r}}^*$ implies the actual end-effector motion for the desired motion of $\delta\hat{\boldsymbol{r}}$. In Fig. 9.7, $\|\delta\hat{\boldsymbol{\theta}}\|$ and $\|\delta\hat{\boldsymbol{r}} - \delta\hat{\boldsymbol{r}}^*\|$ for the variable scale factors are plotted under the same conditions with $\hat{w}_0 = \hat{w}(10°)$. We observe from the figures that,

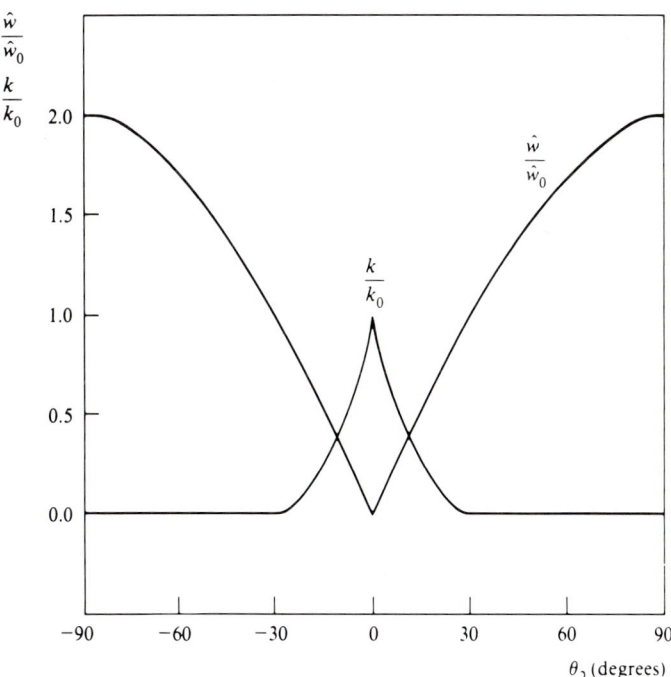

Figure 9.5 Plot of $\hat{w}/\hat{w}_0$ and k/k_0 versus θ_2.

Chapter 9 Singularity-Robust Inverse of Jacobian Matrix

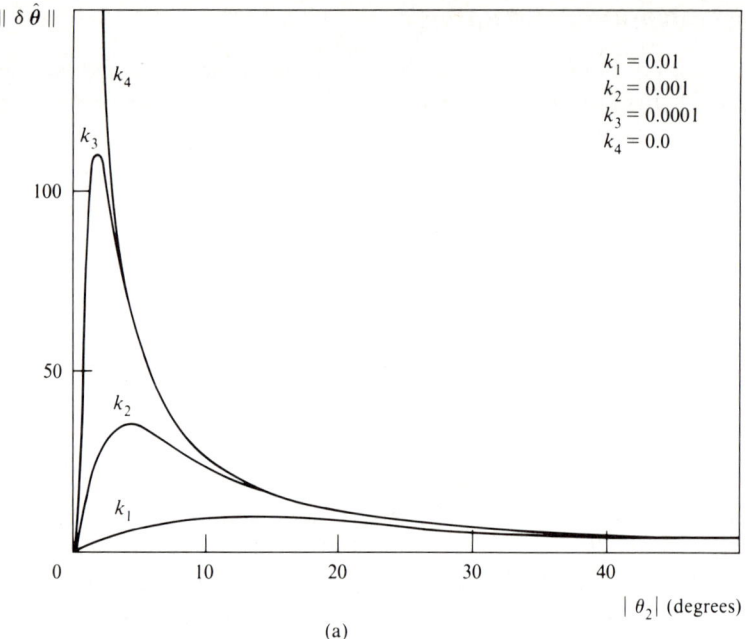

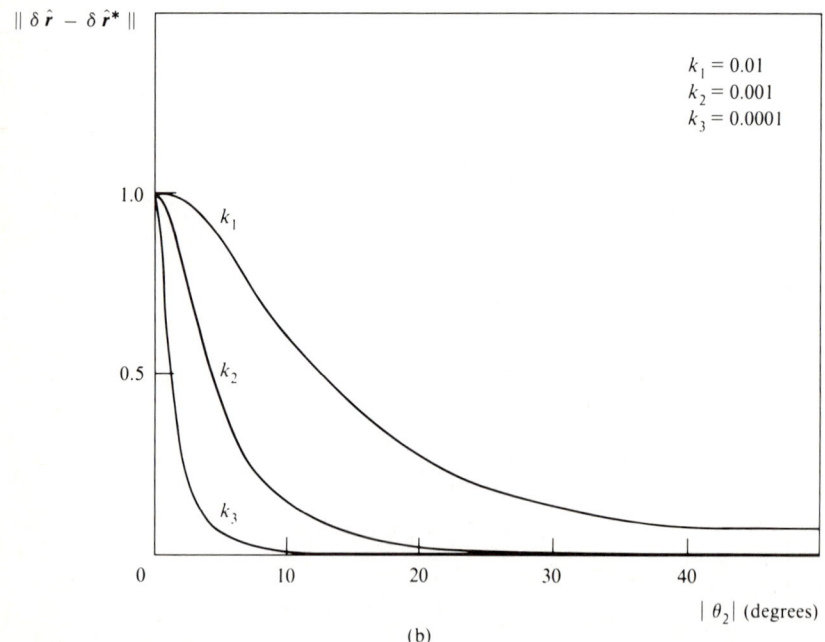

Figure 9.6 Plot of (a) $\| \delta \hat{\boldsymbol{\theta}} \|$ and (b) $\| \delta \hat{\boldsymbol{r}} - \delta \hat{\boldsymbol{r}}^* \|$ for the constant scale factors.

9.6 Variable Scale Factor

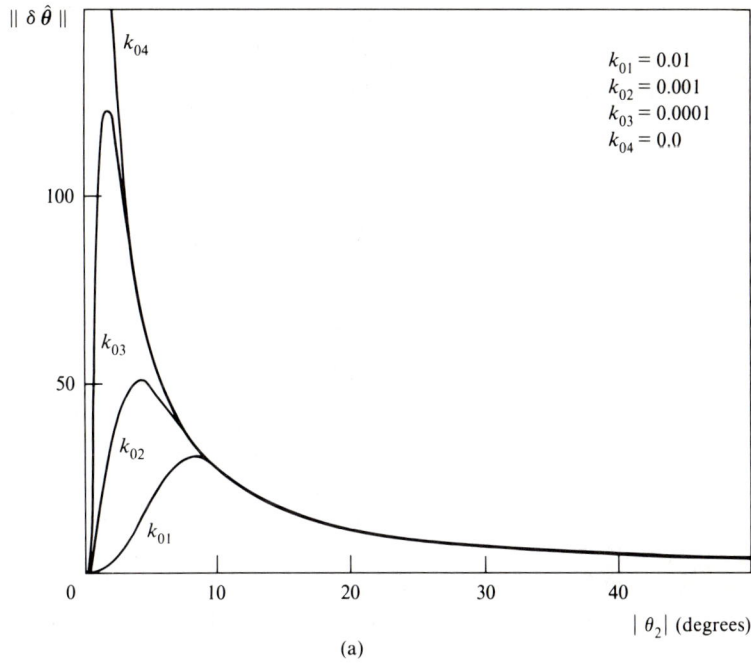

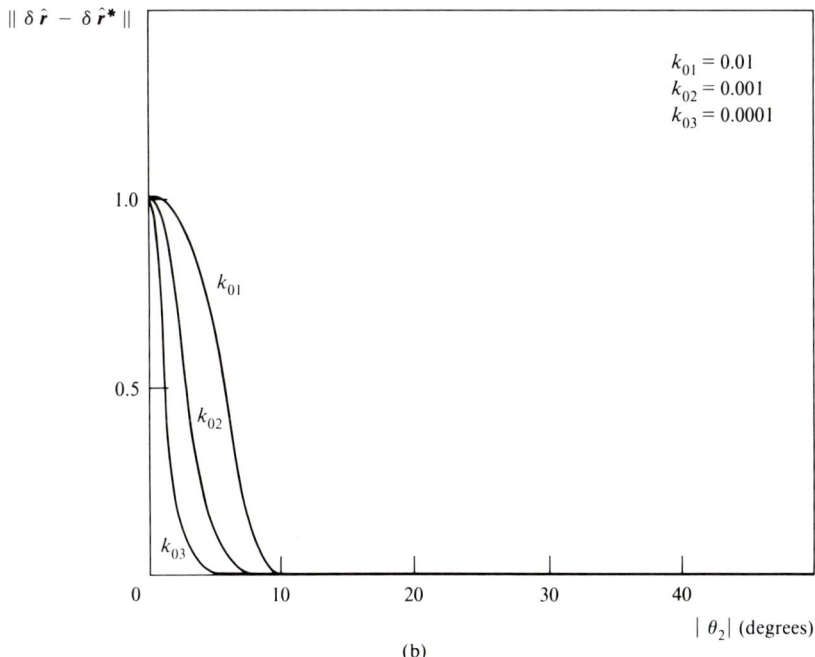

Figure 9.7 Plot of (a) $\| \delta \hat{\boldsymbol{\theta}} \|$ and (b) $\| \delta \hat{\boldsymbol{r}} - \delta \hat{\boldsymbol{r}}^* \|$ for the variable scale factors.

272 Chapter 9 Singularity-Robust Inverse of Jacobian Matrix

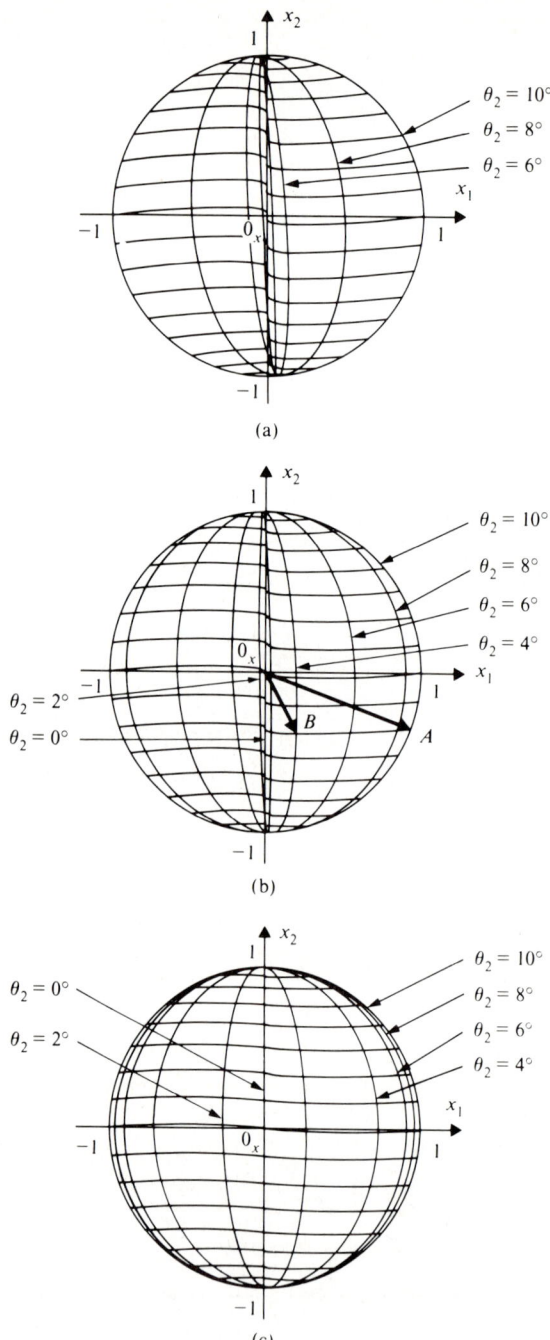

Figure 9.8 End-effector motions obtained by the SR-inverse of the Jacobian matrix with the variable scale factor. (a) $k_0 = 0.1$. (b) $k_0 = 0.01$. (c) $k_0 = 0.001$.

although $k = 0.001$ in Fig. 9.6 and $k_0 = 0.01$ in Fig. 9.7 take the similar magnitude of solution in the neighborhood of the singular point, $k_0 = 0.01$ is preferable, since the end-effector motion error, $\| \delta\hat{r} - \delta\hat{r}^* \|$, is zero outside the neighborhood. Consequently, we can conclude that the variable scale factor of Eq. 9.79 is a better compromise than is the constant scale factor for both requirements of feasibility and exactness of inverse kinematic solutions.

Figure 9.8 shows the direction of error vector $\delta\hat{r} - \delta\hat{r}^*$, where $\hat{w}_0 = \hat{w}(10°)$ and $\theta_1 = 0°$ are assumed, and θ_2 is a variable. When unit end-effector motions are required in all directions, the set of endpoints of vector δr makes a unit circle, which we call the *desired circle*. The three contour circles of Fig. 9.8 are the desired circles. If $\theta_2 \geq 10°$, then $\delta\hat{r}^*$ is equal to $\delta\hat{r}$. Hence, the set of endpoints of $\delta\hat{r}^*$ makes another unit circle. However, if $0° \leq \theta_2 \leq 10°$, then the set of endpoints of $\delta\hat{r}^*$ becomes an ellipsoid, which we call the *modified ellipsoid*. In Fig. 9.8, the lines that look like the lines of longitude indicate the modified ellipsoids for $\theta_2 = 8, 6, 4, 2$, and $0°$. The lines that look like the parallels of latitude represent the traces of $\delta\hat{r}^*$ as θ_2 changes continuously from $0°$ to $10°$, with $\delta\hat{r}$ fixed. For example, in Fig. 9.8(b), if $\delta\hat{r}$ is given by OA, then $\delta\hat{r}^*$ becomes OB when $\theta_2 = 4°$.

The manipulable space at the singular point of $\theta_2 = 0°$ coincides with the x_2 axis. Hence, we observe from Fig. 9.8 that the SR-inverse reduces $\| \delta\hat{\theta} \|$ by modifying the component in the degenerating direction of $\delta\hat{r}$. Note that, if we use $\hat{J}^\#(\theta)$ in place of $\hat{J}^*(\theta)$, then $\delta\hat{r}^\# \triangleq J(\theta)J^\#(\theta)\delta\hat{r}$ always is equal to $\delta\hat{r}$ except at the singular point, where $\delta\hat{r}^\#$ lies in the x_2 axis. This fact implies that the SR-inverse interpolates smoothly the discontinuity of the pseudoinverse at singular points.

9.7 Simulations

9.7.1 A Model of Manipulator Control System

To investigate the effectiveness of the SR-inverse, we shall perform numerical simulations using $J^{-1}(\theta)$ and $J^*(\theta)$ for a desired trajectory passing through a singular point. Figure 9.9 shows the numerical model of a 3-DOF manipulator. The dynamics of an articulated robot manipulator involves various kinds of nonlinearity in the manipulator inertia matrix, the gravity effect, and the effects of Coriolis and centrifugal forces. However, to focus our discussion on the inverse kinematic problem, we assume that the dynamics of the joints are represented by the following decoupled linear equations:

$$0.01\,\ddot{\theta}_i + 0.1\,\dot{\theta}_i = u_i \qquad \text{for } i = 1, 2, 3 \qquad 9.82$$

Equation 9.82 means that closing the position feedback loops with unity gains results in the second-order system behavior with $\zeta = 0.5$ and $\omega_n = 10$ rad/sec. When the nonlinear control using inverse dynamics (Freund 1982) is used, the joint dynamics are usually compensated to have a linear decoupled dynamics similar to one assumed by Eq. 9.82. Therefore, the simplified dynamics of Eq. 9.82 would not lose generality.

The manipulation variable is now chosen as

$$\boldsymbol{r} \triangleq \begin{pmatrix} x_1 \\ x_2 \\ \theta_1 + \theta_2 + \theta_3 \end{pmatrix} \in R^3 \qquad 9.83$$

where x_1 and x_2 represent the position of the end-effector, and $\theta_1 + \theta_2 + \theta_3$ denote the end-effector orientation. The normalizing matrices are chosen as $\boldsymbol{N}_r = \mathrm{diag}\,(l/2, l/2, 1/\pi)$ and $\boldsymbol{N}_\theta = \boldsymbol{E}$. The desired trajectory of the manipulation variable is given by

$$\boldsymbol{r}_0(t) = \begin{pmatrix} x_1(t_0) \\ x_2(t_0) - V_0 t \\ 45° \end{pmatrix} \qquad 9.84$$

where $t_0 = 0 \le t \le 6.5$ sec and $V_0 = l/5$ m/sec. The initial joint angles are chosen as $\boldsymbol{\theta}(t_0) = (90° \; -90° \; 45°)^T$ and $\dot{\boldsymbol{\theta}}(t_0) = \boldsymbol{O}°$/sec. Equation 9.84 means a trajectory passing through the singular point, $\boldsymbol{\theta} = (0° \; -180° \; 225°)^T$ at $t = 5$ sec. Figure 9.10 shows the block diagram of the robot manipulator control system.

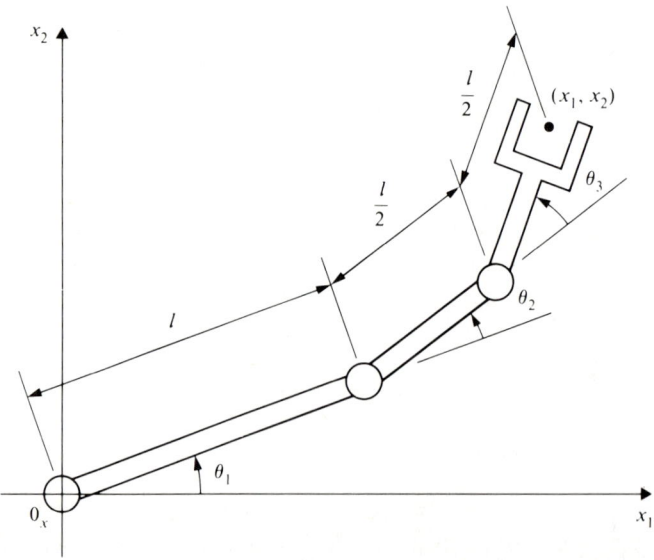

Figure 9.9 A 3-DOF manipulator.

9.7 Simulations 275

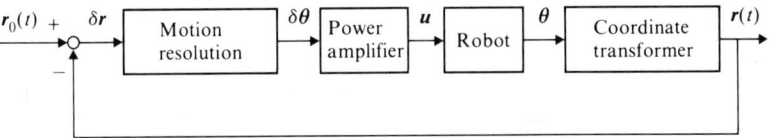

Figure 9.10 Block diagram of the robot manipulator control system.

9.7.2 Results by $J^{-1}(\theta)$

Figure 9.11 shows the result of the simulation performed by using $J^{-1}(\theta)$ and E in the motion-resolution block and the power-amplifier block in Fig. 9.10, respectively. In Fig. 9.11(a), the curves of $\|\delta\hat{\theta}\|$ and $\|\delta\hat{r}\|$ are plotted versus time. Figure 9.11(b) shows the motion of the robot manipulator. Since the desired trajectory $r_0(t)$ is in the manipulable space even at the singular point, the robot manipulator passes through the singular point. We approximated $\det J(\theta)$ by $10^{-10} \text{sgn}\{\det J(\theta)\}$ when $|\det J(\theta)| \leq 10^{-10}$. Due to this approximation, the results would be the same even if we used $J^\#(\theta)$.

In the real situation, a power amplifier should have saturation. Figure 9.12 shows the result of simulation when we assume the following saturation in the power-amplifier block in Fig. 9.10:

$$u_i = \begin{cases} \delta\theta_i & \text{if } |\delta\theta_i| \leq u_{\max} \\ u_{\max} \text{sgn}(\delta\theta_i) & \text{if } |\delta\theta_i| \geq u_{\max} \end{cases} \qquad 9.85$$

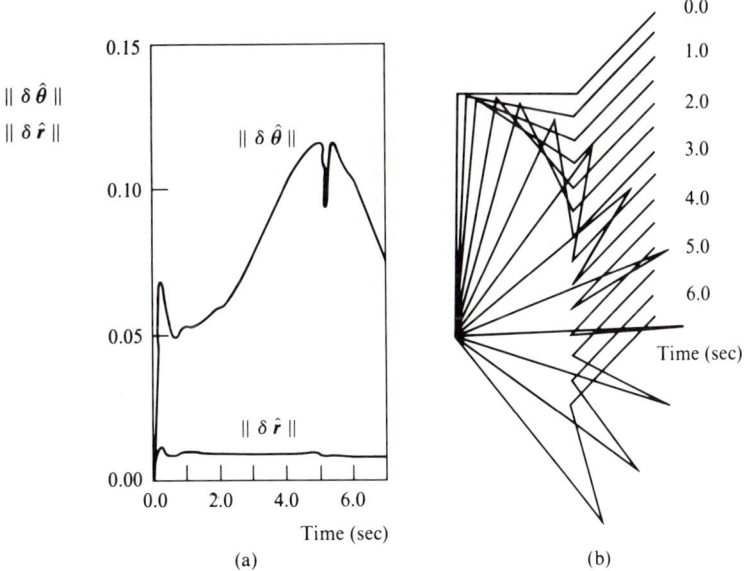

Figure 9.11 Simulation result by $J^{-1}(\theta)$ with no saturation in the power amplifier.

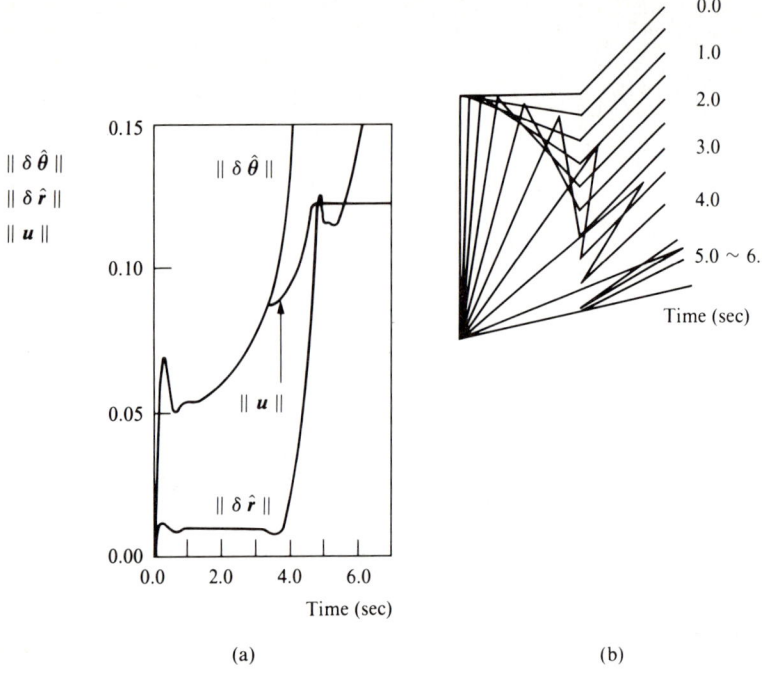

Figure 9.12 Simulation result by $J^{-1}(\theta)$ with saturation in the power amplifier.

where $i = 1, 2, 3$, and $u_{\max} = 4°$. In Fig. 9.12(a), the curves of $\| \delta \hat{\theta} \|$, $\| \delta \hat{r} \|$, and $\| u \|$ are plotted versus time. Figure 9.12(b) is the motion of the robot manipulator. The manipulator starts to drift out of the desired trajectory at $t = 4$ sec and arrives at the singular point at $t = 5$ sec. After that, the manipulator is stuck there and cannot get out of the singular point.

Figure 9.13 indicates the curves of θ, $\dot{\theta}$, and u for the period of $5 \leq t \leq 6.5$ sec. Around the singular point, u_i, $i = 1, 2, 3$, are oscillating between $u_{\max}$ and $-u_{\max}$. We might be able to interpret that the robot manipulator was stuck at the singular point due to a phenomenon like the sliding mode (Young 1978) caused by the discontinuous switching of u_i. This behavior did not appear in Fig. 9.11. This is because, in Fig. 9.12, the position error was caused by the saturation of power amplifier, and it required large u_i to correct the error since the error was no longer in the manipulable space at the singular point, whereas the error was always in the manipulable space in Fig. 9.11.

9.7.3 Results by $J^*(\theta)$

Figures 9.14 and 9.15 show the results of simulation performed using $J^*(\theta)$, where $k_0 = 0.01$ and $\hat{w}_0 = \hat{w}(\theta_w)$, $\theta_w = (0° \ -150° \ 150°)^T$ were used.

9.7 Simulations 277

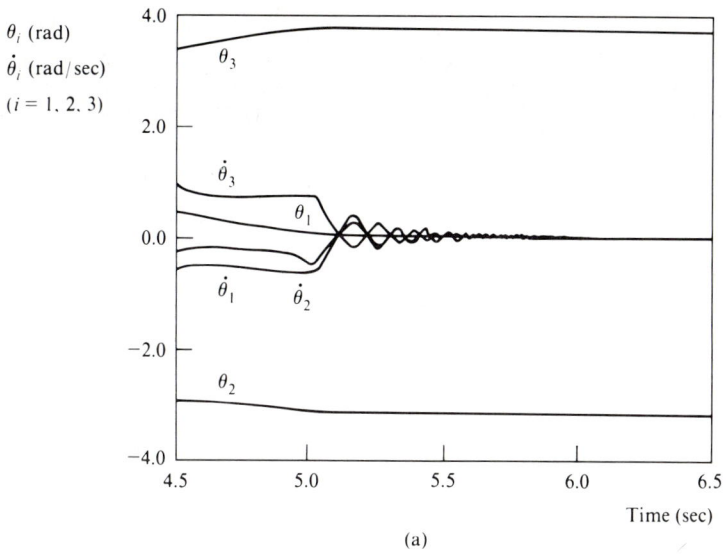

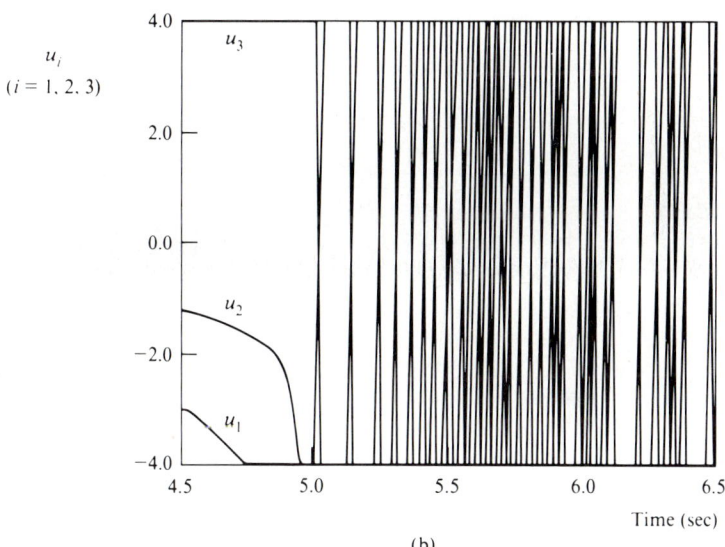

Figure 9.13 Plot of (a) θ, $\dot{\theta}$, and (b) u for the motion of Fig. 9.12 ($4.5 \leq t \leq 6.5$ sec).

278 Chapter 9 Singularity-Robust Inverse of Jacobian Matrix

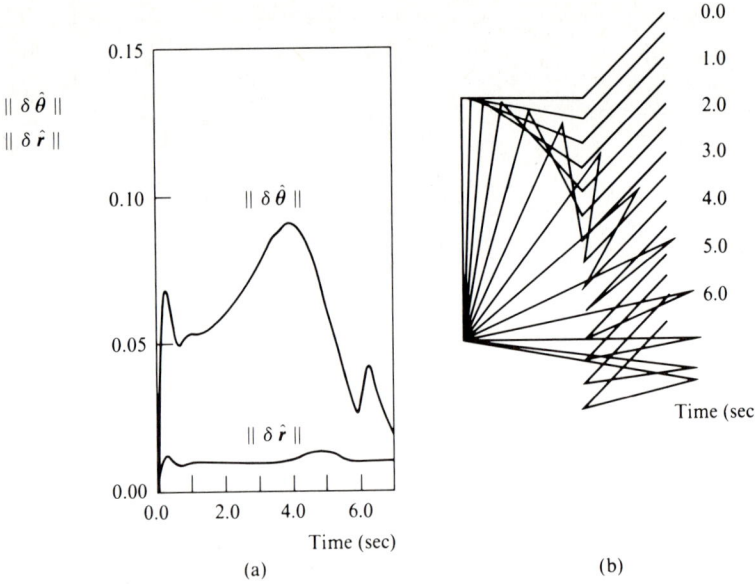

Figure 9.14 Simulation result by $J^*(\theta)$ with no saturation in the power amplifier.

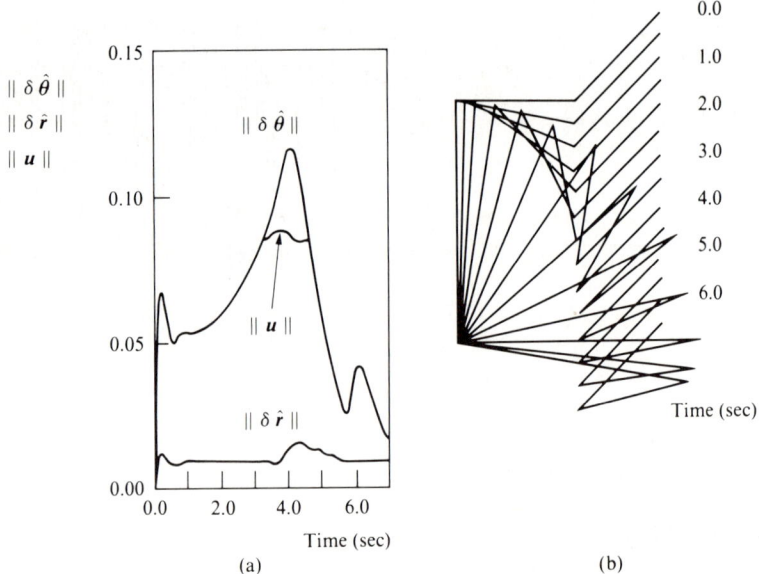

Figure 9.15 Simulation result by $J^*(\theta)$ with saturation in the power amplifier.

Figure 9.14 shows the case of the power amplifier without saturation. Although a small error was generated at $4.5 \leq t \leq 5.5$ sec in Fig. 9.14(b), the magnitude of solution $\| \delta \hat{\boldsymbol{\theta}} \|$ remained small compared with that in Fig. 9.11.

Figure 9.15 shows the case of the power amplifier with saturation. Although the resultant end-effector motion drifted slightly out of the desired trajectory at $4.0 \leq t \leq 5.5$ sec, the robot manipulator passed successfully through the neighborhood of the singular point.

These results indicate clearly that the SR-inverse of the Jacobian matrix is effective in overcoming the singularity problem of robot manipulators.

Summary

The singularity problem of robot manipulators was discussed. The singularity-robust inverse (SR-inverse) of the Jacobian matrix was introduced. The SR-inverse provides a feasible inverse kinematic solution even at or in the neighborhood of singular points.

The fundamental properties of the SR-inverse were investigated with SVD. The computational complexity was discussed comparing the SR-inverse with the inverse and the pseudoinverse. We saw that the SR-inverse requires little more computation than does the pseudoinverse. The computational amount increases linearly as the degrees of redundancy increase. It amounts to approximately one-third of the computation for inverse dynamics by the Newton–Euler recursive method, even when we utilize kinematic redundancy.

Simulation results showed clearly the effectiveness of the SR-inverse as a remedy for the singularity problem of robot manipulators. The inverse kinematics by the SR-inverse of the Jacobian matrix can be applied to any kind of robot manipulator, whether it be redundant or nonredundant.

It is still an open question how we should systematically select k_0 and w_0 in Eq. 9.79. The tradeoff shown in Fig. 9.7 is useful in choosing k_0. Practically speaking, we can determine the w_0 of a specific manipulator by seeing the configuration and judging the boundary of the neighborhood of singular points. However, it would also be useful to take account of the characteristics of power amplifiers and the dynamics of manipulators in order to determine w_0.

Finally, the author would like to make a comment on the related research. Wampler (1986), independently of the author's group, proposed the same idea under the name *damped least-squares method*, where the adaptive change of scale factor was not considered. Wampler and Leifer (1988) applied the methods to the resolved-rate and resolved-acceleration control of manipulators.

References

Atkerson, C. G., An, C. H., and Hollerbach, J. M. 1986. Estimation of inertial parameters of manipulator loads and links. *Int. J. Robotics Res.* 5(3): 101–119.

Boullion, T. L., and Odell, P. L. 1971. *Generalized inverse matrices.* New York: Wiley-Interscience.

Freund, E. 1982. Fast nonlinear control with arbitrary pole-placement for industrial robots and manipulators. *Int. J. Robotics Res.* 1(1): 65–78.

Golub, G., and Van Loan, C. 1983. *Matrix computations.* Baltimore, MD: Johns Hopkins University Press.

Hollerbach, J. M. 1980. A recursive formulation of Lagrangian manipulator dynamics. *IEEE Trans. Sys., Man, Cyber.* 10(11): 730–736.

Hollerbach, J. M. 1985. Optimum kinematic design for a seven degree of freedom manipulator. In *Robotics research 2*, eds. H. Hanafusa and H. Inoue, pp. 215–222. Cambridge, MA: MIT Press.

Klein, C. A., and Huang, C. H. 1983. Review of pseudoinverse control for use with kinematically redundant manipulators. *IEEE Trans. Sys., Man, Cyber.* SMC-13 (3): 245–250.

Lang, S. 1968. *Analysis I.* Reading, MA: Addison-Wesley.

Nakamura, Y., and Hanafusa, H. 1984. Singularity low-sensitive motion resolution of articulated robot arms. *Trans. Society of Instrument and Control Engineers* 20(5): 453–459 *(in Japanese).*

Nakamura, Y., and Hanafusa, H. 1985 (Miami Beach). Inverse kinematic solutions with singularity robustness for robot manipulator control. *Robotics and Manufacturing Automation-PED-.* Vol. 15 (Book No. G00321), eds. M. Donath and M. Leu, 1985 ASME Winter Annual Meeting, ASME, pp. 193–204.

Nakamura, Y., and Hanafusa, H. 1986. Inverse kinematic solutions with singularity robustness for robot manipulator control. *J. Dyn. Sys., Meas., Contr.* 108 (3): 163–171.

Orin, D. E., and Schrader, W. W. 1984. Efficient computation of the Jacobian for robot manipulators. *Int. J. Robotics Res.* 3 (4): 66–75.

Pieper, D. L. 1968. The kinematics of manipulators under computer control. Ph.D. thesis, Stanford University, Dept. of Computer Science.

Tihonov, A. N. 1964. Solution of incorrectly formulated problems and the regularization method. *Sov. Math.*, 4: 1035–1038.

Twomey, S. 1965. The application of numerical filtering to the solution of integral equations encountered in indirect sensing measurements. *Journal of the Franklin Institute,* 279(2): 95–109.

Uchiyama, M. 1979. Study on dynamic control of artificial arms—part 1. *Trans. Japanese Society of Mechanical Engineers.* C-45(391): 314–322 *(in Japanese).*

Waldron, K. J., Wang, S. L., and Bolin, S.J. 1984. A study of the Jacobian matrix of serial manipulators. *Design Engineering Technical Conference*, ASME paper No. 84-DET-109.

Wampler, C. W. 1986. Manipulator inverse kinematic solutions based on vector formulations and damped least-squares methods. *IEEE Trans. Sys., Man, Cyber.* 16 (1): 93–101.

Wampler, C. W., and Leifer, L. J. 1988. Applications of damped least-squares methods to resolved rate and resolved-acceleration control of manipulators. *J. Dyn. Sys., Meas., Contr.* 110 (1): 31–38.

Whitney, D. E. 1972. The mathematics of coordinated control of prostheses and manipulators. *J. Dyn. Sys., Meas., Contr.* 94 (4): 303–309.

Yoshikawa, T. 1984. Control of robots with redundancy. *J. Robotics Society of Japan*, 2(6): 587–592 *(in Japanese)*

Yoshikawa, T. 1985. Manipulability of robotic mechanisms. In *Robotics research 2*, eds. H. Hanafusa and H. Inoue, pp. 439–446. Cambridge, MA: MIT Press.

Young, K. -K. D. 1978. Controller design for a manipulator using theory of variable structure systems. *IEEE Trans. Sys., Man, Cyber.* 8(2): 101–109.

CHAPTER

10

Redundancy in Multiaxis Force Sensing

10.1 Introduction

Force control has been discussed from the beginning of the robotics research (Scheinman 1969; Inoue 1971), and has been recognized as an important control scheme for using robots in advanced applications (Goto, Inoyama, and Takeyasu 1986; Hatamura 1986).

Concerning end-effector force sensing, researchers have proposed that it be computed from the measurement of the strain of driving joint axis (Nakano, Ozaki, Ishida, and Kato 1974), and from the measurement of the driving current of the direct-drive manipulators (Arai and Tachi 1986). However, these methods have the drawback that the friction and inertial forces must be eliminated to get the meaningful information (Uchiyama, Yokota, and Hakomori 1985). A direct and simple method to measure the end-effector force is to place a force sensor close to the end-effector of a robot.

This chapter is adapted from, by permission, Nakamura, Y., Yoshikawa, T., and Futamata, I. 1988. "Design and signal processing of six-axis force sensor," in *Robotics research 4*, eds. R. Bolles and B. Roth, pp. 75–80, Cambridge, MA: MIT Press; and Nakamura, Y., Yoshikawa, T., and Futamata, I. 1987. "Elastic component design criteria and signal processing of force sensors," *Trans. Society of Instrument and Control Engineers*, Vol. 23, No. 5 pp. 433–439 (*in Japanese*).

Force sensing has as long a research history as does force control, and many force sensors have been developed (Loewen, Marshall, and Shaw 1951; Kinoshita 1984; Flatau 1976; Watson and Drake 1975; Kasai, Takeyasu, Uno, and Muraoka 1981; Asakawa 1985; De Fazio, Seltzer, and Whitney 1986; Uchiyama 1986; Ono et al. 1985). Several six-axis force sensors are commercially available at present (Ono 1985; LORD Corporation 1985). A major problem in developing force sensors is the design of the elastic component, which has been done heuristically, dependent on the experience of designers. The theoretical evaluation of the designed force sensors and their comparative evaluation have not been done sufficiently. Uchiyama and Hakomori (1985) proposed that we evaluate the structure of elastic component of force sensor by the condition number of a sensor compliance matrix. This evaluation method made the first contribution not just to the evaluation of the developed force sensors, but also to the theoretical principles of force-sensor design.

In this chapter, first we discuss evaluation of the structure of the elastic component of force sensors by the following three criteria: the *strain-gauge sensitivity*, the *force sensitivity*, and the *minimum stiffness*. The *strain gauge sensitivity* is the sensitivity of each strain gauge to forces and moments within the measurement range given by the design specifications. The *force sensitivity* evaluates the force information of the whole sensor and is used to check whether the output signal can include sufficient signals in all six axes. The mechanical stiffness of the force sensor is liable to be small due to the elastic component. The evaluation of the mechanical stiffness is required at the design stage. The *minimum stiffness* implies the mechanical stiffness in the most compliant direction.

Next, we discuss the utilization of redundancy in force sensing. To calculate from the obtained strain gauge signals the force and moment exerted on the end-effector, we commonly compute the inverse matrix or the pseudoinverse of the sensor compliance matrix and multiply it by the strain-gauge signal vector (Uchiyama and Hakomori 1985). The drawback of this method is that it requires a large amount of computation, which in turn limits the sampling time of commercial six-axis force sensors to roughly 2 to 4 msec. Ono proposed that we design the elastic component such that a strain gauge would be sensitive to only one of the orthogonal axes (Ono 1985). This method does not require the computation of inverse transformation. Its drawback lies in the complexity of the structure of the elastic component. In Section 10.4, we propose a method to determine the structure of elastic component and the location of strain gauges such that the Wheatstone bridges, which are usually used for thermal compensation, work as a part of the inverse transformation circuit to compute the force and moment exerted on the end-effector. This method allows us to extract the decoupled force information as analog signals without making the mechanical structure complex.

Finally, a prototype six-axis force sensor is designed as an example to show the design procedure based on the principles presented.

10.2 Force Sensing

10.2.1 Basic Equations

Let O_h-xyz represent the hand coordinates fixed at the end-effector of a robot manipulator, and O_s-xyz be the sensor coordinates fixed at the force sensor, as shown in Fig. 10.1. We represent a vector from O_s to O_h described in the sensor coordinates by r_o; A_o is an orthogonal matrix with unit vectors in the x, y, and z directions of the hand coordinates described in terms of the sensor coordinates as the first, second, and third column vectors, respectively. The force and moment applied to the end-effector are represented by F_h and N_h at the origin of the hand coordinates. F_h and N_h can be transformed to the force and moment at the origin of the sensor coordinates F_s and N_s by

$$F_s = A_o\, F_h \qquad\qquad 10.1$$

$$N_s = A_o\, N_h + r_o \times (A_o\, F_h) \qquad\qquad 10.2$$

Assuming elastic strain for the elastic component of force sensor, the strain ϵ_i ($i = 1, \ldots, m$) of the strain gauges can be represented by the following linear equation of F_s and N_s (Uchiyama and Hakomori 1985):

$$\epsilon = C\, f \qquad\qquad 10.3$$

$$\epsilon = \mathrm{col}(\epsilon_i) \in R^m$$

$$f = (F_s{}^T\ N_s{}^T)^T \in R^6$$

where m is the number of strain gauges and $C \in R^{m \times 6}$ is called the *sensor compliance matrix*.

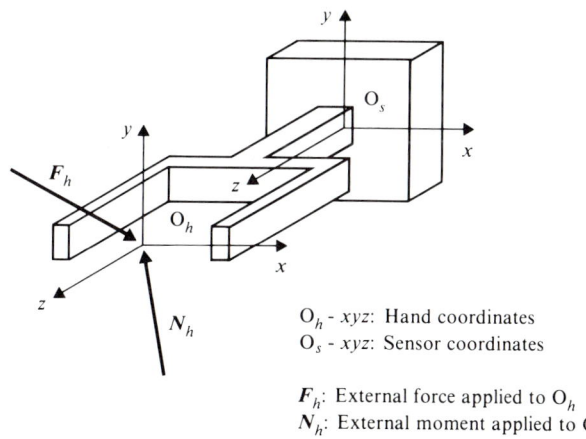

Figure 10.1 Principle of force and moment measurement.

Force sensing is the problem of obtaining $\boldsymbol{F}_h$ and $\boldsymbol{N}_h$ from measured ϵ and Eqs. 10.1 through 10.3. Therefore, it would be desirable to design force sensors based on the relationship of Eqs. 10.1 through 10.3. However, since $\boldsymbol{A}_o$ and $\boldsymbol{r}_o$ in Eqs. 10.1 and 10.2 depend on tasks, force sensors cannot be designed without tasks being specified. In this chapter, we discuss the design of force sensors by simplifying the problem to that of obtaining $\boldsymbol{F}_s$ and $\boldsymbol{N}_s$ from measured ϵ and Eq. 10.3. To generalize the discussion, we use the following as the basic equation:

$$\bar{\epsilon} = \bar{C}\bar{f} \qquad 10.4$$

$\bar{\epsilon} = \epsilon/\epsilon_{\max}$
$\bar{f} = (\boldsymbol{F}_s^T/F_{\max}\ \boldsymbol{N}_s^T/N_{\max})^T$
$\bar{C} = 1/\epsilon_{\max}\ \boldsymbol{CH}$
$\boldsymbol{H} = \mathrm{diag}(F_{\max}, F_{\max}, F_{\max}, N_{\max}, N_{\max}, N_{\max})$

which we obtain by normalizing Eq. 10.3 by means of maximum force $F_{\max}$ and maximum moment $N_{\max}$ given as the design specification, and maximum strain $\epsilon_{\max}$, which is determined by the proportional limit of the elastic material. Note that, although the proportional limit of steel is 0.12 percent, for example, $\epsilon_{\max}$ should be determined rather conservatively such that the strain at the places where strain gauges are not fixed will not exceed the proportional limit.

10.2.2 Structure Evaluation by Condition Number

For $\bar{f}$ to be computed from $\bar{\epsilon}$ for an arbitrary force, $m \geq 6$ and rank $\bar{C} = 6$ are required (Uchiyama and Hakomori 1985). In this case, the singular value decomposition (Section 2.3) of $\bar{C}$ is represented by

$$\bar{C} = U\Sigma V^T \qquad 10.5$$

$$\Sigma = \begin{pmatrix} \sigma_1 & 0 & \cdots & 0 \\ 0 & \sigma_2 & \cdots & 0 \\ \vdots & \vdots & \ddots & \vdots \\ 0 & 0 & \cdots & \sigma_6 \\ 0 & 0 & \cdots & 0 \\ \vdots & \vdots & \ddots & \vdots \\ 0 & 0 & \cdots & 0 \end{pmatrix} \in R^{m \times 6}$$

where $U \in R^{m \times m}$ and $V \in R^{6 \times 6}$ are orthogonal matrices and the σ_is satisfy $\sigma_1 \geq \sigma_2 \geq \ldots \geq \sigma_6 > 0$. For $m = 6$, $\bar{f}$ is computed from measured $\bar{\epsilon}$ by

$$\bar{f} = \bar{C}^{-1}\bar{\epsilon} \qquad 10.6$$

Uchiyama and Hakomori (1985) showed that, if the computation error $\Delta\bar{f}$ included in $\bar{f}$ is caused only by measurement error $\Delta\bar{\epsilon}$ contained in $\bar{\epsilon}$, then $\Delta\bar{\epsilon}$ and $\Delta\bar{f}$ satisfy the following relationship (see Section 2.3.3):

10.2 Force Sensing

$$(\text{cond } \bar{C})^{-1} \leq \frac{\|\Delta \bar{f}\|/\|\bar{f}\|}{\|\Delta \bar{\epsilon}\|/\|\bar{\epsilon}\|} \leq \text{cond } \bar{C} \qquad 10.7$$

where $\|\cdot\|$ represents the Euclidean norm and $\text{cond } \bar{C} \triangleq \sigma_1/\sigma_6$ is called the *condition number*. Equation 10.7 implies that $\text{cond } \bar{C}$ and $(\text{cond } \bar{C})^{-1}$ are respectively the maximum and minimum magnification ratios of the relative error of $\bar{\epsilon}$ to the relative error of $\bar{f}$. Based on this result, Uchiyama and Hakomori proposed that we use the condition number of the sensor compliance matrix as the index for the structural evaluation of force sensors, and made it a goal to design a force sensor with $\text{cond } \bar{C} = 1$. This evaluation method provided the designers of force sensors with a theoretical principle for force sensor design.

The structure evaluation by condition number, however, has the following problems:

1. Suppose that two force sensors have sensor compliance matrices $\bar{C}^1$ and $\bar{C}^2$, respectively. If $\bar{C}^1 = \frac{1}{2}\bar{C}^2$, then the condition numbers of both force sensors are equal to each other. Although $\bar{C}^2$ seems better than $\bar{C}^1$ from the viewpoint of so-called *sensor sensitivity*, the condition number cannot distinguish between the two.

2. Assume
$$\bar{C}^1 = \begin{pmatrix} 1 & 0 \\ 0 & 1 \end{pmatrix}$$

Now, we put another strain gauge at the same location as the first one. Then, the sensor compliance matrix becomes
$$\bar{C}^2 = \begin{pmatrix} 1 & 0 \\ 1 & 0 \\ 0 & 1 \end{pmatrix}$$

Since the condition numbers become $\text{cond } \bar{C}^1 = 1$ and $\text{cond } \bar{C}^2 = \sqrt{2}$, $\bar{C}^1$ will be evaluated to be better than $\bar{C}^2$. Although $\bar{C}^2$ may have a redundant strain gauge, $\bar{C}^2$ does not seem inferior to $\bar{C}^1$.

3. In case of $m > 6$, Eq. 10.6 can be replaced by
$$\bar{f} = \bar{C}^{\#}\bar{\epsilon} \qquad 10.8$$
$$\bar{C}^{\#} = V\Sigma^{\#}U^T$$
$$\Sigma^{\#} = \begin{pmatrix} 1/\sigma_1 & 0 & \cdots & 0 & 0 & \cdots & 0 \\ 0 & 1/\sigma_2 & \cdots & 0 & 0 & \cdots & 0 \\ \vdots & \vdots & \ddots & \vdots & \vdots & \ddots & \vdots \\ 0 & 0 & \cdots & 1/\sigma_6 & 0 & \cdots & 0 \end{pmatrix} \in R^{6\times m}$$

where $\bar{C}^{\#} \in R^{6 \times m}$ is the pseudoinverse of $\bar{C}$. The $\bar{f}$ and $\bar{\epsilon}$ satisfy the following relationship:

$$\begin{aligned}\|\bar{f}\|^2 &= \bar{\epsilon}^T (\bar{C}^{\#})^T \bar{C}^{\#} \bar{\epsilon} \\ &= \bar{\epsilon}^T U(\Sigma^{\#})^T \Sigma^{\#} U^T \bar{\epsilon}\end{aligned} \qquad 10.9$$

which yields the following equation for nonzero $\bar{\epsilon}$:

$$0 \le \frac{\|\bar{f}\|}{\|\bar{\epsilon}\|} \le \frac{1}{\sigma_6} \qquad 10.10$$

The same relationship can be obtained for $\Delta \bar{f}$ and nonzero $\Delta \bar{\epsilon}$:

$$0 \le \frac{\|\Delta \bar{f}\|}{\|\Delta \bar{\epsilon}\|} \le \frac{1}{\sigma_6} \qquad 10.11$$

From Eqs. 10.10 and 10.11, the relationship between the relative errors is represented as follows:

$$0 \le \frac{\|\Delta \bar{f}\|/\|\bar{f}\|}{\|\Delta \bar{\epsilon}\|/\|\bar{\epsilon}\|} < \infty \qquad 10.12$$

Therefore, the condition number offers no information about the relative errors if $m > 6$.

4. If $m > 6$ and rank $\bar{C}$ = rank $[\bar{C}, \bar{\epsilon}]$ are fulfilled, the inverse transformation of Eq. 10.4 is given not only by Eq. 10.8, but also more generally by the following equation:

$$\bar{f} = \bar{C}^- \bar{\epsilon} \qquad 10.13$$

where $\bar{C}^- \in R^{6 \times m}$ is the generalized inverse matrix of $\bar{C}$. In case of rank $\bar{C} = 6$, $\bar{C}^-$ becomes the reflexive generalized inverse matrix and rank $\bar{C}^- = 6$ (Section 2.4.2, generalized inverse, (2)). Unlike $\bar{C}^{\#}$, $\bar{C}^-$ is not unique. Choosing the pseudoinverse as a generalized inverse in Eq. 2.134, the general form of $\bar{C}^-$ is represented by

$$\bar{C}^- = \bar{C}^{\#} + (Z - \bar{C}^{\#} \bar{C} Z \bar{C} \bar{C}^{\#}) \qquad 10.14$$

where $Z \in R^{6 \times m}$ is an arbitrary matrix. The condition number does not account for the variety of the inverse transformations.

In the following section, taking these problems into consideration, we propose a new structure-evaluation method based on the three principles we mentioned: the *strain-gauge sensitivity*, the *force sensitivity*, and the *minimum stiffness*.

10.3 Structure Evaluation of Elastic Components

10.3.1 Strain-gauge Sensitivity

We assume that the specification of measurement range is given by

$$\|\bar{f}\| \leq 1 \qquad 10.15$$

Then, $\bar{\epsilon}_i$ satisfies the following inequality:

$$|\bar{\epsilon}_i| \leq \|\bar{C}_i\| \qquad 10.16$$

where $\bar{C}_i$ means the ith row vector of $\bar{C}$. Therefore, for $|\epsilon_i|$ not to exceed $\epsilon_{\max}$, the following equation must be satisfied:

$$\|\bar{C}_i\| \leq 1 \qquad 10.17$$

Since $\|\bar{C}_i\|$ can be considered as the maximum sensitivity of the ith strain gauge for all of $\bar{f}$ that satisfy Eq. 10.15, we call $\|\bar{C}_i\|$ the *strain-gauge sensitivity*. So that the S/N (signal to noise) ratio of sensor signals can be increased, it is desirable to design the elastic component such that it provides large strain readings. Consequently, the first design principle is to bring the strain-gauge sensitivity to one, but not to allow it to exceed one.

10.3.2 Force Sensitivity

When all of the forces that satisfy $\|\bar{f}\| = 1$ are applied, the minimum singular value, σ_6, represents, from Eqs. 10.4 and 10.5, the minimum magnitude of $\bar{\epsilon}$. Accordingly, in the conservative sense, we call minimum singular value σ_6 the *force sensitivity*. Since it is desirable that a force sensor be able to generate enough output signals in all directions, the second design principle is to make the force sensitivity as large as possible.

Now, we investigate the relationship between the force sensitivity and the strain-gauge sensitivity. From Eq. 10.5, the following equation is derived:

$$\begin{aligned}
\text{trace } \bar{C}\bar{C}^T &= \text{trace } U\Sigma\Sigma^T U^T \\
&= \text{trace } \Sigma\Sigma^T U^T U \\
&= \text{trace } \Sigma\Sigma^T \qquad 10.18 \\
&= \sum_{i=1}^{6} \sigma_i^2
\end{aligned}$$

On the other hand, the following equation also holds:

$$\text{trace } \bar{C}\bar{C}^T = \sum_{i=1}^{m} \|\bar{C}_i\|^2 \qquad 10.19$$

From Eqs. 10.18 and 10.19, the strain-gauge sensitivity and the singular values satisfy the following relationship:

$$\sum_{i=1}^{6} \sigma_i^2 = \sum_{i=1}^{m} \|\bar{C}_i\|^2 \qquad 10.20$$

Therefore, when the strain-gauge sensitivities are fixed, the force sensitivity takes the following maximum value at $\sigma_1 = \sigma_2 = \ldots = \sigma_6$ (cond $\bar{C} = 1$):

$$\sigma_{6\,\max} = \sqrt{\frac{1}{6}\sum_{i=1}^{m} \|\bar{C}_i\|^2} \qquad 10.21$$

If the strain-gauge sensitivities are optimized—that is, $\|\bar{C}_i\| = 1$ ($i = 1, \ldots, m$)—the possible maximum value of force sensitivity is

$$\sigma_{6\,\max} = \sqrt{\frac{m}{6}} \qquad 10.22$$

This is the theoretical maximum value of force sensitivity. Note that Eqs. 10.20 and 10.21 imply that, when the strain-gauge sensitivities are fixed, the force-sensitivity criterion is equivalent to the condition-number criterion.

Next, we discuss the relationship between the force sensitivity and the magnification of measurement error. We assume the inverse transformation by the generalized inverse matrix of the sensor compliance matrix as shown in Eq. 10.13. Suppose that the strain-gauge signal is represented as the sum of the true value $\bar{\epsilon}$ and the measurement error $\Delta\bar{\epsilon}$. Then, the force information is computed as the sum of the true value $\bar{f}$ and the error $\Delta\bar{f}$:

$$\bar{f} + \Delta\bar{f} = \bar{C}^{-}(\bar{\epsilon} + \Delta\bar{\epsilon}) \qquad 10.23$$

Subtracting Eq. 10.13 from Eq. 10.23, we obtain the following equation:

$$\Delta\bar{f} = \bar{C}^{-}\Delta\bar{\epsilon} \qquad 10.24$$

Representing an arbitrary matrix by $Z = V(Z_1\ Z_2)U^T$, $Z_1 \in R^{6\times 6}$, $Z_2 \in R^{6\times(m-6)}$ and substituting Eq. 10.5 and the arbitrary matrix into Eq. 10.14, we obtain the following equation:

$$\bar{C}^{-} = V\Sigma^{-}U^T \qquad 10.25$$

$$\Sigma^{-} = \begin{pmatrix} 1/\sigma_1 & \cdots & 0 \\ \vdots & \ddots & \vdots & Z_2 \\ 0 & \cdots & 1/\sigma_6 \end{pmatrix} \in R^{6\times m}$$

If the magnification ratio of measurement error is defined for nonzero $\Delta\bar{\epsilon}$ by

$$\alpha = \max \frac{\|\Delta\bar{f}\|}{\|\Delta\bar{\epsilon}\|} \qquad 10.26$$

then we observe from Eqs. 10.24 and 10.25 that α depends on $\boldsymbol{Z}_2$ and becomes equal to $1/\sigma_6$ in case of $\boldsymbol{Z}_2 = 0$—that is, $\bar{\boldsymbol{C}}^- = \bar{\boldsymbol{C}}^\#$—and that α never becomes smaller than $1/\sigma_6$. In other words, the measurement-error magnification ratio of inverse transformation by the pseudoinverse is the lower limit of the measurement-error magnification ratio by the generalized inverse matrix, and it is equal to the reciprocal of the force sensitivity.

10.3.3 Minimum Stiffness

The measurement principle for a force sensor is to measure the strain produced in the elastic component, and to compute the applied force and moment from it. Since low stiffness of the elastic component generally means high force sensitivity, the elastic component tends to have low stiffness. However, low stiffness of the force sensor makes the total stiffness of the robotic mechanism low. Therefore, it is one of the important issues in elastic-component design to address the contradictory requirement that both the force sensitivity and the mechanical stiffness be maximized simultaneously. In this subsection, an approximate method for computing the mechanical stiffness of a force sensor is proposed. This method will enable stiffness evaluation at the design stage.

As shown in Fig. 10.2, we assume that the elastic component is approximated by a curve that passes through the center of the elastic component (the *axis of the elastic component*) and that all the stress is caused only by bending moments about axes defined at every point on the axis of the elastic component and perpendicular to it (the *axis of elastic strain*). The axis of the elastic component is chosen as the x-axis, and a vector from the point where $\boldsymbol{F}_s$ and $\boldsymbol{N}_s$ are exerted to x is represented by $\boldsymbol{r}_x$. If we represent by $\boldsymbol{e}_x \in R^3$ the unit vector in the direction of the axis of elastic strain at x, the bending moment at x is calculated by

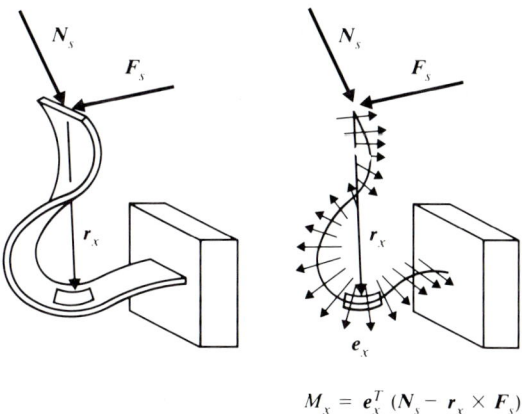

$$M_x = \boldsymbol{e}_x^T (\boldsymbol{N}_s - \boldsymbol{r}_x \times \boldsymbol{F}_s)$$

Figure 10.2 Computation of elastic strain energy.

$$M_x = e_x^T(N_s - r_x \times F_s) \qquad 10.27$$

The elastic strain energy of the whole elastic component is computed by the following equation (Timoshenko and Young 1965):

$$U = \int_x \frac{M_x^2}{2EI_x} dx \qquad 10.28$$

where E means the modulus of elasticity of the material and I_x is the second moment of area at x. Substituting Eq. 10.27 into Eq. 10.28, we obtain the following equation:

$$U = \frac{1}{2} f^T G f \qquad 10.29$$

$$G = \int_x \frac{1}{EI_x} h_x h_x^T dt$$

$$h_x = \begin{pmatrix} -R_x^T e_x \\ e_x \end{pmatrix}$$

$$R_x = \begin{pmatrix} 0 & -r_{x3} & r_{x2} \\ r_{x3} & 0 & -r_{x1} \\ -r_{x2} & r_{x1} & 0 \end{pmatrix}$$

$$r_x = \begin{pmatrix} r_{x1} & r_{x2} & r_{x3} \end{pmatrix}^T$$

Displacement u at the point where $f = (F_s^T \ N_s^T)^T$ is applied is calculated based on *Castigliano's theorem* (Timoshenko and Young 1965) as follows:

$$u = \left(\frac{\partial U}{\partial f}\right)^T = Gf \qquad 10.30$$

Equation 10.30 is normalized as follows:

$$\bar{u} = \bar{G}\bar{f} \qquad 10.31$$

$$\bar{u} = \frac{1}{N_{max}} Hu$$

$$\bar{G} = \frac{1}{N_{max}} H^T G H$$

where H is as defined in Eq. 10.4 and $\bar{G}$ denotes the normalized compliance matrix at the point where the force is applied. If we represent the singular values of $\bar{G}$ by $\sigma_{G1} \geq \sigma_{G2} \geq \ldots \geq \sigma_{G6} > 0$, then $1/\sigma_{G6}$ denotes the stiffness in the stiffest direction and $1/\sigma_{G1}$ denotes the stiffness in the most compliant direction. Therefore, we call $1/\sigma_{G6}$ the *maximum stiffness* and $1/\sigma_{G1}$

the *minimum stiffness*. The third design principle is to keep the minimum stiffness large.

10.4 Use of Redundancy in Force Sensing

Theoretically, the minimum number of strain gauges required to measure six elements of force and moment is six. However, it is common to use redundant strain gauges. Scheinman used 16 strain gauges on all the surfaces of the four prisms that make a cross-shaped force sensor (Ono et al. 1985). The advantages are (1) compensating thermal drift, (2) reducing the effect of strain-gauge location error, and (3) simplifying the inverse transformation. On the other hand, Ono et al. (1985) and Ono (1985) designed the structure of elastic component so that the sensor compliance matrix becomes a diagonal matrix.

In this section, using the fact that the generalized inverse matrix allows a variety of inverse transformations in the case of $m > 6$, as shown in Eqs. 10.13 and 10.14, we use redundancy in force sensing to simplify signal processing. We design an elastic component and determine the strain-gauge locations such that the Wheatstone bridges work as a part of the analog inverse transformation circuit.

Figure 10.3 shows a schematic drawing of a Wheatstone bridge. The equilibrium condition for the Wheatstone bridge to give $V = 0$ is described by (Smith 1976)

$$\frac{r_2}{r_1} = \frac{r_3}{r_4} \qquad 10.32$$

If the initial resistance values are $r_1 = r_2 = r_3 = r_4$ and each resistance value changes from r_i to $r_i + \Delta r_i$, the bridge output becomes (Shiota and Taniguchi 1983)

$$V \propto \Delta r_1 - \Delta r_2 + \Delta r_3 - \Delta r_4 \qquad 10.33$$

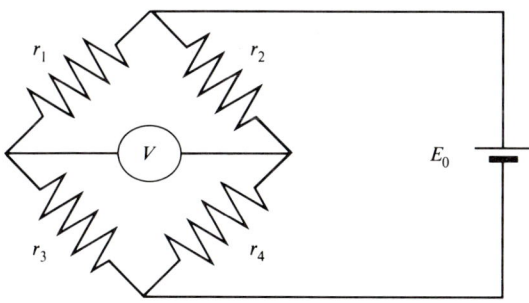

E_0 : Power source
r_i : Resisters
V : Voltage meter

Figure 10.3 Wheatstone bridge.

We use strain gauges for all of $r_1, \ldots, r_4$ and assume that resistance change is proportional to strain. Then, we have the following equation:

$$V = k(\epsilon_{r1} - \epsilon_{r2} + \epsilon_{r3} - \epsilon_{r4}) \qquad 10.34$$

where k is a constant and ϵ_{ri} denotes the strain that corresponds to Δr_i. When 24 strain gauges and six Wheatstone bridges are used, the transfer function $B \in R^{6 \times 24}$ of the circuit becomes

$$B = k \begin{pmatrix} 1 & -1 & 1 & -1 & 0 & 0 & 0 & 0 & \cdots & 0 & 0 & 0 & 0 \\ 0 & 0 & 0 & 0 & 1 & -1 & 1 & -1 & \cdots & 0 & 0 & 0 & 0 \\ 0 & 0 & 0 & 0 & 0 & 0 & 0 & 0 & \cdots & 0 & 0 & 0 & 0 \\ 0 & 0 & 0 & 0 & 0 & 0 & 0 & 0 & \cdots & 0 & 0 & 0 & 0 \\ 0 & 0 & 0 & 0 & 0 & 0 & 0 & 0 & \cdots & 0 & 0 & 0 & 0 \\ 0 & 0 & 0 & 0 & 0 & 0 & 0 & 0 & \cdots & 1 & -1 & 1 & -1 \end{pmatrix} \qquad 10.35$$

Figure 10.4 shows a block diagram of the circuit including six Wheatstone bridges and the corresponding gain-adjustment amplifiers w_i ($i = 1, \ldots, 6$). If $\bar{C}$ is designed such that the transfer function of the circuit could satisfy the following equation by adjustment of gain matrix $W = \text{diag}(w_i) \in R^{6 \times 6}$:

$$WB = \frac{1}{\epsilon_{max}} \bar{C}^- \qquad 10.36$$

then output $\bar{f}^*$ of the circuit becomes equal to $\bar{f}$.

It is difficult to give the general structure of elastic component that satisfies Eq. 10.36. However, it is comparitively easy to make $\bar{C}$ satisfy Eq. 10.36 by designing a symmetric structure and adjusting the strain-gauge location. In the next section, we shall look at an example of such a structure. Note that symmetric structure of the elastic component simplifies analysis and is useful in analytically evaluating the strain-gauge sensitivity, the force sensitivity, and the minimum stiffness.

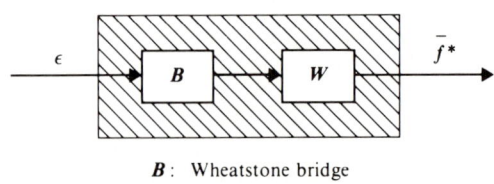

B: Wheatstone bridge
$W = \text{diagonal}(w_i)$

Figure 10.4 Analog inversion circuit.

A drawback of this method lies in the fact that, however faithfully $\bar{C}$ may fulfill Eq. 10.36 in the theoretical analysis, it is difficult for $\bar{C}$ to satisfy Eq. 10.36 accurately in practice because of the machining error, the strain-gauge location error, and the effect of neglected higher order strain, which was not considered in the theoretical analysis. Therefore, this method may inherently contain some measurement error. In spite of this drawback, since the inverse transformation can be done by analog circuit, this method will be effective in the case where force feedback control must be done by high sampling rate.

10.5 An Example of Force-Sensor Design

Figure 10.5 shows an example of elastic-component design. The design specifications were $F_{max} = 10\,\text{N}$ and $N_{max} = 100\,\text{N/mm}$. Steel was chosen as the elastic material. Considering safety coefficient 3, the proportional limit was determined as $\epsilon_{max} = 0.02$ percent, since the proportional limit of steel is 0.12 percent and the maximum strain at the place where the strain gauges are not fixed is approximately twice as large as the maximum strain at the place where the strain gauges are fixed. The sensor compliance matrix was calculated analytically and became

$$\bar{C} = \begin{pmatrix}
-a_1 & -a_2 & 0 & 0 & 0 & -a_3 \\
a_1 & a_2 & 0 & 0 & 0 & -a_3 \\
-a_1 & a_2 & 0 & 0 & 0 & a_3 \\
a_1 & -a_2 & 0 & 0 & 0 & a_3 \\
a_4 & a_5 & 0 & 0 & 0 & a_6 \\
a_4 & -a_5 & 0 & 0 & 0 & -a_6 \\
-a_4 & a_5 & 0 & 0 & 0 & -a_6 \\
-a_4 & -a_5 & 0 & 0 & 0 & a_6 \\
0 & 0 & a_7 & 0 & -a_8 & 0 \\
0 & 0 & -a_7 & 0 & a_8 & 0 \\
0 & 0 & a_7 & 0 & a_8 & 0 \\
0 & 0 & -a_7 & 0 & -a_8 & 0 \\
0 & 0 & a_9 & -a_{10} & 0 & 0 \\
0 & 0 & -a_9 & a_{10} & 0 & 0 \\
0 & 0 & -a_9 & -a_{10} & 0 & 0 \\
0 & 0 & a_9 & a_{10} & 0 & 0 \\
0 & 0 & a_{11} & 0 & -a_{12} & 0 \\
0 & 0 & -a_{11} & 0 & a_{12} & 0 \\
0 & 0 & -a_{11} & 0 & -a_{12} & 0 \\
0 & 0 & a_{11} & 0 & a_{12} & 0 \\
-a_{13} & -a_{14} & 0 & 0 & 0 & -a_{15} \\
-a_{13} & a_{14} & 0 & 0 & 0 & a_{15} \\
a_{13} & a_{14} & 0 & 0 & 0 & -a_{15} \\
a_{13} & -a_{14} & 0 & 0 & 0 & a_{15}
\end{pmatrix} \quad 10.37$$

296 Chapter 10 Redundancy in Multiaxis Force Sensing

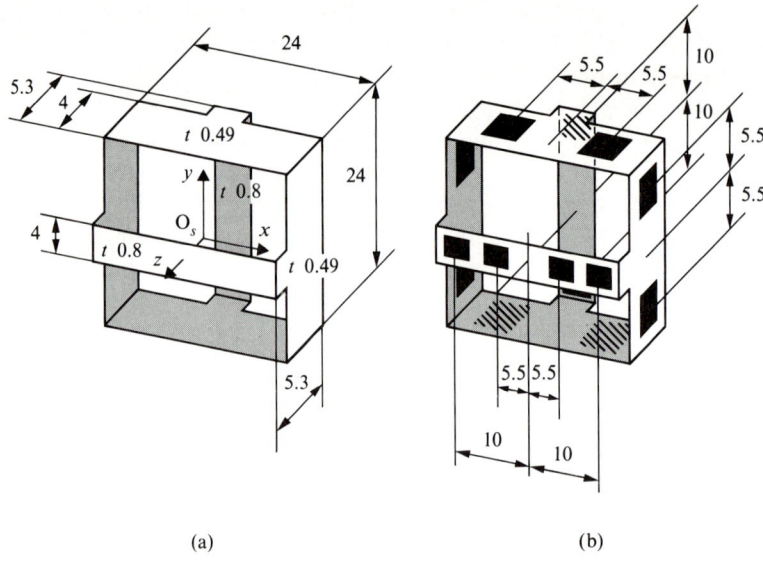

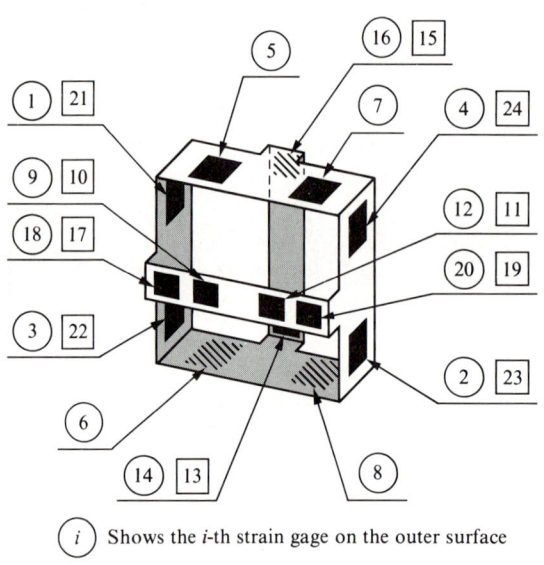

Figure 10.5 Designed six-axis force sensor. (a) Designed elastic component. (b) Strain-gauge location. (c) Strain-gauge numbers.

10.5 An Example of Force-Sensor Design

$$a_1 = 0.930 \qquad a_2 = -0.0372 \qquad a_3 = 0.391$$
$$a_4 = 0.0372 \qquad a_5 = 0.930 \qquad a_6 = -0.391$$
$$a_7 = 0.419 \qquad a_8 = -0.917 \qquad a_9 = -0.837$$
$$a_{10} = -0.614 \qquad a_{11} = -0.837 \qquad a_{12} = 0.614$$
$$a_{13} = -0.930 \qquad a_{14} = 0.0372 \qquad a_{15} = -0.391$$

where $E = 2.1 \times 10^5$ N/mm^2 was used as the modulus of elasticity of steel. Accordingly, the strain-gauge sensitivities were calculated as follows:

$$\|\bar{C}_i\| = \begin{cases} 1.01 & (i = 1, \ldots, 12, 21, 22, 23, 24) \\ 1.04 & (i = 13, \ldots, 20) \end{cases} \qquad 10.38$$

The singular values of $\bar{C}$ were obtained as follows:

$$\text{col}(\sigma_i) = (2.63 \; 2.51 \; 2.21 \; 1.86 \; 1.35 \; 1.23)^T \qquad 10.39$$

Therefore, the force sensitivity was 1.23. The condition number was 2.14. Compliance matrix $\bar{G}$ was also computed analytically and obtained as follows:

$$\bar{G} = 10^{-4} \text{diag}(2.02, 2.02, 1.16, 1.01, 1.01, 0.686) \qquad 10.40$$

The maximum stiffness and the minimum stiffness became 1.46×10^4 and 4.95×10^3, respectively. Note that these are the normalized stiffnesses. The value of the minimum stiffness means that, if $F_{\max} = 10$N is applied in the weakest direction, it will produce 2.02×10^{-3}mm of displacement, which seems a fairly high stiffness. On the other hand, gain matrix W of the inverse transformation circuit using the Wheatstone bridges was obtained as follows:

$$W = \frac{1}{k \, \epsilon_{\max}} \text{diag}(-0.269, 0.269, 0.597, 0.406, -0.406, 0.640) \qquad 10.41$$

and the singular values of generalized inverse matrix $\bar{C}^-$ became

$$\text{col}(\sigma_i^-) = (1.28 \; 1.20 \; 0.811 \; 0.811 \; 0.538 \; 0.538)^T \qquad 10.42$$

where the maximum singular value was 1.28 and the condition number was 2.38. Although these values are a little greater than the maximum singular value $1/\sigma_6 = 0.813$ and condition number 2.14 of pseudoinverse $\bar{C}^\#$, the measurement error magnification by these values is not serious and, therefore, the Wheatstone bridges and the gain matrix W would be acceptable for the analog inverse transformation circuit of the force sensor. The fabricated six-axis force sensor is shown in Fig. 10.6.

Figure 10.6 Fabricated six-axis force sensor for robotic fingers and a 3.5 inch floppy disk.

Summary

The design of elastic component of force sensor can be done based on the following three evaluation principles:

1. *The strain-gauge sensitivity*: $\|\bar{C}_i\|$ $(i = 1, \ldots, m)$ is called the strain gauge sensitivity. Condition $\|\bar{C}_i\| \leq 1$ prevents the material from undergoing plastic deformation. The corresponding design principle is to bring the strain-gauge sensitivity close to one, but not to allow it to exceed one.

2. *The force sensitivity*: The minimum singular value of $\bar{C}$ is called the force sensitivity. The corresponding design principle is to make the force sensitivity as large as possible, because this guarantees that the force sensor can generate enough output signals in all force directions. When the strain-gauge sensitivities are optimized—that is, $\|\bar{C}_i\| = 1$ $(i = 1, \ldots, m)$—the possible maximum value of force sensitivity is equal to $\sqrt{m/6}$, which is the theoretical maximum value of the force sensitivity. It also implies that the possible maximum value of the force sensitivity increases as the number of strain gauges increases. Since the measurement-error magnification ratio of inverse transformation by the pseudoinverse is equal to the reciprocal of the force sensitivity and is the lower limit of the measurement-error magnification ratio by the generalized inverse matrix, higher force sensitivity is better for reducing the measurement-error magnification.

3. ***The minimum stiffness***: The compliance matrix at the point where forces are applied is obtained based on *Castigliano's theorem*. The reciprocal of the maximum singular value of the compliance matrix is called the minimum stiffness, which is the stiffness in the most compliant direction. The corresponding design principle is to keep the minimum stiffness large.

There is a large number of DOF in the design of elastic component of force sensors. The three principles are not sufficient to identify a unique structure of the elastic component. We have also to consider practical requirements, such as the light weight of the elastic component, its compact shape for packaging, its simple structure (which would ease machining the elastic components and putting strain gauges on them), and so on. However, we are convinced that the three design principles will indicate a clear design direction of the elastic components of force sensors.

We showed that a variety of inverse transformations of the force signals exist when there is redundancy in force sensing. We proposed designing an elastic component and determining the strain-gauge locations such that the Wheatstone bridges work as a part of the analog inverse transformation circuit. Since the inverse transformation can be done by an analog circuit, this method will be especially effective in the case where force feedback control must be done by high sampling rate.

An example of six-axis force-sensor design was shown to illustrate the design procedure based on the proposed design principles and the inverse transformation method. In the example, the strain-gauge sensitivity, the force sensitivity, and the minimum stiffness were computed analytically with several assumptions. The use of the finite element method will enable more accurate evaluation of these three values and make it possible to realize an interactive computer-assisted design system for force sensors.

References

Arai, H., and Tachi, S. 1986. Operational force measurement and active force assistance in human operation of direct drive manipulators. *J. Japan Robotics Society* 4 (3): 209–219. *(in Japanese)*.

Asakawa, K. 1985. Realization of highly accurate tasks by sensor feedback. *J. Precision Engineering* 51 (11): 2034–2039 *(in Japanese)*.

De Fazio, T. L., Seltzer, D. S., and Whitney, D. E. 1986. The IRCC instrumented remote center compliance. In *Robot sensors*, Vol. 2, ed. A. Pugh, pp. 33–44. Bedford, UK: IFS Ltd.

Flatau, C. R. 1976 (Warsaw). Force sensing in robots and manipulators. *Proc. 2nd Int. CISM-IFToMM Symp. Theory and Practice of Robots and Manipulators*, pp.294–306.

Goto, T., Inoyama, T., and Takeyasu, K. 1986. Precise insert operation by tactile-controlled robot. In *Robot sensors*, Vol. 2, ed. A. Pugh, pp. 45–52. Bedford, UK: IFS Ltd.

Hatamura, Y. 1986. Force torque sensor. *J. Japan Society of Mechanical Engineers* 89 (814): 1055–1058 *(in Japanese)*.

Inoue, H. 1971. Computer controlled bilateral manipulator. *Bulletin of Japan Society of Mechanical Engineers* 14 (69): 199–207.

Iri, M., Kodama, S., and Suda, N. 1982. Singular value decomposition and its application to system control. *J. Society of Instrument and Control Engineers* 21(8): 763-772 *(in Japanese)*.

Kasai, M., Takeyasu, Y., Uno, M., and Muraoka, K. 1981 (Tokyo). Trainable assembly system with an active sensory table possessing six axes. *Proc. 11th Int. Symp. Industrial Robots*, pp. 393–404.

Kinoshita, G. 1984. A survey of tactile sensor development. *J. Japan Robotics Society* 2 (5): 430–437 *(in Japanese)*.

Loewen, E. G., Marshall, E. R., and Shaw, M. C. 1951. Electric strain gauge tool dynamometers. *Proc. SESA*, 8 (2): 1–16.

LORD Corporation. 1985. Force/torque wrist sensing systems. Technical Note F/T Series 6/85.

Nakamura, Y., Yoshikawa, T., and Futamata, I. 1987. Elastic component design criteria and signal processing of force sensors. *Trans. Society of Instrument and Control Engineers* 23(5): 433–439 *(in Japanese)*.

Nakamura, Y., Yoshikawa, T., and Futamata, I. 1988. Design and signal processing of six-axis force sensors. In *Robotics research 4*, eds. R. Bolles and B. Roth, pp. 75–80. Cambridge, MA: MIT Press.

Nakano, E., Ozaki, S., Ishida, T., and Kato, I. 1974 (Tokyo). Cooperative control of a pair of anthropomorphous manipulators: MELARM. *Proc. 4th Int. Symp. Industrial Robots*, pp. 251–260.

Ono, K., Hatamura, Y., Ogata, K., Takada, R., and Kusaki, T. 1985 (Osaka). Development of 6 axis force sensor LSA6000. *Proc. 3rd Annual Conf. Japan Robotics Society*, pp. 19–20 *(in Japanese)*.

Ono, K. 1985. Six axis force sensor for high performance robot manipulators. *Automation* 30 (2): 48–52 *(in Japanese)*.

Scheinman, V. D. 1969. Design of a computer controlled manipulator. Stanford Artificial Intelligence Project Memo AIM-92, Stanford University.

Shiota, Y., and Taniguchi, Y. 1983. *Sensors*. Tokyo: Sangyo Tosyo Ltd. *(in Japanese)*.

Smith, R. J. 1976. *Circuits, devices, and systems*, third edition. New York: John Wiley & Sons.

Timoshenko, S., and Young, D.H. 1965. *Theory of structures*. New York: McGraw-Hill.

Uchiyama, M., Yokota, M., and Hakomori, K. 1985 (Tokyo). Kalman filtering the six-axis robot wrist force sensor signal. *Proc. '85 Int. Conf. Advanced Robotics*, pp. 153–160.

Uchiyama, M., and Hakomori, K. 1985. A few considerations on structure design of force sensors. *Proc. 3rd Annual Conf. Japan Robotics Society*, pp. 17–18 *(in Japanese)*.

Uchiyama, M. 1986. Robot sensors. In the text of *Seminar on Fundamental Robotics*. (Tokyo). Society of Instrument and Control Engineers, pp. 89–113 *(in Japanese)*.

Watson, P. C., and Drake, S. H. 1975. Pedestal and wrist sensors for automatic assembly. *Proc. 5th Int. Symp. Industrial Robots*, pp. 501–511.

CHAPTER 11

Geometric Optimization for Sensor Fusion

11.1 Introduction

Advanced applications require robotic systems to have various kinds of external sensors, including force sensors, tactile sensors, proximity sensors, ultra sonic sensors, range sensors, vision sensors, and so on, in addition to the basic internal sensors, such as encoders and tachometers. Industrial applications tend to need robotic systems that have more adaptability. Nonindustrial tasks in uncertain environments, such as in outer space and under water, cannot be done without high-level adaptability based on sensing. Robotic systems will naturally have to equip more and more sensors to identify their environments and to determine their actions.

Sensor fusion is the technology to extract the better information from multiple sensors. The primary aim is to pick up more accurate and less uncertain information by actively utilizing redundant information. For instance, a robot manipulator with a vision sensor and a set of joint sensors can identify its end-effector location by either of them. The conventional approach is

---

This chapter is adapted from, by permission, Nakamura, Y., and Xu, Y. 1989 (Scottsdale, AZ). "Geometrical Fusion Method for Multi-Sensor Robotic Systems," *Proc. 1989 IEEE International Conference on Robotics and Automation*, pp. 668–673.

to select the one that looks more appropriate for the situation than the other. This selection procedure must be written explicitly as a program by a person with a knowledge of the details of the measuring function of sensors and the function of robotic mechanisms. Although it is not too difficult when the degree of redundancy of sensors is relatively low, the load of the programmer gets heavier and heavier as the degree of redundancy increases. More important, this strategy takes no advantage of the possibility of improving accuracy and suppressing uncertainty that could be done by utilizing redundant sensory information. If we could appropriately combine both of vision and joint information, we would get better information than we do from either one of them. The improvement in quality of information becomes even more significant as the degree of sensing redundancy increases.

The goal of this chapter is to provide a geometrically motivated mathematical method of multisensor fusion for robotic systems that increases the accuracy and reduces the uncertainty by combining redundant sensory information.

11.2 Background

Although research in the area of sensor fusion has a relatively short history, many papers have been published in journals and conferences recently. Henderson and Shilcrat (1984) discussed the framework of *logical sensor systems* to treat multisensor systems in a coherent and efficient way, and Henderson, Fai, and Hansen (1984) developed a *multisensor kernel system*. Bajcsy and Allen (1985) proposed fusing disparate sensory data for object recognition. The modeling of measurement uncertainty was addressed by Brooks (1985), who defined the *uncertainty manifold* for mobile robots and proposed back reasoning based on land mark information.

Statistical-uncertainty modeling for multisensor systems was initiated by Durrant-Whyte (1985), who assumed the Gaussian distribution and applied Bayesian inference with minimum variance estimate to fuse linearly structured multisensor systems. This method was applied to a system that includes a stereo camera and a tactile array sensor mounted on a PUMA 560 (Durrant-Whyte 1986). Luo, Lin, and Scherp (1987) defined the confidence distance measure using linear Gaussian models, proposed a hypothesis test to reject sensory data obtained by malfunction, and discussed an iterative computational method of fusion. Hashimoto and Paul (1987) applied the approach by Durrant-Whyte to integrating encoder data and tachometer data for control of manipulators.

It is interesting that similar statistical approaches have been used, independently, in the field of mobile robots. Chatila and Laumond (1986) proposed averaging scalar data by weighting with variances. Smith and Cheeseman (1986) discussed the uncertainty of mobile robot location caused by the uncertain coordinate transformation, where a nonlinear model associated with the orientation of coordinate frames was used. Although the fusion method was derived using the Kalman filter theory, it is essentially equivalent to Bayesian inference with minimum variance estimate. Matthies and Shaper (1987) discussed the error modeling of stereo vision for mobile robots, where the fusion method was obtained by the weighted least-squares estimate with the inverse of the covariance matrix as the weight. It is well known (Bryson and Ho 1975) that, with this choice of weighting matrices and assuming the Gaussian distribution, the weighted least-squares estimate is identical to Bayesian inference with minimal variance estimate. The method to consider nonlinearity was discussed in several textbooks on optimal estimation, among which Bryson and Ho (1975) gave an example of estimating the location of a point object by sight angle measurements from several fixed points (pp. 352–355). This is similar to the fusion of stereo vision by Matthies and Shaper.

The effectiveness of parallel processing in sensory data fusion was discussed by Chiu, Morley, and Martin (1986). Harmon, Bianchini, and Pinz (1986) proposed a distributed blackboard for sensory data fusion. Ruokangas, Black, Martin, and Schoenwald (1986) discussed the implementation of a vision sensor, an acoustic ranging sensor, and a force-torque sensor. Shekhar, Khatib, and Shimojo (1986) discussed fusion for object localization using tactile sensory information. Multiview integration for three-dimensional object modeling was studied by Wang and Aggarwal (1987).

It is an important property of robotic sensors that the data from a set of sensors are used to obtain various kinds of information. For instance, the joint sensors of robotic manipulators may be used to acquire the position and orientation of the end-effector for operational space motion, to compute the position of the elbow for obstacle-avoidance motion, and to obtain joint angles for joint motion. Therefore the nonlinearity between the sensory data, which is the low-level data from specific sensors, and the sensory information, which is the high-level information to be obtained by processing of the sensory data, comes from both the inherent structural nonlinearity related to the mechanism to generate the sensory data and the computational nonlinearity related to the processing to obtain the desired sensory information. The generalization of nonlinearity should include the case where sensors contribute only part of the desired information. This would make it possible to fuse, for example, joint sensors and a single-camera vision sensor to obtain the end-effector position.

In this chapter, we discuss a general statistical fusion method for multiple sensor systems that is directly motivated by the geometry of uncertainties. The treatment of nonlinearity is generalized so as to include both the structural nonlinearity and the computational nonlinearity. First, assuming the Gaussian noise additive to the sensory data, the uncertainty ellipsoid is defined associated with the covariance matrix of the error of the sensory information. Second, the optimal fusion is defined as one that minimize the geometric volume of the ellipsoid among all of the possible linear combinations of sensory information. It is shown that the optimal fusion method results in an algorithm similar to those obtained by Bayesian inference by minimum variance estimate, by the Kalman filter theory, and by the weighted least-squares estimate. Finally, the method is extended to include the fusion of partial information.

11.3 Sensing Model and Uncertainty Ellipsoid

11.3.1 Nonlinear Sensing Model

We define $\boldsymbol{\theta}^i \in \mathcal{R}^{m_i}, i = 1, \ldots, p$, as sensory data from sensor unit i, where sensory data are the low-level measurements inherent to a specific physical sensor, m_i denotes the number of the independent measurements, and p is the number of the sensor units. A sensor unit is not necessarily a single physical sensor, but rather may represent a set of sensors whose data usually are used simultaneously. *Sensory information*, $\boldsymbol{x}^i \in \mathcal{R}^n, i = 1, \ldots, p$, is computed from sensory data $\boldsymbol{\theta}^i$; n is the dimension of the sensory information. Therefore, $\boldsymbol{x}^i$ can generally be represented as a nonlinear vector function of $\boldsymbol{\theta}^i$ as follows:

$$\boldsymbol{x}^i = \boldsymbol{f}^i(\boldsymbol{\theta}^i) \qquad 11.1$$

where we assume $n \leq m_i$. When $n > m_i$, the relationship between the sensory data and the sensory information cannot generally be represented as Eq. 11.1, because sensory data have only part of the required sensory information. Modeling and fusion of partial information will be discussed in Section 11.5.

Equation 11.1 is used as a general model of sensors in this chapter. Three typical examples follow:

1. *joint angles $\rightarrow$ end-effector location*. When the position and orientation of the end-effector are computed from joint angles as shown in Fig. 11.1, the joint angles are $\boldsymbol{\theta}^i \in \mathcal{R}^{m_i}$ and the position and the orientation of the end-effector are $\boldsymbol{x}^i \in \mathcal{R}^n$. Here, m_i denotes the number of joints, and $n = 6$. Note that $\boldsymbol{x}^i$ can be computed from $\boldsymbol{\theta}^i$ even when $n > m_i$, because $\boldsymbol{\theta}^i$ limits the end-effector motion and, therefore, includes all the information of $\boldsymbol{x}^i$. This is an exceptional case.

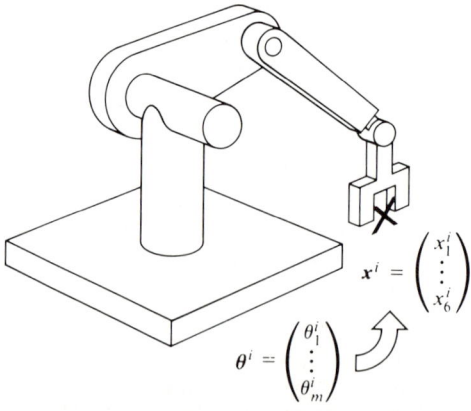

Figure 11.1 Sensing model example 1: Joint angles $\rightarrow$ end-effector location.

2. *stereo vision*. Figure 11.2 is the sensing model of stereo vision. The position of the reference point P in the absolute coordinates is $\boldsymbol{x}^i = (p_x\ p_y\ p_z)^T \in \mathcal{R}^3$. The $\boldsymbol{x}^i$ is computed from $\boldsymbol{\theta}^i = (x_R\ y_R\ x_L\ y_L)^T \in \mathcal{R}^4$, where $(x_R\ y_R)^T$ and $(x_L\ y_L)^T$ are the position of the image of P in the image frames of Camera R and Camera L, respectively. In this case, $n(=3) < m_i(=4)$ is satisfied. If only Camera L is available, $n(=3) > m_i(=2)$ and, therefore, $\boldsymbol{x}^i$ cannot be computed.

3. *range sensor*. Figure 11.3 is the sensing model of a range sensor. The low-level measurements are the sight angles α and β and the distance

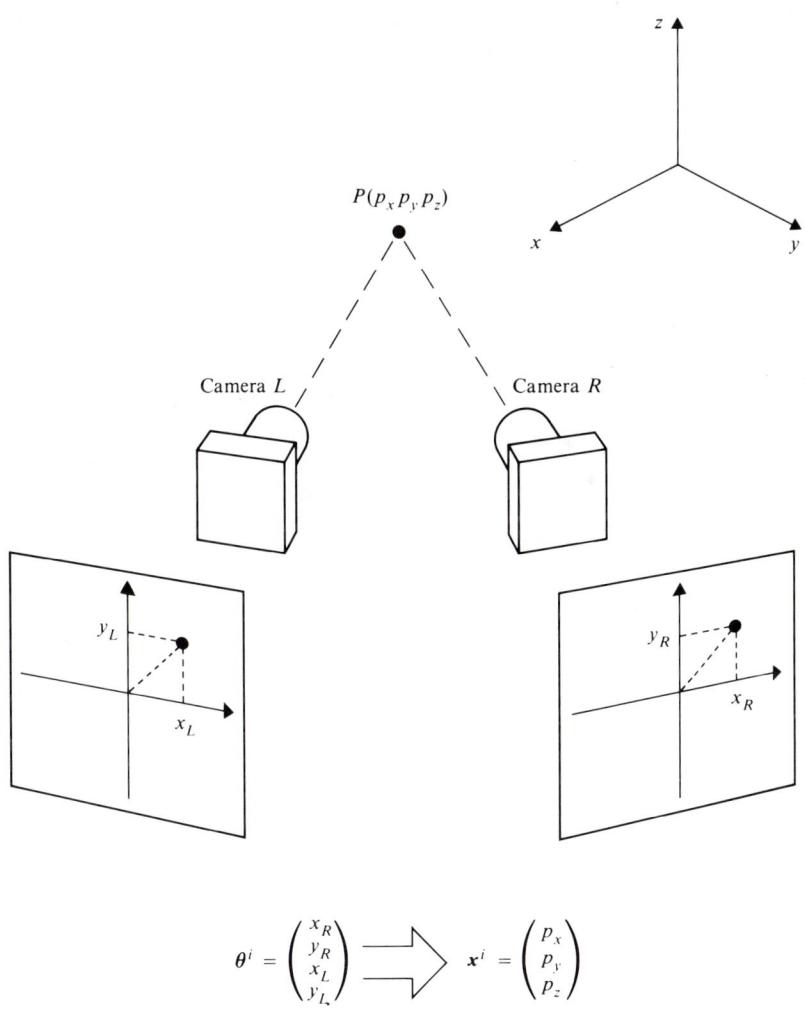

Figure 11.2 Sensing model example 2: Stereo vision.

d from the origin of the sensing frame $O - xyz$ to the reference point $P(p_x\ p_y\ p_z)$. Accordingly, the sensory data and the sensory information become $\boldsymbol{\theta}^i = (\alpha\ \beta\ d)^T \in \mathcal{R}^3$ and $\boldsymbol{x}^i = (p_x\ p_y\ p_z)^T \in \mathcal{R}^3$, respectively, and satisfy $n = m_i\ (= 3)$.

A nonlinear sensing model was used by Smith and Cheeseman (1986) to discuss the sensing uncertainty of mobile robots where the nonlinearity between the positions and orientations of neighboring coordinate frames is considered. Matthies and Shaper (1987) also used a nonlinear sensing model that is the same as example 2. In these problems, the models were used to find a consensus from the same type of information; therefore, only one nonlinear model was considered at a time. The main interest of sensor fusion is to make a consensus from various different types of information. The nonlinear model of Eq. 11.1 is intended to model various different types of nonlinear sensing structures in a systematic manner. Sensor fusion has mostly been discussed using linear models (Durrant-Whyte 1985, 1986; Luo, Lin, and Scherp 1987; Hashimoto and Paul 1987).

Another distinctive point is that $\boldsymbol{\theta}^i$ is defined as low-level information whose physical meaning is determined by the inherent physical structure of

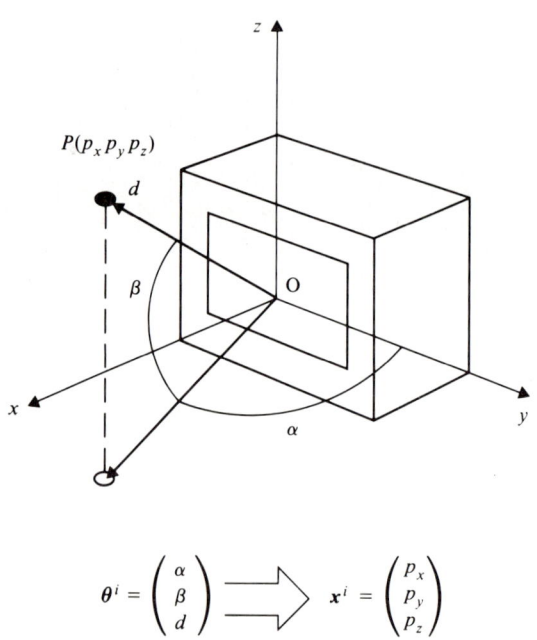

Figure 11.3 Sensing model example 3: Range sensor.

a specific sensor. Later in the section, we shall consider Gaussian statistical disturbances added to the low-level sensory data. It enables simple mathematical formulation. Although this simple assumption can cover the random noise added to the output signals of the sensors, it is not sufficient to model strictly disturbances caused by many kinds of sources. However, it is our idea that we can consider some of non-Gaussian disturbances on the high-level sensory information of x^i by the disturbances that appear as the result of mapping of Gaussian disturbances on θ^i by Eq. 11.1. This is more realistic than assuming Gaussian disturbances for high-level information such as the coordinate frames (Smith and Cheeseman 1986).

The third point to be noted is that a sensor unit and sensory data should not necessarily be coupled with a single kind of sensory information. For example, the joint angle sensors of a manipulator can be used to measure not only the end-effector location, but also the position of the elbow, which is required to avoid obstacles, and, of course, the joint angles for joint control.

So that intelligence can be integrated fully into the motion control of robotic systems in order that the robots can determine their own motion, the motion controller of a robotic system should be able to identify the necessary information and to request the latter of the sensor system. The sensor system will compose and return the requested information if available by processing the low-level sensory signals; otherwise, it will report the information's unavailability. This suggests the separation of the motion controller and the sensor-management system and bidirectional communication between them, whereas the conventional sensor system can return only the fixed sensory data to the controller.

11.3.2 Uncertainty Ellipsoid

The disturbance or uncertainty included in the sensory data is assumed additive and is represented as follows:

$$\theta^i = \underline{\theta^i} + \delta\theta^i \qquad 11.2$$

where $\underline{\theta^i} \in \mathcal{R}^{m_i}$ $(i = 1, \ldots, p)$ is the undisturbed data or the true value, and $\delta\theta^i \in \mathcal{R}^{m_i}$ $(i = 1, \ldots, p)$ is the disturbance. Now, we assume a Gaussian distribution for $\delta\theta^i$; that is,

$$E[\delta\theta^i] \triangleq \overline{\delta\theta^i} = O \in \mathcal{R}^{m_i} \qquad 11.3$$

$$V[\delta\theta^i] \triangleq E[(\delta\theta^i - \overline{\delta\theta^i})(\delta\theta^i - \overline{\delta\theta^i})^T]$$

$$= Q^i$$

$$= \text{diag}(\sigma_1^{i\,2}, \ldots, \sigma_{m_i}^{i\,2}) \in \mathcal{R}^{m_i \times m_i} \qquad 11.4$$

where $E[*]$ means the expectation of $*$, and it is also assumed that $\delta\theta_j^i$ ($j = 1, \ldots, m_i$), which is the jth element of $\delta\boldsymbol{\theta}^i$, is not correlated and $\sigma_j^{i\,2}$ is the variance of $\delta\theta_j^i$. Here, $\boldsymbol{Q}^i$ is the covariance matrix of $\delta\boldsymbol{\theta}^i$.

Substituting Eq. 11.2 into Eq. 11.1 provides

$$\boldsymbol{x}^i = \boldsymbol{f}^i(\underline{\boldsymbol{\theta}^i} + \delta\boldsymbol{\theta}^i) \qquad 11.5$$

Now, if we assume $\delta\boldsymbol{\theta}^i$ is small enough, Eq. 11.5 is approximated by

$$\boldsymbol{x}^i = \boldsymbol{f}^i(\underline{\boldsymbol{\theta}^i}) + \boldsymbol{J}^i(\underline{\boldsymbol{\theta}^i})\delta\boldsymbol{\theta}^i \qquad 11.6$$

$$\boldsymbol{J}^i(\underline{\boldsymbol{\theta}^i}) = \frac{\partial \boldsymbol{f}^i}{\partial \boldsymbol{\theta}^i} \in \mathcal{R}^{n \times m_i} \qquad 11.7$$

where $\boldsymbol{J}^i(\underline{\boldsymbol{\theta}^i})$ is the Jacobian matrix of $\boldsymbol{f}^i$ with respect to $\boldsymbol{\theta}^i$. From Eqs. 11.3 and 11.6, the mean and the covariance matrix of $\boldsymbol{x}^i$ become

$$E[\boldsymbol{x}^i] \triangleq \overline{\boldsymbol{x}^i} = \boldsymbol{f}^i(\underline{\boldsymbol{\theta}^i}) \qquad 11.8$$

$$V[\boldsymbol{x}^i] \triangleq E[(\boldsymbol{x}^i - \overline{\boldsymbol{x}^i})(\boldsymbol{x}^i - \overline{\boldsymbol{x}^i})^T]$$

$$= E[\boldsymbol{J}^i \, \delta\boldsymbol{\theta}^i \delta\boldsymbol{\theta}^{i\,T} \boldsymbol{J}^{i\,T}]$$

$$= \boldsymbol{J}^i \boldsymbol{Q}^i \boldsymbol{J}^{i\,T} \qquad 11.9$$

Equation 11.8 implies that, if we repeat the infinite number of measurements and compute the $\boldsymbol{x}^i$s, their average will converge to the true value of $\boldsymbol{x}^i$. This fact is a natural result of the neglect of the global error that can be identified by careful calibration. The calibration error is global and deterministic, whereas the disturbances discussed here are local and statistical. Although both are sources of uncertainty, they should be treated separately. In this chapter, we focus on the statistical uncertainty and assume that the calibration error is compensated beforehand. This assumption was also used by Smith and Cheeseman (1986).

Equation 11.9 means that the covariance matrix of $\boldsymbol{x}^i$ is no longer diagonal, since the Jacobian matrix is not diagonal in general. In other words, the correlation of x_j^i ($j = 1, \ldots, n$), that is the jth element of $\boldsymbol{x}^i$, is included in the model although $\delta\theta_j^i$ ($j = 1, \ldots, m_i$) are assumed uncorrelated. Note that, for a full rank $\boldsymbol{J}^i$, $\boldsymbol{J}^i\boldsymbol{Q}^i\boldsymbol{J}^{i\,T}$ becomes positive definite because $\boldsymbol{Q}^i$ is positive definite from Eq. 11.4.

Since $\boldsymbol{J}^i\boldsymbol{Q}^i\boldsymbol{J}^{i\,T}$ is symmetric, its singular value decomposition (Chapter 2.3) is represented by

$$J^i Q^i J^{iT} = U^i D^i U^{iT} \qquad 11.10$$

$$U^i = (e_1^i \cdots e_n^i) \in \mathcal{R}^{n \times n} \qquad 11.11$$
$$e_j^i \in \mathcal{R}^n$$
$$e_j^{iT} e_k^i = \begin{cases} 1 & \text{for } j = k \\ 0 & \text{for } j \neq k \end{cases}$$

$$D^i = \text{diag}(d_1^i, \ldots, d_n^i) \qquad 11.12$$
$$d_1^i \geq \cdots \geq d_n^i \geq 0$$

where U^i is an orthogonal matrix and d_j^i ($j = 1, \ldots, n$) are the singular values of $J^i Q^i J^{iT}$. The scalar variance in the direction indicated by a unit vector e_j^i is given by

$$\begin{aligned} V[e_j^{iT} x^i] &= e_j^{iT} J^i Q^i J^{iT} e_j^i \\ &= e_j^{iT} U^i D^i U^{iT} e_j^i \\ &= d_j^i \end{aligned} \qquad 11.13$$

Therefore, $\sqrt{d_j^i}$ represents the uncertainty of x^i in the direction of e_j^i. If we check the scalar variance in all the directions using unit vectors, the collection of the vectors whose directions are those of the unit vectors and magnitudes are the corresponding uncertainties forms an ellipsoid with e_j^i as the directions of principal axes and $2\sqrt{d_j^i}$ as their lengths, as shown in Fig. 11.4 for the three-dimensional case. This ellipsoid is called the *uncertainty ellipsoid*. Here, e_1^i and $\sqrt{d_1^i}$ correspond to the most uncertain direction and e_n^i and $\sqrt{d_n^i}$ correspond to the least uncertain direction.

11.4 Geometric Fusion Method

11.4.1 Uncertainty of Fused Information

Fusion combines the multiple sensory information x^i in a systematic manner to get a good consensus x. We discuss this problem in the scope of the linear combination; that is,

$$x = \sum_{i=1}^{p} W^i x^i \qquad 11.14$$

where $W^i \in \mathcal{R}^{n \times n}$ is the weighting matrix. The main problem is to determine W^i.

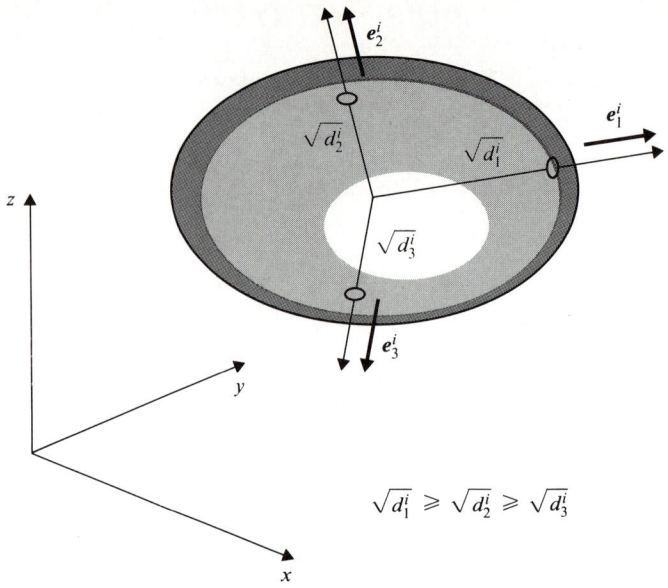

Figure 11.4 Uncertainty ellipsoid: three-dimensional case.

The mean of the fused information x is computed from Eqs. 11.8 and 11.14 as follows:

$$E[\boldsymbol{x}] = \sum_{i=1}^{p} \boldsymbol{W}^i \, E[\boldsymbol{x}^i] = \sum_{i=1}^{p} \boldsymbol{W}^i \, \overline{\boldsymbol{x}^i} \qquad 11.15$$

Due to the assumption that the global calibration errors have been compensated, $\overline{\boldsymbol{x}^i} = \overline{\boldsymbol{x}}$ for all i, where $\overline{\boldsymbol{x}}$ is the true value of $\boldsymbol{x}$. Hence, we have

$$E[\boldsymbol{x}] = (\sum_{i=1}^{p} \boldsymbol{W}^i) \, \overline{\boldsymbol{x}} \qquad 11.16$$

Since $E[\boldsymbol{x}]$ should satisfy

$$E[\boldsymbol{x}] = \overline{\boldsymbol{x}} \qquad 11.17$$

we have the following constraint on the weighting matrices:

$$\sum_{i=1}^{p} \boldsymbol{W}^i = \boldsymbol{E}_n \qquad 11.18$$

where $\boldsymbol{E}_n \in \mathcal{R}^{n \times n}$ is an identity matrix.

On the other hand, using $\overline{x^i} = \overline{x}$ and Eqs. 11.6, 11.8, 11.9, 11.14, 11.15, and 11.17, the covariance matrix of x is given by

$$V[x] \triangleq E[(x - \overline{x})(x - \overline{x})^T]$$

$$= E[(\sum_{i=1}^{p} W^i J^i \delta\theta^i)(\sum_{i=1}^{p} W^i J^i \delta\theta^i)^T]$$

$$= \sum_{i=1}^{p} W^i J^i Q^i J^{iT} W^{iT}$$

$$= WQW^T \in \mathcal{R}^{n \times n} \qquad 11.19$$

$$W \triangleq (W^1 \; W^2 \cdots W^p) \in \mathcal{R}^{n \times pn} \qquad 11.20$$

$$Q \triangleq \begin{pmatrix} J^1 Q^1 J^{1T} & \cdots & O \\ \vdots & \ddots & \vdots \\ O & \cdots & J^p Q^p J^{pT} \end{pmatrix} \in \mathcal{R}^{pn \times pn} \qquad 11.21$$

In deriving Eq. 11.19 we assumed that $\delta\theta^i$ and $\delta\theta^j$ $(i \neq j)$ are uncorrelated. Equation 11.19 implies that the shape and the size of the uncertain ellipsoid of x depends on the choice of the weighting matrices. In the next section, we shall discuss how to obtain the W^i that minimizes the volume of the uncertainty ellipsoid of x. The geometric meaning of Eq. 11.19 is conceptually represented in Fig. 11.5 for the case of $n = 2$ and $p = 2$.

11.4.2 Minimizing the Volume of the Uncertainty Ellipsoid

To get more accurate and less uncertain information, we determine the weighting matrix W such that it minimizes the volume of the uncertainty ellipsoid. Similar to Eq. 11.10, the singular value decomposition of the covariance matrix of x becomes

$$WQW^T = UDU^T \qquad 11.22$$

$$U = (e_1 \cdots e_n) \in \mathcal{R}^{n \times n}, \quad e_j \in \mathcal{R}^n$$

$$D = \text{diag}(d_1, \ldots, d_n), \quad d_1 \geq \cdots \geq d_n > 0$$

where $2\sqrt{d_i}$ gives the length of the ith longest principal axis of the uncertainty ellipsoid of the fused information x, and e_i represents its direction. The geometric volume of an ellipsoid whose principal axes have $2\sqrt{d_i}$ as their lengths is computed from Eq. 2.101 as follows:

$$\text{volume} = \frac{\pi^{n/2}}{\Gamma(1 + n/2)} (\prod_{i=1}^{n} d_i)^{1/2} \qquad 11.23$$

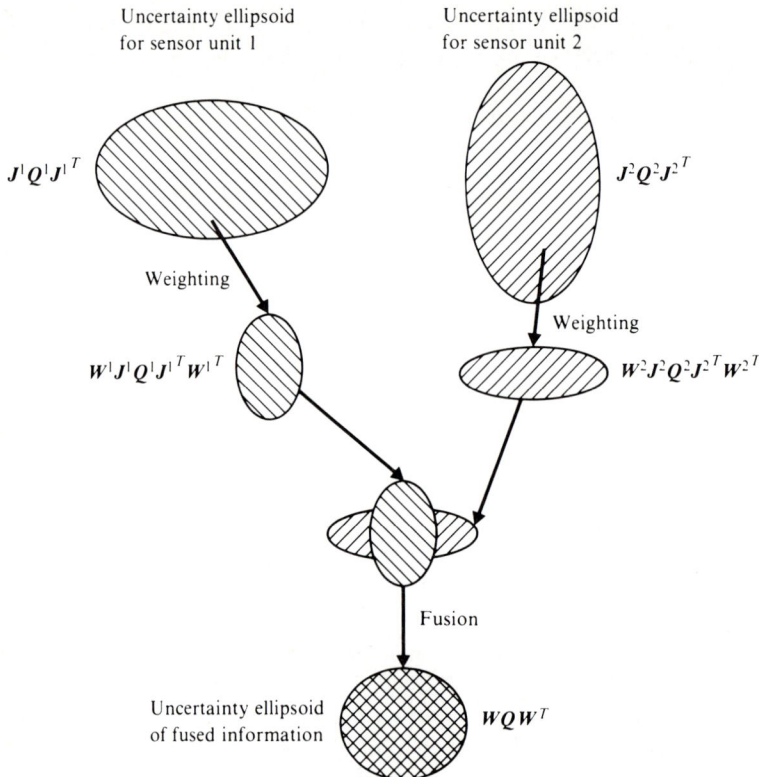

Figure 11.5 Geometric interpretation of Eq. 11.19.

where $\Gamma(*)$ is the gamma function. On the other hand, the determinant of a matrix is computed as the product of the singular values. That is,

$$\det(\boldsymbol{WQW}^T) = \det(\boldsymbol{UDU}^T)$$
$$= \prod_{i=1}^{n} d_i \qquad 11.24$$

Hence, from Eqs. 11.23 and 11.24, we have

$$\text{volume} = \frac{\pi^{n/2}}{\Gamma(1+n/2)}\sqrt{\det(\boldsymbol{WQW}^T)} \qquad 11.25$$

We find that minimizing the volume of the uncertainty ellipsoid is equivalent to minimizing the determinant of the covariance matrix of $\boldsymbol{x}$.

Now, our problem is to minimize

$$\text{P.I.} \triangleq \det(\boldsymbol{WQW}^T) \qquad 11.26$$

subject to the constraint of Eq. 11.18. This problem can be solved using the Lagrange multiplier theorem (Theorem 2.8). Considering the constraint, the criterion of Eq. 11.26 is replaced with the following one:

$$\text{P.I.}^* \triangleq \det(\boldsymbol{W}\boldsymbol{Q}\boldsymbol{W}^T) + P \qquad 11.27$$

$$P = \sum_{i=1}^{n}\sum_{j=1}^{n}\lambda_{ij}\left(\sum_{k=1}^{p}W_{ij}^k - \delta_{ij}\right) \qquad 11.28$$

$$\delta_{ij} = \begin{cases} 1 & \text{for } i=j \\ 0 & \text{for } i\neq j \end{cases}$$

$$\boldsymbol{\Lambda} \triangleq \begin{pmatrix} \lambda_{11} & \cdots & \lambda_{1n} \\ \vdots & \ddots & \vdots \\ \lambda_{n1} & \cdots & \lambda_{nn} \end{pmatrix} \qquad 11.29$$

where λ_{ij} are the Lagrange multipliers, W_{ij}^k is the (i, j) element of $\boldsymbol{W}^k$, and $(\sum_{k=1}^{p}W_{ij}^k - \delta_{ij})$ is the (i, j) element of $(\sum_{i=1}^{p}\boldsymbol{W}^i - \boldsymbol{E}_n)$, which should be equal to zero from Eq. 11.18. For P.I.* to be minimized, $\boldsymbol{W}$ must satisfy

$$\frac{\partial \text{P.I.}^*}{\partial \boldsymbol{W}} = \begin{pmatrix} \partial \text{P.I.}^*/\partial \boldsymbol{W}^1 \\ \vdots \\ \partial \text{P.I.}^*/\partial \boldsymbol{W}^p \end{pmatrix} = \boldsymbol{O} \qquad 11.30$$

where $\partial \text{P.I.}^*/\partial \boldsymbol{W}^i \in \mathcal{R}^{n\times n}$ has an entry $\partial \text{P.I.}^*/\partial W_{jk}^i$ as the (k, j) element. We compute $\partial \text{P.I.}^*/\partial \boldsymbol{W}^i$ using Eq. 11.27 as follows:

$$\frac{\partial \text{P.I.}^*}{\partial \boldsymbol{W}^i} = \frac{\partial}{\partial \boldsymbol{W}^i}\{\det(\boldsymbol{W}\boldsymbol{Q}\boldsymbol{W}^T)\} + \frac{\partial P}{\partial \boldsymbol{W}^i} \qquad 11.31$$

The second term, $\partial P/\partial \boldsymbol{W}^i$, becomes, from Eqs. 11.28 and 11.29,

$$\frac{\partial P}{\partial \boldsymbol{W}^i} = \boldsymbol{\Lambda}^T \qquad 11.32$$

Let the (j, k) entry of the first term, $\partial/\partial \boldsymbol{W}^i\{\det(\boldsymbol{W}\boldsymbol{Q}\boldsymbol{W}^T)\}$, be represented by α_{jk}^i. Then, α_{jk}^i is computed from Eqs. 2.59 and 2.60 as follows

$$\begin{aligned}\alpha_{jk}^i &= \frac{\partial}{\partial W_{kj}^i}\{\det(\boldsymbol{W}\boldsymbol{Q}\boldsymbol{W}^T)\} \\ &= \text{trace}[\{\frac{\partial}{\partial W_{kj}^i}(\boldsymbol{W}\boldsymbol{Q}\boldsymbol{W}^T)\}\,\text{adj}(\boldsymbol{W}\boldsymbol{Q}\boldsymbol{W}^T)] \\ &= \det(\boldsymbol{W}\boldsymbol{Q}\boldsymbol{W}^T)\,\text{trace}[\{\frac{\partial}{\partial W_{kj}^i}(\boldsymbol{W}\boldsymbol{Q}\boldsymbol{W}^T)\}(\boldsymbol{W}\boldsymbol{Q}\boldsymbol{W}^T)^{-1}]\end{aligned} \qquad 11.33$$

where trace(*) is the trace of matrix * and adj(*) is the adjoint matrix of *. Now, the trace in Eq. 11.33 is calculated (the derivation is given in Section 11.4.3) by

$$\text{trace}[\{\frac{\partial}{\partial W^i_{kj}}(WQW^T)\}(WQW^T)^{-1}]$$
$$= 2\{J^i Q^i J^{iT} W^{iT}(WQW^T)^{-1}\}_{jk} \quad 11.34$$

where $\{*\}_{jk}$ represents the (j, k) element of *. From Eqs. 11.31 through 11.34, Eq. 11.30 yields the following equation:

$$2\det(WQW^T)J^i Q^i J^{iT} W^{iT}(WQW^T)^{-1} + \Lambda^T = O \quad 11.35$$
$$\text{for } i = 1, \cdots, p$$

Therefore,

$$W^i = -\frac{1}{2\det(WQW^T)}(WQW^T)\Lambda(J^i Q^i J^{iT})^{-1} \quad 11.36$$
$$\text{for } i = 1, \cdots, p$$

where the symmetry of Q is used. By substituting Eq. 11.36 into Eq. 11.18, we obtain the Lagrange multipliers as follows:

$$\Lambda = -2\det(WQW^T)(WQW^T)^{-1}\{\sum_{i=1}^{p}(J^i Q^i J^{iT})^{-1}\}^{-1} \quad 11.37$$

Finally, by substituting Eq. 11.37 into Eq. 11.36, we have the optimal weighting matrices as follows:

$$W^i = \{\sum_{i=1}^{p}(J^i Q^i J^{iT})^{-1}\}^{-1}(J^i Q^i J^{iT})^{-1} \quad 11.38$$

In addition, we can compute the covariance matrix of x by substituting Eq. 11.38 into Eq. 11.19 as follows:

$$V[x] = \{\sum_{i=1}^{p}(J^i Q^i J^{iT})^{-1}\}^{-1} \quad 11.39$$

11.4.3 Derivation of Eq. 11.34

In this section, we derive Eq. 11.34; that is,

$$\text{trace}[\{\frac{\partial}{\partial W^i_{kj}}(WQW^T)\}(WQW^T)^{-1}] \quad 11.40$$
$$= 2\{J^i Q^i J^{iT} W^{iT}(WQW^T)^{-1}\}_{jk}$$

From Eq. 11.19, the following equation holds:

$$\frac{\partial}{\partial W^i_{kj}}(WQW^T) = \frac{\partial}{\partial W^i_{kj}}(W^i J^i Q^i J^{iT} W^{iT}) \qquad 11.41$$

$$= A_{kj} J^i Q^i J^{iT} W^{iT} + W^i J^i Q^i J^{iT} A_{kj}{}^T$$

where A_{kj} is a matrix with one as the (k, j) entry and zeros as the other entries. Therefore,

$$\mathrm{trace}[\{\frac{\partial}{\partial W^i_{kj}}(WQW^T)\}(WQW^T)^{-1}] \qquad 11.42$$

$$= \mathrm{trace}\{A_{kj} J^i Q^i J^{iT} W^{iT} (WQW^T)^{-1}\}$$
$$+ \mathrm{trace}\{W^i J^i Q^i J^{iT} A_{kj}{}^T (WQW^T)^{-1}\}$$

Using $\mathrm{trace}(AB) = \mathrm{trace}(BA)$ (for $A \in \mathcal{R}^{m \times n}$ and $B \in \mathcal{R}^{n \times m}$), $\mathrm{trace}(A) = \mathrm{trace}(A^T)$, and the symmetry of $(WQW^T)^{-1}$ and Q^i, the second term of Eq. 11.42 becomes

$$\mathrm{trace}\{W^i J^i Q^i J^{iT} A_{kj}{}^T (WQW^T)^{-1}\} \qquad 11.43$$
$$= \mathrm{trace}\{A_{kj} J^i Q^i J^{iT} W^{iT} (WQW^T)^{-1}\}$$

Substituting Eq. 11.43 into Eq. 11.42 provides

$$\mathrm{trace}[\{\frac{\partial}{\partial W^i_{kj}}(WQW^T)\}(WQW^T)^{-1}] \qquad 11.44$$

$$= 2\,\mathrm{trace}(A_{kj} C^i)$$
$$= 2\,C^i_{jk}$$

where

$$C^i \triangleq J^i Q^i J^{iT} W^{iT} (WQW^T)^{-1} \qquad 11.45$$

and C^i_{jk} is the (j, k) entry of C^i. Equation 11.44 yields Eqs. 11.40 and 11.34.

11.4.4 Relationship to Previous Work

Equation 11.38 has the same form as do the results of Smith and Cheeseman (1986) and Matthies and Shaper (1987), although the former used the Kalman filter theory, which is equivalent to Bayesian inference with minimum variance estimate, and the latter applied the weighted least-squares estimate with the inverse of the covariance matrix.

The Kalman filter theory and Bayesian inference determine the weighting matrix W^i of Eq. 11.14 that minimizes

$$a^T V[x] a = a^T W Q W^T a \qquad 11.46$$

for any constant vector $a \in \mathcal{R}^n$.

In contrast, the weighted least-squares estimate with the inverse of covariance matrix suggests determining the weighting matrix that minimizes the following scalar function:

$$\frac{1}{p}\sum_{i=1}^{p}(\boldsymbol{x}-\boldsymbol{x}^i)^T V[\boldsymbol{x}^i]^{-1}(\boldsymbol{x}-\boldsymbol{x}^i)$$

$$=\frac{1}{p}\sum_{i=1}^{p}(\boldsymbol{x}-\boldsymbol{x}^i)^T(\boldsymbol{J}^i\boldsymbol{Q}^i\boldsymbol{J}^{iT})^{-1}(\boldsymbol{x}-\boldsymbol{x}^i) \qquad 11.47$$

It is known that the weighted least-squares estimate with the inverse of covariance matrix as the weighting matrix is identical to Bayesian inference with minimum variance estimate assuming a Gaussian distribution of disturbances (Bryson and Ho 1975). The formulation we developed in Section 11.4.2 is based on the geometric motivation that the volume of the uncertainty ellipsoid should be minimized by the choice of a weighting matrix, which is a new approach to this problem. The coincidence of the resultant weighting matrices provides one more rationale for using Eq. 11.38 as the weighting matrices for multisensor fusion, and a geometric insight to the fusion method.

11.4.5 Comments on the Structure of Computation

Consider the problem of fusing $\boldsymbol{x}^{p+1}$ into $\boldsymbol{x}$ that has been obtained fusing $\boldsymbol{x}^1, \ldots, \boldsymbol{x}^p$. If we define the covariance matrix of $\boldsymbol{x}^i$ by

$$\boldsymbol{H}^i \triangleq \boldsymbol{J}^i\boldsymbol{Q}^i\boldsymbol{J}^{iT} \qquad 11.48$$

the covariance matrix of $\boldsymbol{x}$ becomes

$$\boldsymbol{H} = \{\sum_{i=1}^{p}(\boldsymbol{J}^i\boldsymbol{Q}^i\boldsymbol{J}^{iT})^{-1}\}^{-1}$$

$$= \{\sum_{i=1}^{p}(\boldsymbol{H}^i)^{-1}\}^{-1} \qquad 11.49$$

From Eqs. 11.14 and 11.38 the fusion of $\boldsymbol{x}$ and $\boldsymbol{x}^{p+1}$ is given by

$$\boldsymbol{x}^* = \{\boldsymbol{H}^{-1}+(\boldsymbol{H}^{p+1})^{-1}\}^{-1}\boldsymbol{H}^{-1}\boldsymbol{x}$$
$$+ \{\boldsymbol{H}^{-1}+(\boldsymbol{H}^{p+1})^{-1}\}^{-1}(\boldsymbol{H}^{p+1})^{-1}\boldsymbol{x}^{p+1} \qquad 11.50$$

Substituting Eq. 11.49 into Eq. 11.50 yields

$$\boldsymbol{x}^* = \sum_{i=1}^{p+1}\{\sum_{j=1}^{p+1}(\boldsymbol{H}^j)^{-1}\}^{-1}(\boldsymbol{H}^i)^{-1}\boldsymbol{x}^i \qquad 11.51$$

Equation 11.51 implies that the fusion of x and x^{p+1} is identical to the fusion of $x^1, \ldots, x^p$, and x^{p+1}. This fact suggests the following recursive computation of multisensor fusion.

Algorithm 11.1

1. Initialize; $x = 0$, $H^{-1} = 0$, choose the first i.
2. $x^a = x$, $(H^a)^{-1} = H^{-1}$,
   $x^b = x^i$, $(H^b)^{-1} = (H^i)^{-1}$.
3. $x = \{(H^a)^{-1} + (H^b)^{-1}\}^{-1}((H^a)^{-1}x^a + (H^b)^{-1}x^b)$,
   $H^{-1} = (H^a)^{-1} + (H^b)^{-1}$.
4. If all of x^j $(j = 1, \ldots, p)$ have been fused, then stop. End.
5. Go to step 2 for next i.

Algorithm 11.1 possesses the following significant structural characteristics of the fusion method:

1. The order of i does not matter in the loop from step 2 to step 5.

2. At every stage, x is the best estimate that can be made using fused x^is.

3. Since steps 2 and 3 require only x^i and $(H^i)^{-1}$, except for x and H^{-1}, the computation to fuse x^i can be done locally—that is, even in the sensor unit i.

Characteristics 1 and 2 imply that a sensor unit with faster response can be fused earlier than can one with slower response, and that a change in the number of sensor units or the order of fusion does not cause any inconsistency. This fact allows asynchronous computation of fusion. Characteristic 3 suggests the distributed structure of computation. These structural characteristics show that the computation of fusion is suitable for implementation in intelligent systems with blackboards (Harmon, Bianchini, and Pinz 1986; Nii 1986; Englemore and Moregan 1987) or whiteboards (Shafer, Stentz, and Thorpe 1986).

11.5 Fusion of Partial Information

11.5.1 Computation of Covariance Matrix

When sensor unit i measures only partial information, Eq. 11.1 cannot describe the relationship between θ^i and x^i since θ^i may not have sufficient

320   Chapter 11   Geometric Optimization for Sensor Fusion

information to recover $\boldsymbol{x}^i$. In this section, we discuss the fusion of partial information. The relationship between $\boldsymbol{\theta}^i$ and $\boldsymbol{x}^i$ for partial information should be described by

$$\boldsymbol{\theta}^i = \boldsymbol{g}^i(\boldsymbol{x}^i) \qquad 11.52$$

where $\boldsymbol{\theta}^i \in \mathcal{R}^{m_i}$, $\boldsymbol{x}^i \in \mathcal{R}^n$, and $m_i < n$. For example, when $\boldsymbol{x}^i$ is a point in three-dimensional space and $\boldsymbol{\theta}^i$ is the location in the image frame of a single-camera vision, as shown in Fig. 11.6, the relationship between $\boldsymbol{\theta}^i$ and $\boldsymbol{x}^i$ is governed not by Eq. 11.1, but rather by Eq. 11.52.

Let $\underline{\boldsymbol{\theta}}^i$ and $\underline{\boldsymbol{x}}^i$ represent the true values of $\boldsymbol{\theta}^i$ and $\boldsymbol{x}^i$. Also let $\delta\boldsymbol{\theta}^i$ be the additive disturbance of $\boldsymbol{\theta}^i$. Here, $\delta\boldsymbol{x}^i$ is the error of $\boldsymbol{x}^i$ caused by $\delta\boldsymbol{\theta}^i$. By substituting $\boldsymbol{\theta}^i = \underline{\boldsymbol{\theta}}^i + \delta\boldsymbol{\theta}^i$ and $\boldsymbol{x}^i = \underline{\boldsymbol{x}}^i + \delta\boldsymbol{x}^i$, Eq. 11.52 becomes

$$\underline{\boldsymbol{\theta}}^i + \delta\boldsymbol{\theta}^i = \boldsymbol{g}^i(\underline{\boldsymbol{x}}^i + \delta\boldsymbol{x}^i) \qquad 11.53$$

When $\delta\boldsymbol{x}^i$ is assumed to be small, subtracting $\underline{\boldsymbol{\theta}}^i = \boldsymbol{g}^i(\underline{\boldsymbol{x}}^i)$ from Eq. 11.53 yields the following equation:

$$\delta\boldsymbol{\theta}^i = \boldsymbol{K}^i\, \delta\boldsymbol{x}^i \qquad 11.54$$

$$\boldsymbol{K}^i \triangleq \frac{\partial \boldsymbol{g}^i}{\partial \boldsymbol{x}^i} \in \mathcal{R}^{m_i \times n} \qquad 11.55$$

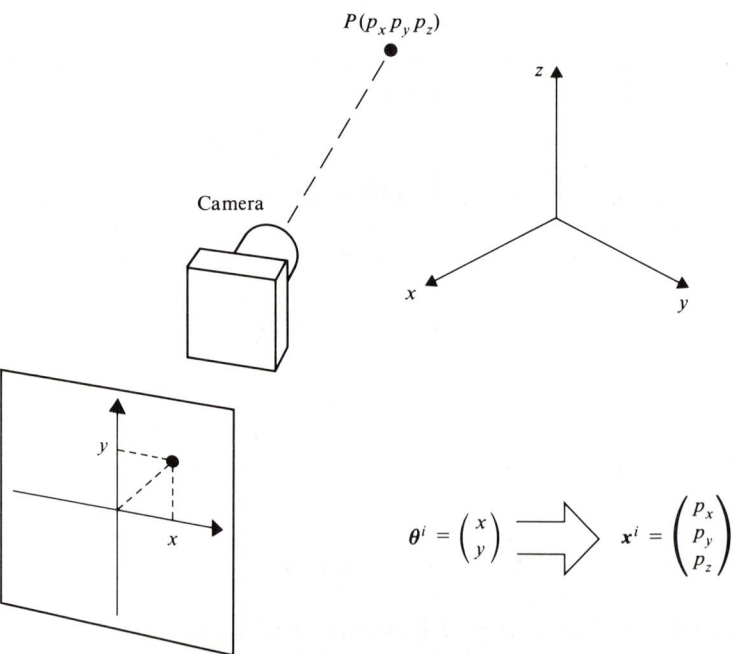

Figure 11.6 Sensing of partial information.

where K^i is the Jacobian matrix of g^i with respect to x^i and is a function of x^i.

The covariance matrix of $\delta\theta^i$ is calculated as follows:

$$V[\delta\theta^i] = E[\delta\theta^i \delta\theta^{iT}]$$
$$= K^i \; E[\delta x^i \delta x^{iT}] \; K^{iT}$$
$$= K^i \; V[x^i] \; K^{iT} = Q^i \quad \quad 11.56$$

To apply the fusion method defined by Eqs. 11.14 and 11.38, we need $V[x^i]^{-1}$—that is, $(J^i Q^i J^{iT})^{-1}$—in Eq. 11.38. Since K^i is an $m_i \times n$ ($m_i < n$) matrix, $V[x^i]$ cannot be obtained from Eq. 11.56 in a straightforward manner.

Now, we add dummy measurements $\theta_d^i \in \mathcal{R}^{n-m_i}$ to θ^i so that K^i becomes a nonsingular matrix. Equations 11.52, 11.54, and 11.55 are modified as follows:

$$\theta^{*i} = g^{*i}(x^i) \quad \quad 11.57$$

$$g^{*i}(x^i) \triangleq \begin{pmatrix} g^i(x^i) \\ g_d^i(x^i) \end{pmatrix}$$

$$\theta_d^i = g_d^i(x^i)$$

$$\delta\theta^{*i} = K^{*i} \; \delta x^i \quad \quad 11.58$$

$$K^{*i} \triangleq \begin{pmatrix} K^i \\ \partial g_d^i / \partial x^i \end{pmatrix} \in \mathcal{R}^{n \times n} \quad \quad 11.59$$

where K^{*i} is nonsingular. Suppose that the additional dummy measurements are extremely uncertain; Eq. 11.56 becomes

$$V[\delta\theta^{*i}] = K^{*i} \; V[x^i] \; K^{*iT}$$
$$= Q^{*i} \quad \quad 11.60$$

$$Q^{*i} = \begin{pmatrix} Q^i & & O & \\ & +\infty & \cdots & 0 \\ O & \vdots & \ddots & \vdots \\ & 0 & \cdots & +\infty \end{pmatrix}$$

Hence, the following equation is obtained:

$$V[x^i]^{-1} = K^{*iT} \; (Q^{*i})^{-1} \; K^{*i} \quad \quad 11.61$$

Substituting

$$(Q^{*i})^{-1} = \begin{pmatrix} (Q^i)^{-1} & & O & \\ & +0 & \cdots & 0 \\ O & \vdots & \ddots & \vdots \\ & 0 & \cdots & +0 \end{pmatrix} \quad 11.62$$

into Eq. 11.61 yields the following equation:

$$V[x^i]^{-1} = K^{iT} (Q^i)^{-1} K^i \quad 11.63$$

Equation 11.63 can be used in the place of $(J^i Q^i J^{iT})^{-1}$ in the fusion method of Eqs. 11.14 and 11.38, when the ith sensor unit measures only the partial information.

The geometric meaning of Eq. 11.63 is explained as follows. Considering that $K^{iT}(Q^i)^{-1}K^i$ is a singular symmetric matrix, the singular value decomposition of $V[x^i]^{-1}$ becomes

$$V[x^i]^{-1} = U^i \widehat{D}^i U^{iT} \quad 11.64$$

$U^i \in \mathcal{R}^{n \times n}$; an orthogonal matrix

$\widehat{D}^i = \text{diag}(\hat{d}_1^i, \ldots, \hat{d}_{m_i}^i, 0, \ldots, 0)$

$\hat{d}_1^i \geq \cdots \geq \hat{d}_{m_i}^i > 0.$

From Eq. 11.64, $V[x^i]$ can be considered as

$$V[x^i] = U^i (\widehat{D}^i)^{-1} U^{iT} \quad 11.65$$
$$(\widehat{D}^i)^{-1} = \text{diag}(1/\hat{d}_1^i, \ldots, 1/\hat{d}_{m_i}^i, +\infty, \ldots, +\infty).$$

We find that the sensory information is extremely uncertain in the directions of the $(m_i + 1)$th through nth column vectors of U^i in the space of x. The corresponding uncertainty ellipsoid has infinitely long principle axes in these directions.

In Fig. 11.6, this direction coincides with the line from the center of the lens of the camera to point P. We call this line the sight line of P. The uncertainty is physically illustrated by the fact that the image of P on the image frame stays still even if P moves on the sight line of P.

11.5.2 Computation of x^i

We require x^i not only to evaluate Eq. 11.14, but also to compute K^i of Eq. 11.55. However, we cannot solve x^i from Eq. 11.52, because it is underdeterminate. In this subsection, an approximating computation method for x^i is proposed.

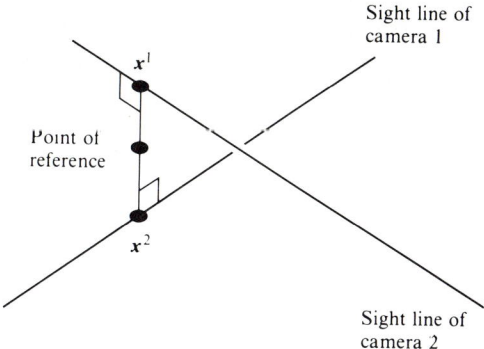

Figure 11.7 Computation of x^i for two nonstereo cameras.

When the recursive computational method given in Section 11.4.5 is used, x^i can be computed taking advantage of x, which is the consensus obtained using the previous sensory data. In this case, x^i can be approximated by the point on the manifold determined by Eq. 11.52 that has the shortest distance from x. In Fig. 11.6, this means obtaining the intersection of the sight line of P and its normal passing through x.

If the previous consensus of x is not available and only several kinds of partial information are available, first a point of reference should be computed by the least-squares method. The point of reference is generally obtained as the point that minimizes the sum of the squared distances from the point to the manifolds of partial information. Then, using the point of reference as a previous consensus, x^i is obtained as the point on the manifold that has the shortest distance from this point.

When two cameras are considered not as a stereo-vision sensor but as two sensor units that provide partial information, the point of reference is given as the centroid of the common normal between two sight lines of P. We compute x^i as the point of intersection of the corresponding sight line and the common normal. This case is shown in Fig. 11.7.

Summary

A general geometric fusion method was proposed for multisensor robotic systems. This method is based on the geometric analysis of the sensing uncertainty, and is motivated by the geometric idea that the volume of the uncertainty ellipsoid should be minimized. The resultant fusing equation coincides with those obtained by Bayesian inference, by the Kalman filter theory, and by weighted least-squares estimation. This fact provides another geometric rationale for using the equation in multisensor fusion, and a geometric insight to the fusion method.

Since the developed nonlinear sensing model is general and covers most robotic sensors, the proposed fusion method works as a general fusion method for multisensor robotic systems.

The fusion method was also extended to handle partial sensory information. This generalization enables robotic systems to utilize all of the sensors to obtain more accurate and less uncertain information.

Although the proposed fusion method provides a good consensus even if it is combined with a conventional robot controller, it will be more efficient when used with an intelligent robot controller that can decide and request by itself the necessary information. One of the most important future research problems is to develop a robotic system that includes an intelligent controller and a sensor-management system with the function of sensor fusion. A future research problem of multisensor fusion theory is to establish a method to relate and fuse the sensory data with different physical dimensions.

References

Bajcsy, R., and Allen, P. 1985. Converging disparate sensory data. In *Robotics research 2*, eds. H. Hanafusa and H. Inoue, pp. 81–86. Cambridge, MA: MIT Press.

Brooks, R. A. 1985 (St. Louis). Visual map making for a mobile robot. *Proc. IEEE Int. Conf. Robotics and Automation*, pp. 1947–1953.

Bryson Jr., A. E., and Ho, Y. C. 1975. *Applied optimal control: Optimization, estimation, and control.* Washington, D.C.: Hemisphere.

Chatila, R., and Laumond, J. P. 1985 (St. Louis). Position referencing and consistent world modeling for mobile robots. *Proc. IEEE Int. Conf. Robotics and Automation*, pp. 138–145.

Chiu, S. L., Morley, D. J., and Martin, J. F. 1986 (San Francisco). Sensor data fusion on a parallel processor. *Proc. IEEE Int. Conf. Robotics and Automation*, pp. 1629–1633.

Durrant-Whyte, H. F. 1985. Integrating distributed sensor information, an application to a robot system coordinator. *Proc. IEEE Int. Conf. Syst., Man Cybernet.*, pp. 415–419.

Durrant-Whyte, H. F. 1986 (San Francisco). Consistent integration and propagation of disparate sensor observations. *Proc. IEEE Int. Conf. Robotics and Automation*, pp. 1464–1469.

Englemore, R., and Moregan, A. 1987. *Blackboard systems: Application and framework.* Reading, MA: Addison-Wesley.

Harmon, S. Y., Bianchini, B. E., and Pinz, B. E. 1986 (San Francisco). Sensor data fusion through a distributed blackboard. *Proc. IEEE Int. Conf. Robotics and Automation*, pp. 1449–1454.

Hashimoto, M., Paul, R. P. 1987 (Tsukuba, Japan). Integration of multi-sensor manipulator actuator information for robust robot control systems. *Proc. Annual Conf. Japan Robotics Society*, pp. 393–396 *(in Japanese)*.

Henderson, T. C., Fai, W. S., and Hansen, C. 1984. MKS: A multisensor kernel system. *IEEE Trans. Syst., Man Cybernet.* SMC-14 (5): 784–791.

Henderson, T. C., and Shilcrat, E. 1984. Logical sensor systems. *J. Robotics Systems* 1 (2): 169–193.

Klema, V. C., and Laub, A. T. 1980. The singular value decomposition: its computation and some applications. *IEEE Trans. Automatic Control* 25 (2): 164–176.

Kodama, S., and Suda, N. 1978. *Matrix theory for systems control.* Tokyo: Society of Instruments and Control Engineers *(in Japanese)*.

Luo, R. C., Lin, M. H., and Scherp, R. S. 1987 (Raleigh, NC). The issues and approaches of a robot multi-sensor integration. *Proc. IEEE Int. Conf. Robotic and Automation*, pp. 1941–1946.

Matthies, L., and Shafer, S. A. 1987. Error modeling in stereo navigation. *IEEE J. Robotics Automat.* 3 (3): 239–248.

Nakamura, Y., and Xu, Y. 1989 (Scottsdale, AZ). Geometrical fusion method for multi-sensor robotic systems. *Proc. 1989 IEEE Int. Conf. Robotics and Automation*, pp. 668–673.

Nii, H. P. 1986. Blackboard systems: Blackboard application systems, blackboard systems from a knowledge engineering perspective. *AI MAGAZINE*, 7(3): 82–106.

Ruokangas, C. C., Black, M. S., Martin, J. F., and Schoenwald, J. S. 1986 (San Francisco). Integration of multiple sensors to provide flexible control strategies. *Proc. IEEE Int. Conf. Robotics and Automation*, pp. 1947–1953.

Shafer, S. A., Stentz, A., and Thorpe, C. E. 1986. An architecture for sensor fusion in a mobile robot. Technical Report 86-9, Carnegie–Mellon University, Robotics Institute.

Shekhar, S., Khatib, O., and Shimojo, M. 1986 (San Francisco). Sensor fusion and object localization. *Proc. IEEE Int. Conf. Robotic and Automation*, pp. 1623–1628.

Smith, R. C., and Cheeseman, P. 1986. On the representation and estimation of spatial uncertainty. *Int. J. Robotics Res.* 5 (4): 56–68.

Wang, Y. F., and Aggarwal, J. K. 1987 (Raleigh, NC). On modeling 3-D objects using multiple sensory data. *Proc. IEEE Int. Conf. Robotics and Automation*, pp. 1098–1103.

Index

Absolute coordinates, 176
Absolute values of eigenvalues, 32
Acceleration control, 255
Acceleration of the contact point, 181
Accelerations of the object, 181, 192
Active constraints, 79
Active utilization of redundancy, 126
Actuation redundancy, 3, 7–8, 207, 230, 236–238, 242, 244, 246, 249
Adaptive change of scale factor, 279
Additional actuators, 232
Adjoint matrix, 29, 316
Adjoint variables, 170
Adjoint vector, 83–84, 99, 157, 162
Admissible control, 82, 84, 86, 92, 95
Algebraic equations, 175
Algorithmic singularity, 239
Analog inverse transformation circuit, 293, 299
Anthropomorphic 7-DOF manipulator, 126, 162, 230, 248

Approximation at singular points, 256
Arbitrary acceleration, 183
Arbitrary vector, 129, 155, 215, 238
Artificial potential and dissipative functions, 136
Asymmetric control schemes, 175
Asymptotic stability, 179
Asynchronous computation of fusion, 319
ATAN2($\star,\star$), 108
Augmented dynamics equation, 83
Averaging scalar data by weighting with variances, 304
Axis of elastic strain, 291
Axis of the elastic component, 291

Back reasoning, 304
Backward integration, 163, 165
Bang-bang control, 91
Base coordinates, 106, 109

328 Index

Base frame, 113
Basic equations for coordinative manipulation of a rigid object, 176
Basic Jacobian matrix, 110, 113
Bayesian inference with minimum variance estimate, 304–305, 317–318, 323
Bending moment, 291
Bidirectional communication, 309
Bilinear equation, 215, 224, 236
Boundary condition, 72, 85, 98–99, 101, 156, 158, 167, 169–170
Boundary of the constraints, 189–190
Boundary of the friction cone, 189, 193, 199
Boundary of the neighborhood of singular points, 268, 279
Boundary of the workspace, 254

Calculus of variations, 170
Calibration error, 310
Canonical equations of Hamilton, 84
Cartesian coordinates, 142
Castigliano's theorem, 292
Centrifugal and Coriolis forces, 161
Centrifugal forces, 24
Centroid of the common normal, 323
Circular cone, 195
Classical variational method, 11, 15, 82, 170
Class of disturbing forces, 191
Closed-form solution, 105
Closed-link mechanisms with actuation redundancy, 207, 225
Closed-link mechanisms, 3, 205, 232, 238
Closed-link finger mechanism, 217
Closed-link mechanism with kinematic redundancy, 230
Closed set, 82, 170
Closest components to the singularity, 260
Closing the position feedback loops, 273
CMU DD Arm Robot model I, 230
Coefficient of maximum static friction, 185
Cofactor expansion, 28
Combined inertia matrix of the object and the robotic mechanisms, 195

Common normal, 323
Compliance matrix, 179
Computational complexity for the inverse kinematic solutions, 266
Computational complexity of the closed-link mechanism, 221
Computational methods of pseudoinverse, 34, 62
Computational nonlinearity, 305
Computation error, 286
Computation of the Lagrange multipliers, 210
Computation of the minimizing internal forces, 188, 199
Computation of the optimal internal forces, 199
Computation of the SR-inverse, 258
Computation of two-point boundary value problems, 101
Computer-assisted design system for force sensors, 299
Concavity, 188
Condition number, 32, 37, 121, 287–288, 290, 297
Condition number of a sensor compliance matrix, 284
Condition number of the Jacobian matrix, 121
Condition of a free right-hand endpoint, 156
Conditions for admissible control, 82
Confidence distance measure, 304
Constant scale factors, 269, 273
Constraint equation, 73
Constraint-free stationary value problem, 78
Constraint surfaces, 180
Contact condition, 174
Contact points, 3, 174, 176–178, 191, 193
Contact stability, 174–175, 185, 191, 193, 195, 200
Contact-stability cone, 195
Contact-stability ellipsoid, 194
Control of actuation redundancy, 241
Controllable, 208, 213
Conventional mechanical-design, 125
Convex function, 187
Convexity of the quadratic function, 188

Index   329

Convex quadratic function, 216
Coordination of multiple robotic mechanisms, 173
Coordinative manipulability, 183, 185
Coordinative manipulation by the resultant force, 175
Coriolis and centrifugal forces, 111, 133, 142, 273
Coriolis forces, 25
Cost function, 84, 95
Covariance matrix, 305, 310, 313, 316, 321

D'Alembert's principle, 181, 206, 210–211, 224
Damped least-squares method, 279
Decoupled and configuration-invariant, 206
Decoupled dynamics, 217
Decoupled force information, 284
Degenerated block of the Jacobian matrix, 254
Degenerated directions, 115
Degree of manipulability, 115, 117, 119
Degree of redundancy, 116–117, 154, 162, 279
Degree of redundancy of sensors, 304
Denavit–Hartenberg convention, 113, 217
Derivative of a scalar with respect to a vector, 17
Derivative of determinants, 28
Derivatives of matrices, 19
Derivatives of vectors, 16
Design of elastic component of force sensors, 299
Design of force sensors, 286
Determination of the initial joint variable, 155
Difference between the pseudoinverse and the SR-inverse, 260
Differential equations of the adjoint variables, 96
Differential kinematics, 105–106, 116, 122, 255
Differential kinematics and statics, 122
Direct-drive actuation, 229
Direct-drive robots, 205

Discontinuity of the pseudoinverse at singular points, 273
Distance from singularity, 122
Distance from singular points, 137, 268
Distributed blackboard for sensory data fusion, 305
Distributed structure of computation, 319
Distribution of the resultant force into individual end-effector forces, 180
Disturbing forces, 175, 191–192, 195
DOM, 116–117, 119
DOR, 116–117
DP, 102
Duality, 122
Dummy measurements, 321
Dynamic coordination of multiple robotic mechanisms, 175
Dynamic load distribution between actuators in the closed loop, 215
Dynamic manipulation of the object, 179
Dynamic programming, 102
Dynamic stability, 179
Dynamics of an object, 175
Dynamics of robot manipulators, 161
Dynamics of the tree-structure mechanism, 211

Effective inertia, 40
Effective point of the end-effector, 113
Efficient computation of the Jacobian matrix, 109, 254
Efficient recursive computational algorithm, 114
Eigenvalues, 25, 33, 90
Elastic component of force sensor, 285
Elastic fingers, 174
Elastic strain energy, 292
Ellipsoid, 39, 120, 193
Elliptic cone, 195
Elliptic geometry, 39, 120
End-effector acceleration, 106
End-effector coordinates, 106
End-effector force sensing, 283
Energy minimization, 127
Equality constraints, 78–79
Equivalent angle-axis representation, 112

Error magnification ratio, 37
Estimates of unknown left-hand endpoint boundary values, 157
Euclidean norms, 51, 54, 57, 75, 165, 178, 256, 287
Euler angles, 162
Euler–Lagrange equation, 70–71
Evaluation method of contact stability, 191
Exactness of the solution, 256, 268
Exact solution, 52–53, 62, 254, 256
Example of elastic-component design, 295
Existence of the minimizing internal forces, 185
Extended regular point, 79
External forces, 177
External sensors, 303
Extreme value, 74, 224

Feasibility, 257, 264, 267, 273
Feasible inverse kinematic solution, 279
Feasible span, 216, 224
Feedback control, 111, 133
First manipulation variable, 128–129, 135, 137, 154, 160, 162
First-order differential equation, 170
First priority, 128, 132
Five-bar closed-link mechanism, 217
Five different types of solutions of inverse kinematics, 264
Fixed-domain problems, 91–92, 94, 99
Fixed-point problems, 91
Fixed-time and free end-state problems, 98, 156
Fixed-time problems, 97, 99
Floating-point arithmetic operations, 266
Follower, 174
Force applicability, 122
Force-closure grasp of polyhedra, 174
Force-closure grasp, 182
Force control, 174, 180–181, 283
Force distribution, 206
Force sensing, 283, 286
Force sensitivity, 284, 289–291, 294, 297–298
Force sensors, 283, 289, 303

Forward dynamics of general closed-link mechanisms, 210
Free end-point problem, 99
Free workspace, 125
Frictional constraints, 174
Frictional point contact, 176
Frictional soft-finger contact model, 174
Friction cone, 191–195
Frictionless enveloping grasping, 174
Fusion of partial information, 305–306

Gamma function, 314
Gaussian distribution, 304–305, 309
Gaussian disturbances for high-level information, 309
Gaussian elimination, 68, 264–265
Gaussian noise, 305
General form of the least P-weighted-norm solutions, 55
General form of the least-squares solutions, 51, 55
Generalized inertia ellipsoid, 41, 194
Generalized inverse, 41–42, 45–46, 48, 288, 293
Generalized inverse matrix of the sensor compliance matrix, 290
Generalized inverse of the Jacobian matrix, 126
Generalized momentum, 84
Generalized total energy, 84
General solution of closed-link actuator torque, 238
General solution of joint velocity, 126
Geodesic line, 78
Geometric fusion method, 323
Geometric volume of the ellipsoid, 305
Geometry of uncertainties, 305
Global criterion for redundancy utilization, 154
Globally optimal control, 127, 154
Globally optimal control of redundancy, 154, 169
Global optimization of joint rates, 170
Global optimization of kinematic redundancy, 97
Global trajectory planning of kinematically redundant manipulators, 99
Gradient, 17, 75

Index   **331**

Gradient vector of a scalar function, 126, 153
Grasping and manipulation, 173
Grasping stability, 174
Gravity torque vector, 142, 161
Great circles, 78
Greville's algorithm, 266

Hamilton form, 83, 86, 99
Hamiltonian, 83–87, 94–95, 156, 159–161
Hard squeezing, 174, 191, 195
Hierarchical structure, 256–257
Home position, 142
Hybrid position/force control, 180
Hyperellipsoid, 38–39
Hypothesis test, 304

Idempotent, 46, 49–50, 129, 131
Ill conditioned, 32, 37
Independent actuators, 238
Independent boundary conditions, 157
Independent parameter of actuation redundancy, 239
Index for the structural evaluation of force sensors, 287
Inequality constraints, 78–79, 189
Inertia ellipsoid, 40
Inertia matrix, 24, 27, 41, 111, 133, 161–162, 191
Inertia matrix of closed-link mechanisms, 206
Inertia tensor, 176, 181–182
Infinitely long principle axes, 322
Infinite number of measurements, 310
Infinitesimal, 71, 75, 211
Infinitesimal motion of the tree-structure mechanism, 209
Initial value adjusting method, 157
Initial value problem, 157
Inner normal direction, 174
Inner unit normal vector, 178
Integrated value of kinetic energy, 154
Internal forces, 174–175, 180, 184–186, 195, 200
Internal sensors, 303
Internal singularities, 255
Intersection of hypersurfaces, 91

ith longest principal axis of the uncertainty ellipsoid, 313
Intrinsic singularity of Z-Y-Z Euler angles, 108, 110
Inverse dynamics, 114, 267, 273
Inverse dynamics of closed-link mechanisms, 206–207, 224
Inverse function, 254
Inverse kinematics, 105, 116, 253–256
Inverse kinematics based on the order of priority, 137
Inverse kinematics of redundant manipulators, 148
Inverse kinematics of robot manipulators, 127
Inverse kinematic solution considering the order of priority, 134
Inverse kinematic solution taking account of the priority of subtasks, 129
Inverse kinematic solution with singularity robustness, 255
Inverse matrix, 29, 284
Inverse of the condition number, 121
Inverse of the Jacobian matrix, 255
Inverse transformation circuit, 284, 297
Invertible, 27

Jacobian matrix, 11, 18, 20, 39–40, 106, 109, 114, 122, 155, 192, 240, 253–254, 269, 310, 321
Jacobian matrix of the ith manipulation variable, 128
Jacobian matrix of the nonlinear mapping, 105
Jacobian matrix of the unactuated joints with respect to the actuated ones, 206, 209, 224
Joint torque of closed-link mechanism, 209, 215
Joint torque of the equivalent open-link tree mechanism, 215
Joint torques of the tree-structure mechanism with the virtual actuators, 207
Joint trajectory farthest from singular points, 154, 169

Kalman filter, 305, 317, 323
Kinematically redundant direct-drive manipulators, 230, 248

332 Index

Kinematically redundant manipulator, 126, 254, 260
Kinematic equivalent of a 7-DOF anthropomorphic manipulator, 230
Kinematic redundancy, 2, 11, 106, 126, 229, 254, 266
Kinematic singularity, 11, 127
Kinetic energy, 23, 27
Kinetic friction cones, 191
Known left-hand endpoint boundary values, 157
Known right-hand endpoint boundary values, 157
Kuhn–Tucker theorem, 79, 81

Lagrange equation, 23, 206
Lagrange formulation of dynamics, 84
Lagrange multipliers, 72, 74, 76, 206, 316
Lagrange multiplier theorem, 75, 78, 315
Lagrangian functions, 206, 211
Large condition number, 32
Large internal forces, 175
Largest and smallest singular values, 31–32
Largest singular value, 40
Leading principal minors, 26, 28
Least P-weighted-norm solutions, 55
Least-squares mapping, 129
Least-squares method, 323
Least-squares solutions, 51, 54–55, 128
Least-squares solution with the minimum norm, 54, 256
Least uncertain direction, 311
Left-hand endpoint, 156
Lengths of the principal axes, 193
Linear algebraic methods, 11, 15
Linearly independent resultant forces, 183–184
Linear programming, 102, 174, 215
Linear time-invariant dynamical systems, 89
Local and global optimization of joint torque, 170
Local least squares solutions of the joint accelerations, 170
Locally optimal control of redundancy, 127, 153, 155

Local maxima, 166
Local minima, 166, 187
Local minimum search, 169
Local redundancy control, 135
Location of strain gauges, 284
Logical sensor systems, 304
Longest principal axis, 38
Longest principal axis of the contact-stability ellipsoid, 195
Low-level sensory data, 309
Low-stability grasping, 175
LP, 102

Magnification ratio of measurement error, 290
Magnification ratio of relative error, 37–38
Magnification ratio of the absolute error, 37
Magnitude of the grasping force, 174
Maintaining force, 174
Manifold, 71, 156, 170
Manipulability, 7, 106
Manipulability ellipsoid, 40, 120–121
Manipulable space, 115, 129, 273, 276
Manipulation variable, 115, 119, 128, 274
Manipulator inertia matrix, 273
Mass distribution of links, 206
Master–slave control, 4
Maximum and minimum magnification ratios of the relative error, 287
Maximum number of unactuated joints in a single closed loop, 208
Maximum sensitivity, 289
Maximum singular value, 297
Maximum static friction, 176, 185
Maximum static frictional coefficient, 199
Maximum stiffness, 292, 297
Measurement-error magnification ratio by the generalized inverse matrix, 291
Measurement-error magnification ratio by the pseudoinverse, 291
Measure of contact stability, 175, 194–195, 200–201
Measure of manipulability, 11, 40, 119, 122, 137, 140, 159, 167, 170, 268

Measure of manipulability for the normalized Jacobian matrix, 269
Mechanical stiffness of the force sensor, 284
MELARM, 4
Minimize the integral of kinetic energy, 126
Minimizing internal forces, 175, 185–186, 189–190, 197, 199–200
Minimizing the determinant of the covariance matrix, 314
Minimizing the volume of the uncertainty ellipsoid, 314
Minimum forces, 186, 189–190, 197, 200
Minimum forces under the static frictional constraints, 185
Minimum net forces under static frictional constraints, 175
Minimum norm force, 185, 196
Minimum norm solution, 51, 238
Minimum potential energy criterion, 126
Minimum Q-weighted-norm solution, 55
Minimum singular value, 289
Minimum stiffness, 284, 293–294, 297, 299
Minimum time control, 86, 91
Minimum-energy motion, 154
Minimum-value search problem, 154, 158, 169
Minimum-value search using random numbers, 154
MIT DD Arm II, 230
Modified Denavit–Hartenberg convention, 217
Modified ellipsoid, 273
Modulus of elasticity, 292, 297
MOM, 119–121
Momentum, 84
Moore–Penrose inverse, 42
Most uncertain direction, 311
Multiaxis force sensor, 12
Multifingered robot hand, 6, 173, 200
Multilimb, 3, 213
Multilimb-type closed-link mechanisms, 8
Multiloop closed-link mechanisms, 206, 208
Multiple joint trajectories, 254
Multiple robot manipulators, 173, 200
Multipoint boundary value problems, 157
Multisensor fusion, 12
Multisensor kernel system, 304

Necessary and sufficient condition for the existence of the solution of a linear equation, 157
Necessary and sufficient condition to manipulate an object kinematically, 174
Necessary condition for optimal control, 97
Necessary condition for optimality, 84
Negative definite, 25
Negative real parts, 112
Negative semidefinite, 25
Neighborhood of singular points, 254, 257, 268, 273, 279
Newton–Euler method, 267, 279
Newton's method, 166, 169
Nonanthropomorphic structure, 4
Nonautonomous systems, 93, 97
Nonholonomic constraint, 177
Nonlinear programming, 78
Nonlinear sensing model, 308, 324
Nonlinear time-varying dynamical system, 155
Nonmultilimb-type closed-link mechanisms, 3, 11
Nonnegative definite, 25–26
Nonpositive definite, 25
Nonredundant actuation system, 242
Normal directions of the constraint surfaces, 180
Normalized compliance matrix, 292
Normalized gradient, 17
Normalized stiffnesses, 297
Normalizing matrices, 258
Normal vectors of the manifold, 92, 157
Null space, 58, 116, 184, 237, 244
Null space of the Jacobian matrix, 116, 129
Number of switchings, 90

Object coordinates, 176, 199
Object dynamics, 179

Object stability, 174–175, 179, 191
Object stiffness, 180
Obstacle avoidance, 5, 127, 135, 149, 154, 159
One-dimensional minimum-value search problem, 159
Open-set input variables, 170
Open sets, 82
Open-link tree-structure mechanisms, 206–208, 210, 221
Operational space control, 254
Optimal control problem with a variable left-hand endpoint and a free right-hand endpoint, 156
Optimal forces, 175, 195–197
Optimal fusion, 305
Optimal internal forces, 174–175, 184, 196–197, 201
Optimal in the sense of the measure of manipulability, 122
Optimal redundancy control, 155, 160, 162
Optimal redundancy control problem considering dynamics, 160, 162
Optimal trajectory, 85, 96, 160, 165–167
Optimal trajectory that minimizes the energy cost, 161
Optimization of actuation redundancy, 225, 249
Optimization of finger forces, 11
Optimization problems with constraints, 72
Order of priority, 127–128, 134, 148
Orthogonal complement of the null space, 186
Orthogonal complement of the range space, 115
Orthogonal complement of the redundant space, 157, 258
Orthogonal complements, 58, 129, 180
Orthogonal matrix, 30–31, 36, 38, 106, 285, 311
Orthogonal projection of the adjoint vector, 160
Orthogonal projection of the arbitrary vector, 130
Orthogonal projections, 58, 60, 164
Orthogonal transformations, 260, 262
Orthonormals of the null space, 186

P-weighted norms, 57
Parallel actuated wrists, 205
Parallel computation, 207
Parallel computational scheme of inverse dynamics, 205
Parallel processing in sensory data fusion, 305
Parameterization of actuation redundancy, 11, 230
Partial sensory information, 324
Payload of a closed-link mechanism, 213
Penrose conditions, 41, 43, 47–48
Performance index, 154, 160–162, 164, 166, 169, 242–243, 245
Physical meaning of the internal force, 184
Physical sensor, 306
Plural solutions, 254
Point contact, 11
Polyhedron of admissible joint torques, 223
Polynomial of finite dimension, 154
Pontryagin's maximum principle, 11, 15, 81, 84, 154–155, 161, 166, 169–170
Position control, 174, 180, 182
Positive definite, 26–27, 182, 188–189
Positive definiteness of nonsymmetric matrix, 27
Positive definiteness of the inertia matrix, 192
Positive semidefinite, 25
Potential energy, 23, 174
Potential function, 17, 131, 149, 159
Potential function for avoiding singularity, 137
Potential functions due to the obstacle and the joint limits, 137
Power amplifier with saturation, 279
Principal axes, 40, 120
Principle of variational methods, 68
Principle of virtual work, 40, 210
Product of singular values, 33, 120, 314
Properties of MOM, 119
Properties of the SR-inverse, 269
Proportional limit of the elastic material, 286
Proximity sensors, 303
Pseudoinverse, 15, 34, 41–42, 48, 51,

63, 147, 160, 164, 186, 238, 254–255, 260, 268, 284, 288
Pseudoinverse approximation, 62
Pseudoinverse control of redundant manipulators, 126
Pseudoinverse of a vector, 66
Pseudoinverse of the Jacobian matrix, 118, 126, 128, 133, 155, 256
Pseudoinverse solution, 62
Pseudokinetic-energy, 164
PUMA-type 6-DOF manipulator, 255

Quadratic function, 22, 24, 81, 111, 188, 215, 242, 257
Quadratic programming, 81, 215, 225, 242
Quantitative measure of manipulability, 106
Q-weighted norms, 57

RAC, 111–112
Range sensors, 303
Range space, 58, 91, 115–116, 190
Range space of the Jacobian matrix, 115, 129
Reachable workspace, 128
Real-time redundancy control, 148
Rectangular parallel-piped domain, 90
Recursive computational schemes, 206–207
Recursive computation of multisensor fusion, 319
Recursive Newton–Euler scheme, 221
Redundancy in determining the individual forces, 180
Redundancy of prosthetic arms, 126
Redundancy of the anthropomorphic manipulator, 163
Redundancy utilization, 128
Redundant actuation, 207–208, 213, 223–224, 232, 238, 242, 248
Redundant information, 303
Redundant manipulator, 116, 122
Redundant space, 116, 129, 160, 262
Redundant space of the first manipulation variable, 164
Redundant strain gauges, 287, 293
Reference configuration, 145–146, 149

Reflexive generalized inverse, 41–42, 45, 48, 288
Regularization of experimental data, 259
Regular point on the constraint manifold, 75
Regular points, 73–74, 79, 91
Relationship between MOM and the condition number, 121
Relationship between the force sensitivity and the strain-gauge sensitivity, 289
Relationship between the manipulable space and the redundant space, 117
Relative errors, 288
Relaxation method, 101
Resolved acceleration control, 106, 111, 114
Resolved motion rate control, 106, 111, 114, 127, 140, 162
Restoring forces, 174, 179
Resultant force, 175, 177, 179–180, 183, 189–191
Right-hand endpoint, 156
RMRC, 111–112
Rolling contact, 11
Rotated ellipsoid, 38
Rotational angle of the elbow, 163–164
Rotation of the coordinate frames, 38
Roundoff error, 37

Sampling time of commercial six-axis force sensors, 284
Saturation of power amplifier, 276
Scalar variance, 311
Scale factor, 264, 268
SCARA, 1
Second manipulation variable, 128–131, 133, 136, 149, 154
Second moment of area, 292
Second priority, 128, 132
Sensing model of a range sensor, 307
Sensing model of stereo vision, 307
Sensing redundancy, 2, 11–12
Sensing uncertainty, 323
Sensitivity analysis, 36
Sensor compliance matrix, 284–285, 287, 293
Sensor coordinates, 285

Sensor fusion, 303–304
Sensor-management system, 309
Sensor sensitivity, 287
Sensor unit, 306, 309
Sequence of Z-Y-Z Euler angles, 108
Shape of the ellipsoid, 121
Shooting method, 101
Sight line, 322–323
Signal to Noise (S/N) ratio of sensor signals, 289
Simplex method, 215
Single-camera vision, 320
Singularity, 3, 11, 126, 137, 237, 254
Singularity avoidance, 122, 128
Singularity-free, 159, 169–170
Singularity-free parameterization, 11, 242, 249
Singularity of robotic mechanisms, 122, 279
Singularity of the parameterization of actuation redundancy, 249
Singularity of wrist mechanisms, 120
Singularity-robust inverse, 11, 255, 257, 279
Singularity robustness, 257
Singular Jacobian matrices, 266
Singular points, 105, 115, 117, 125, 168, 254, 256, 264
Singular value decomposition, 15, 30, 120, 259, 286, 310, 322
Singular value decomposition of the covariance matrix, 313
Singular value decomposition of the Jacobian matrix, 119
Singular values, 31, 36, 38, 120, 259, 290, 292, 297, 311
Singular values of a symmetric matrix, 32
Singular values of a symmetric positive or nonnegative definite matrix, 32
Singular values of generalized inverse matrix, 297
Six-axis force sensors, 284, 299
Six-legged walking machine, 10
Sixth-order equation, 200
Skew symmetric matrix, 110
Sliding mode, 276
Slipping, 175, 185, 191, 200
Smallest singular value, 37, 195

Smooth manifold, 91
Soft contact, 11
Soft-finger contact model, 174
Spherical contact, 177
Spherical coordinates, 243
Spherical set of contact-point accelerations, 193
SR-inverse, 255, 257, 260–261, 268, 273, 279
SR-inverse of the Jacobian matrix, 255, 279
Stability of grasping, 11
Stable grasping, 185
Static frictional constraints, 183, 185, 200
Static restoring force, 175
Stationary value, 69, 74, 80
Statistical approaches, 304
Statistical fusion method, 305
Statistical-uncertainty modeling, 304
Stewart Platform Mechanism, 206
Stiffness control of each finger, 179
Stiffness in the most compliant direction, 292
Stiffness in the stiffest direction, 292
Stiffness matrix, 179
Stiffness of the elastic component, 291
Strain gauges, 285
Strain-gauge sensitivity, 284, 289–290, 294, 298–299
Structure evaluation by condition number, 287
Structure of elastic component, 284
Subtasks, 127
Subtasks with lower priority, 127
Surface contact, 11
SVD (see Singular value decomposition)
Switching, 87, 89
Switching point, 92
Sylvester's Criterion, 26
Symmetric positive definite matrix, 27, 41, 54, 162, 215, 257–258

Tactile sensors, 303
Tangential directions of constraint surfaces, 180
Tangential vector, 71, 92
Task priority, 11, 127, 148

Tasks with the order of priority, 128, 154
Tensor, 20
Theoretical maximum value of force sensitivity, 290
Three-dimensional grasp, 183
Torque distribution, 241
Total stiffness of the robotic mechanism, 291
Trace of matrix, 29, 316
Trajectory-planning problem taking manipulability into account, 167
Transversality condition, 72, 91–92, 94, 97, 156, 163, 166, 169
Tree-structure mechanism, 234, 237
Two-dimensional minimum-value search problem, 162
Two-point boundary value problem, 91, 101, 157–158, 162

Uchiyama's mode switching, 254
UCSB hand, 217
UJIBOT, 4, 127, 140, 146, 256
Ultra sonic sensors, 303
Unactuated joints, 3, 207–208, 213, 234, 237, 239
Unavoidable singularities, 255
Uncertainties, 111, 309, 311
Uncertainty ellipsoid, 305, 311, 322
Uncertainty manifold, 304
Uncertainty of mobile robot, 304
Unconstrained optimization, 72
Uncorrelated, 310, 313
Underdeterminate, 207–208
Unexpected disturbing forces, 175, 191, 200
Unified computation of kinematics and dynamics, 114
Unified recursive computational scheme, 220
Unit vector approach, 211
Upper-bounded set, 84

UTAH/MIT dextrous hand, 9
Utilization of redundancy in force sensing, 284

Variable scale factors, 269
Variation, 68, 71
Variational methods, 15, 70, 170
Vertex of a polyhedron or a cone, 177
Virtual actuators, 207
Virtual displacements, 184, 211
Virtual work, 237
Volume of the manipulability ellipsoid, 120
Volume of the uncertainty ellipsoid, 313, 318, 323
VSLAM, 169

Weighted least-squares estimate with the inverse of covariance matrix as the weighting matrix, 305, 317–318
Weighted least-squares estimation, 323
Weighted norm, 54, 257
Weighted-norm solutions, 56
Weighted pseudoinverse, 55
Weighted pseudoinverse of the full-rank Jacobian matrix, 126
Weighting between the exactness and the feasibility, 258
Well conditioned, 32
Wheatstone bridges, 284, 293, 297, 299
Whitney's approximation, 254
Wolfe's method, 215
Wrist singularities, 255

X-Y-Z Euler angles, 109

Zero internal forces, 200
Z-Y-Z Euler angles, 107–108, 115